Artificial Intelligence for Power Electronics

IEEE Press
445 Hoes Lane
Piscataway, NJ 08854

Artificial Intelligence for Power Electronics

Edited by Ahteshamul Haque, Saad Mekhilef, and Azra Malik

IEEE Press
WILEY

Published by John Wiley & Sons, Inc., Hoboken, New Jersey.
Published simultaneously in Canada.

Library of Congress Cataloging-in-Publication Data applied for:

Hardback ISBN: 9781394270774

Cover Design: Wiley
Cover Image: © Eugene Mymrin/Getty Images

Set in 9.5/12.5pt STIXTwoText by Straive, Chennai, India

The work in this book is dedicated to the last Prophet of Humanity, Prophet Muhammad ﷺ ("peace be upon him"), who introduced Our Creator, Almighty Allah, to all of us.

Ahteshamul Haque

I wish to dedicate this book to my parents, who planted the seed of knowledge in my mind and nurtured it.

Saad Mekhilef

I wish to dedicate this book to my beloved daughter Maryam and my husband Tanweer Aadil, who are my pillars of strength and support system.

Azra Malik

Contents

About the Editors

Prof. Ahteshamul Haque is working as Professor in the Department of Electrical Engineering, Jamia Millia Islamia (A Central University) New Delhi. His area of research is Power Electronics and its application in Renewable Energy, drives, electric control system for artificial lighting, Power quality improvements, smart grids, wireless power transfer, electric vehicles, electric traction, smart cities etc. Prior to Jamia Millia Islamia, he was working in the Power Electronics R&D units of world reputed Multi-National Companies. His inventions are patented and awarded in USA and Europe. He has published and presented around 200 research papers in several international conferences and peer reviewed Journals. He has established state of the art Advance Power Electronics Research Lab in which research work in the area of Solar based Energy conversion system, Electric Vehicle, embedded system control of power electronics converters, electric control system for artificial lighting, reliability of Power electronics converter, AI based control of Power Electronics Converter etc are carried out. Prof. Haque is senior member of IEEE PELS, IAS, Smart Cities Society and Branch Counsellor IEEE -JMI, and actively involved in IEEE activities at Institution and in Delhi Section. Prof. Haque is awarded with Outstanding Engineer award 2019 by IEEE Power & Energy society for his research and development contribution in the area of Power Electronics and Renewable Energy. Prof. Haque has also won design contest called by Switzerland based R&D company Typhoon and received RTD emulator as prize. He also won most well newsletter award from IEEE Smart Cities Society. He has also secured a place in the world's top 2% of Scientist curated by the Meta Research Innovation Centre (METRICS), Standford University, USA for the year 2022, 2023.

Prof. Haque is working as Associate Guest Editor of IEEE Journal of Emerging and Selected Topics in Power Electronics and IET Power Electronics Journal.

Prof. Haque has total work experience of 20+ years of Industry and academics in the area of Power Electronics and its application.

Prof. Dr. Saad Mekhilef is an IEEE and IET Fellow. He is a Distinguished Professor at the School of Engineering, Swinburne University of Technology, Melbourne, Australia, an Honorary Professor at the Department of Electrical Engineering, University of Malaya, and a distinguished visiting professor at the Institute of Sustainable Energy, Universiti Tenaga Nasional, Malaysia. He authored and co-authored more than 800 publications in academic journals and proceedings, as well as five books, which have garnered over 64,000 citations, an H-index of 119, and have supervised more than 85 Ph.D. students. He serves as an editorial board member for several top journals, including IEEE Transactions on Power Electronics, IEEE Open Journal of Industrial Electronics, IET Renewable Power Generation, E-Prime, Journal of Power Electronics, and International Journal of Circuit Theory and Applications.

Prof. Mekhilef has been listed by Thomson Reuters (Clarivate Analytics) as one of the world's Highly Cited (World's Top 1%) engineering researchers. He is also listed among the world's top 2% of scientists by Stanford University, USA. He is actively involved in industrial consultancy for major corporations in Power Electronics and Renewable Energy projects. His research interests include Power Conversion Techniques, Control of Power Converters, Maximum Power Point Tracking (MPPT), Renewable Energy, and Energy Efficiency.

Dr. Azra Malik is currently working as an Institute Post-Doctoral Fellow in the Electrical Engineering department, IIT Roorkee, Uttarakhand, India. She received her Ph.D. from Jamia Millia Islamia, New Delhi, India, where her work focused on fault-tolerant converters for multi-inverter photovoltaic (PV) systems. She completed her B. Tech (Electrical Engineering), and M. Tech (Control, and Instrumentation Systems) from Jamia Millia Islamia, New Delhi, India. She has many publications in peer-reviewed journals and presented her research articles in several international conferences. She has served as a reviewer in many reputed journals including MDPI, IEEE transactions, Elsevier, Springer etc. Her research interests include fault tolerant power electronic converters, electric vehicle (EV) charging infrastructure, grid-connected inverter control, and the application of artificial intelligence (AI)-based methods (such as reinforcement learning) in smart energy systems.

List of Contributors

Mohammad Amir
Advance Power Electronics and Research Laboratory
Jamia Millia Islamia (A Central University)
New Delhi
India

Kurukuru Varaha Satya Bharath
Silicon Austria Labs
Villach
Austria

Ahteshamul Haque
Advance Power Electronics and Research Laboratory
Jamia Millia Islamia (A Central University)
New Delhi
India

Manauwar Hussain
Advance Power Electronics and Research Laboratory
Jamia Millia Islamia (A Central University)
New Delhi
India

Md Zafar Khan
Advance Power Electronics and Research Laboratory
Jamia Millia Islamia (A Central University)
New Delhi
India

Mohammed Ali Khan
Centre of Industrial Electronics
University of Southern Denmark
Sønderborg
Denmark

Mansha Khursheed
Advance Power Electronics and Research Laboratory
Jamia Millia Islamia (A Central University)
New Delhi
India

Azra Malik
Advance Power Electronics and Research Laboratory
Jamia Millia Islamia (A Central University)
New Delhi
India

Suwaiba Mateen
Advance Power Electronics and Research Laboratory
Jamia Millia Islamia (A Central University)
New Delhi
India

Izhar Ahmad Saifi
Advance Power Electronics and Research Laboratory
Jamia Millia Islamia (A Central University)
New Delhi
India

Fatima Shabir Zehgeer
Advance Power Electronics and Research Laboratory
Jamia Millia Islamia (A Central University)
New Delhi
India

Preface

The integration of artificial intelligence (AI) with power electronics is revolutionizing the way energy systems are designed, controlled, and optimized. In addition to propelling technical progress, the combination of these two is creating new opportunities for research and innovation. The main goal of this book, *Artificial Intelligence for Power Electronics*, is to inspire, and encourage students, and researchers to learn more about this exciting multidisciplinary topic. This book starts with an overview of the foundations of AI and power electronics before demonstrating how they work together to solve challenging power electronics-related problems. Case studies and illustrated examples provide a hands-on learning experience, which makes complex concepts more approachable and interesting for students.

Enhancing power converter efficiency, forecasting vital power electronic component performance, or optimizing control techniques are just a few examples of how the subject matter connects theory and practice to provide a strong basis. This book's purpose goes beyond simply teaching; it also seeks to ignite curiosity and foster exploration. Tools such as MATLAB/Simulink and python are utilized to solve the power electronics-related problems using AI techniques. The curriculum aims to stimulate students' interest, develop their analytical thinking, and motivate them to make significant contributions to this evolving field by showcasing real-world problems and presenting innovative solutions.

We are extremely appreciative to the researchers, educators, and colleagues whose insights and wisdom have influenced this work. With the help of this book, we wish to inspire and educate students, enabling them to explore and innovate in the field of AI-driven power electronics.

12 March 2025

Ahteshamul Haque
Delhi, India

Saad Mekhilef
Melbourne, Australia

Azra Malik
Delhi, India

1

Fundamentals of Power Electronics and Key Challenges

Azra Malik and Ahteshamul Haque

Advance Power Electronics and Research Laboratory, Jamia Millia Islamia (A Central University), New Delhi, India

1.1 Introduction

Power electronics is a cornerstone of modern electrical engineering, allowing for efficient control and conversion of electrical power in a wide range of applications. This area combines electronics and power engineering ideas to develop devices and systems capable of handling large amounts of electricity effectively [1]. Power electronics encompasses a wide range of devices, including power converters, inverters, and motor drives, which are all necessary for modern industrial operations, consumer electronics, and renewable energy systems. The history of power electronics began with the introduction of the first semiconductor devices. The silicon-controlled rectifier (SCR) was invented in the late 1950s, and it marked an important milestone in electrical power regulation [2]. Since then, advances in semiconductor technology have resulted in the creation of a variety of power electronic devices, including Insulated Gate Bipolar Transistors (IGBTs) and Metal–Oxide–Semiconductor Field-Effect Transistors (MOSFETs), which provide higher efficiency and switching capabilities. These developments were crucial in improving the performance and lowering the costs of power electronic systems. Despite these advances, the area of power electronics encounters significant challenges that should be addressed in order to fulfil the rising demands of present-day technology. One of the pressing challenges is the requirement for increased efficiency and power density. As applications become increasingly demanding, it is critical to minimize energy losses and reduce the weight, and size of power electronic systems. This necessitates novel materials, enhanced semiconductor devices, and better heat management approaches. Wide bandgap (WBG) semiconductors like silicon carbide (SiC) and gallium nitride (GaN) have

Artificial Intelligence for Power Electronics, First Edition.
Edited by Ahteshamul Haque, Saad Mekhilef, and Azra Malik.

emerged as potential options [3]. These materials provide superior electrical characteristics, including higher breakdown voltage and more rapid switching rates, which may drastically enhance power electronic systems' performance.

Power electronics technology in the 21st century envisions a more distributed energy sector with IoT connectivity, and smart devices. Power electronics is the study of how electrical power is controlled and converted through electronic devices. It covers a wide range of technologies and applications, including small power supply and huge power systems [4]. Understanding basic ideas and terminologies, such as the many types of power electronic devices, converters, modulation techniques, and control methods, is critical for anyone working in this field. These ideas underpin a wide range of applications, including renewable energy systems, electric transportation systems, and diverse range of power suppliers. The swift growth of energy planning technology in smart grids, combining sensors, smart devices, storage devices, and non-conventional energy sources utilizing communication networks, is driving the demand for more sustainable and cleaner energy. It is a well-known fact that burning fossil fuels like coal, oil, and natural gas to create power is harmful to the environment and contributes significantly to climate change. Throughout decades of heavy mining of resources based on fossil fuel, the consumption of energy is enormously increasing and is expected to rise further as the global economy expands. Developing inexpensive and clean electricity is crucial for ensuring global sustainability and combating climate change. Efforts to increase the use of renewable energy (REN) sources are ongoing, with new technologies developing. Among these resources, solar photovoltaic (PV) and wind are now the most favorable, accounting for more than half of the entire worldwide capacity being installed.

Worldwide, the installed PV capacity grew significantly, surpassing 1340 GW in 2023, an increase from around 940 GW at the close of 2022 and 760 GW in 2021. This shows an impressive annual growth rate of nearly 24% and 30% for the respective years, as depicted in Figure 1.1 [5]. It is expected that additional wind and solar PV capacity will be built shortly, as the costs of such technologies continue to fall; hence, large-scale wind and solar PV power generation is just around the corner. However, the key problem is the combination of power electronics with these RENs. Renewable energy sources, such as solar and wind, are inherently variable and intermittent, necessitating complex power electronic systems to assure steady and reliable power conversion and delivery. Advanced control algorithms and grid-tied inverters are critical for regulating fluctuations and ensuring grid stability. Furthermore, power electronics play an important part in energy storage systems, which are critical for balancing supply and demand in renewable energy applications [6]. Power electronics is also having a huge influence on the electrification of transportation. Electric vehicles (EVs) require extremely efficient power converters and inverters that can function in a variety of challenging

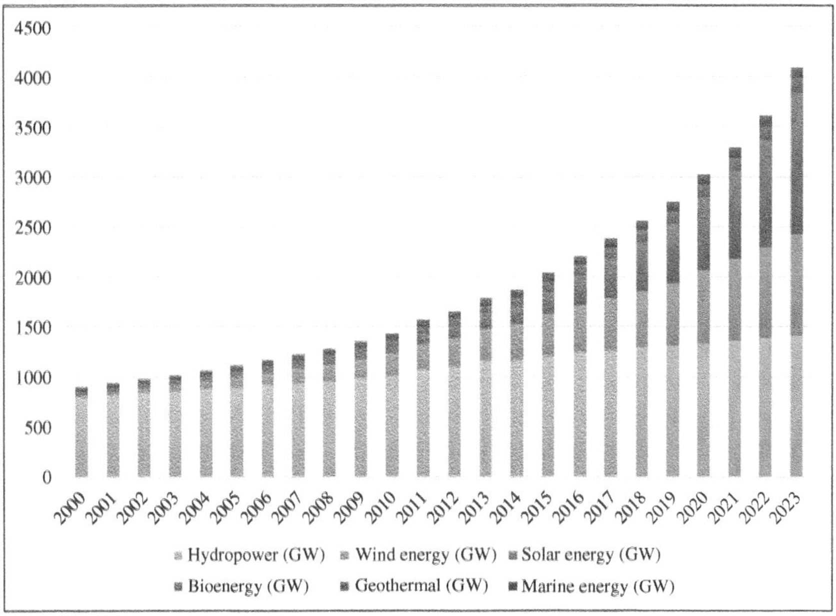

Figure 1.1 Globally installed renewable energy trend.

environments [7]. To fulfil performance and safety regulations, power electronics are widely used in the development of fast-charging infrastructure and advanced battery management systems. Innovations in these areas are vital for the broad adoption of EVs, which is necessary for lowering greenhouse gas emissions and dependency on fossil fuels.

Furthermore, as modern power systems become increasingly complicated, control and reliability issues arise. The integration of smart grid technologies, which rely on power electronics for real-time monitoring and control, is critical for increasing the efficiency and resilience of electricity distribution networks [8]. These technologies allow for improved regulation of electrical flows, the integration of dispersed energy resources, and the general stability of the grid. To summarize, this chapter details the modern power electronics fundamental concepts and related principles. While power electronics has made significant progress over the last few decades, continued research and development is required to solve present problems. Future developments must focus on increasing efficiency, integrating renewable energy sources, and promoting vehicle electrification. By tackling these problems, power electronics can continue to play an important role in increasing energy efficiency and sustainability across a variety of applications.

1.2 Fundamental Concepts and Definitions

Power electronics is an essential field that deals with the control, conversion and delivery of electrical power utilizing different electronic components. Power semiconductor devices are at the heart of power electronics, providing the switching and rectification operations required for effective power control. These semiconductor devices are the foundation of power electronics systems, allowing for efficient conversion and management of electrical power. They enable precise control of required electrical parameters, as per the operation of a wide range of electronic devices and systems, from small consumer electronics to big industrial machinery [9]. Key components of such systems include semiconductor devices like diodes, transistors, and thyristors as discussed here. The fundamental component or devices in power conversion applications are electronic switches, since they have the capability of dealing with high currents and voltages at high-frequency operations. The concept of ideal switch is highly important in this regard, and it can be represented using a three-terminal device as displayed in Figure 1.2 (a) which consists of input, control terminal often pronounced as gate terminal, and output. An "ideal" switch is one, which, when open, conducts zero current and can tolerate infinite voltage. When this ideal switch is closed, the voltage across it is zero, and it may handle infinite current. Furthermore, an ideal switch has the ability to change its state immediately, which means it takes no time to move from ON to OFF or OFF to ON states. An ideal switch may maintain bidirectional voltage, carry bidirectional current, and consume no power. When we draw the graph of the current through the switch in relation to its voltage, we get four quadrants, which are commonly known by the term *vi* plane as depicted in Figure 1.2 (b). As per the description, an ideal switch may function in each of the four quadrants. It is important to understand the concept of ideal switches particularly during the evaluation of various circuit topologies. The presumptions of zero leakage current, zero voltage drop, and instant transitions facilitates the

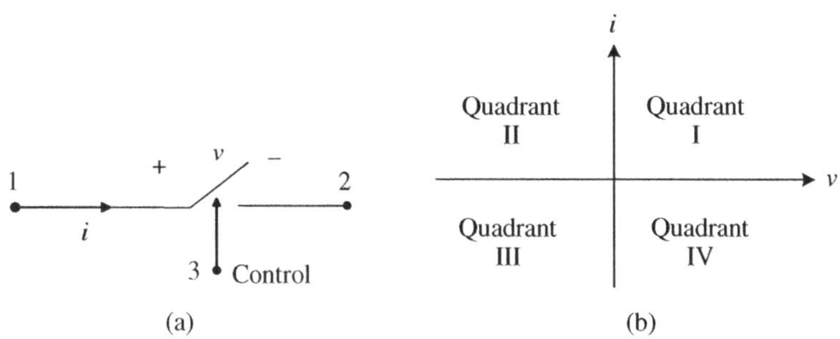

Figure 1.2 (a) Ideal switch and (b) switch *vi* characteristics (4 quadrant).

simulation and modeling of varied electrical devices. Power switches are classified into three types based on the ideal switch characteristics.

1.2.1 Uncontrolled Switches

Uncontrolled switch is a two-terminal device, with the terminals labeled as Cathode (K), and Anode (A), shown in Figure 1.3 (a). Ideally, current flows only from the anode to the cathode, making it unidirectional. The diode voltage is positive when analyzed from the terminal A to terminal K. The ideal diode, displayed in Figure 1.3 (b), conducts current with zero voltage drop when ON and supports a negative voltage with zero current when OFF. The *vi* characteristics of a practical (non-ideal) diode can be illustrated in Figure 1.3 (c) [10]. The diode operates in the ON state in first quadrant, also called the forward-biased region, allowing current flow while maintaining a small voltage. The current increases exponentially with the voltage. In quadrant III, the diode is reverse-biased or in the OFF state, supporting a negative voltage with minimal leakage current. If the reverse voltage exceeds the breakdown voltage, the leakage current immediately increases, which may destroy the diode. To prevent damage, operation beyond this limit should be avoided.

1.2.2 Semi-Controlled Switches

A thyristor, also known as an SCR, is a power semiconductor device that is triggered via its gate terminal to turn ON, but once activated, the gate no longer

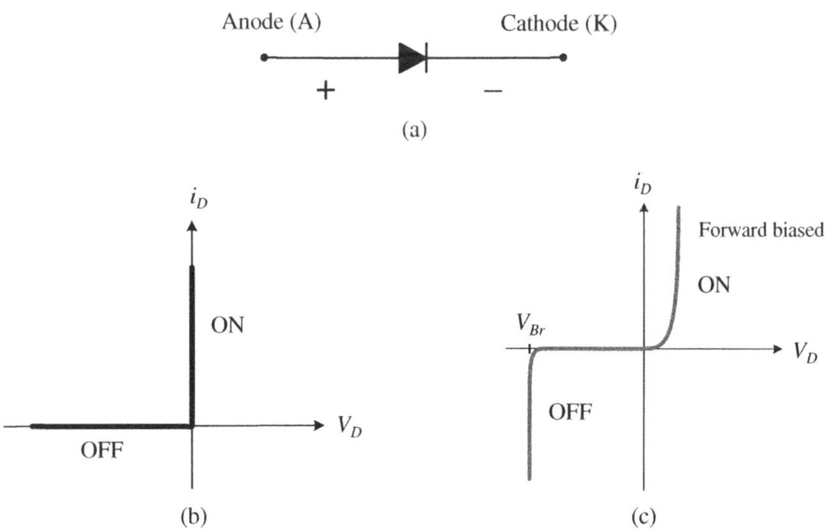

Figure 1.3 (a) Diode symbol, (b) ideal diode characteristics, and (c) *vi* properties of diode.

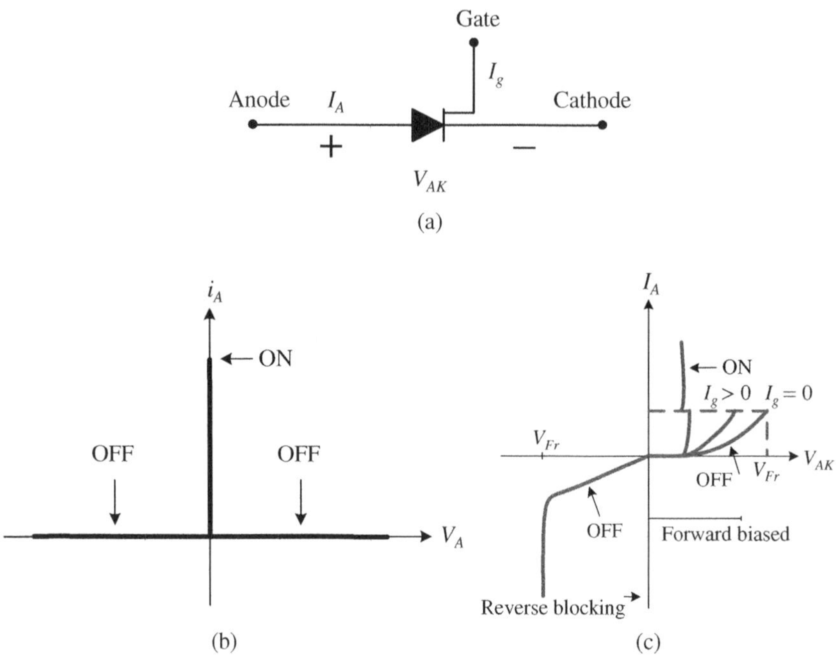

Figure 1.4 (a) Thyristor symbol, (b) ideal characteristics, and (c) *vi* characteristics.

controls it, and the thyristor acts like a diode. For this reason, it is classified as a semi-controlled switch. The term "controlled rectifier" suggests that a thyristor can be seen as a diode whose turn-on is controlled externally. The thyristor's circuit symbol given in Figure 1.4 (a) resembles that of a diode, but their working differs significantly. Current flows from terminal A to terminal K, and the voltage across the thyristor is positive, when the anode is at a greater voltage.

The ideal thyristor characteristics are illustrated in Figure 1.4 (b). In the ON state, it conducts current with no voltage drop, while in the OFF state, it supports either positive or negative voltage. The *vi* characteristics of a real thyristor in Figure 1.4 (c) show that in quadrant I, without a gate signal, the thyristor remains in the forward blocking area, and it is OFF, upholding a positive voltage. Upon receiving a gate signal, the thyristor turns ON, exhibiting diode-like behavior. In the third quadrant, it is in the reverse blocking area, where it remains OFF. Compared to a diode, a thyristor may block both positive and negative voltages when it is turned off, making it useful in applications like AC/AC converters.

A TRIAC shown in Figure 1.5 (a), on the other hand, is another semi-controlled device and can be viewed as two thyristors connected in reverse parallel (Figure 1.5 (b)). Unlike the thyristor, a TRIAC can handle both bidirectional voltage and current, as illustrated by its *vi* characteristics in Figure 1.5 (c). Ideal TRIAC behavior is

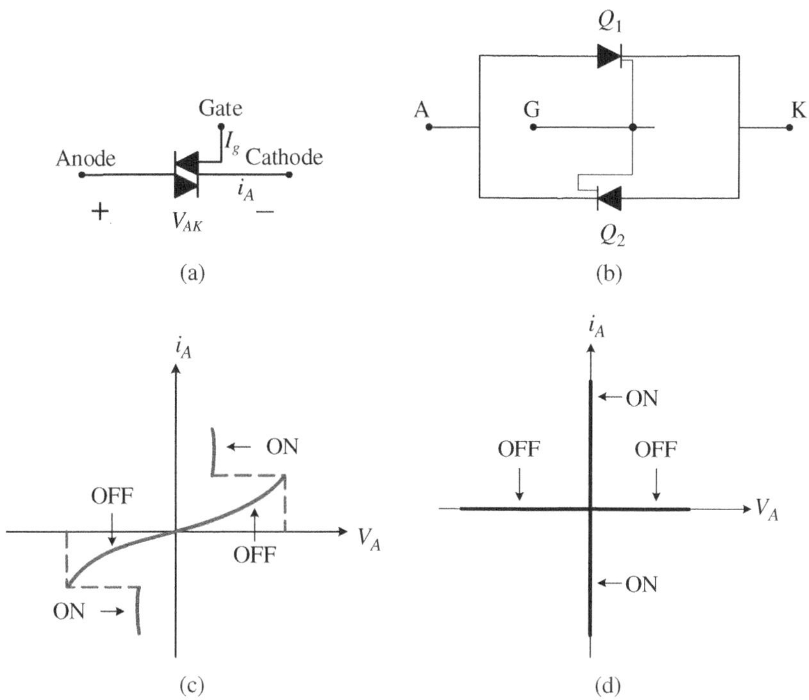

Figure 1.5 (a) TRIAC symbol, (b) two-thyristor equivalent, (c) *vi* properties, and (d) ideal *vi* behavior.

illustrated in Figure 1.5 (d). Due to its less cost and bidirectional switching capability, TRIACs are commonly employed in low-power commercial applications, like control circuits for single-phase motors and light dimmers.

1.2.3 Fully Controlled Switch

In this particular case, the switch turn on and turn off, both are controlled through the gate/control terminal. Examples include bipolar junction transistor (BJT), gate turn off (GTO) thyristor, IGBT, MOSFET, and MOS-controlled thyristor (MCT). A control terminal on a fully controlled switch allows for external activation of the ON and OFF conditions. Controlled switches include a variety of power switches, a number of which are thyristor-based while others use transistors. Some of these devices are described briefly as given below.

1.2.3.1 Bipolar Junction Transistor (BJT)

The circuit symbol that represents a npn-type BJT is displayed in Figure 1.6 (a). The control connector is the base (B), while the emitter (E) and collector (C) are the

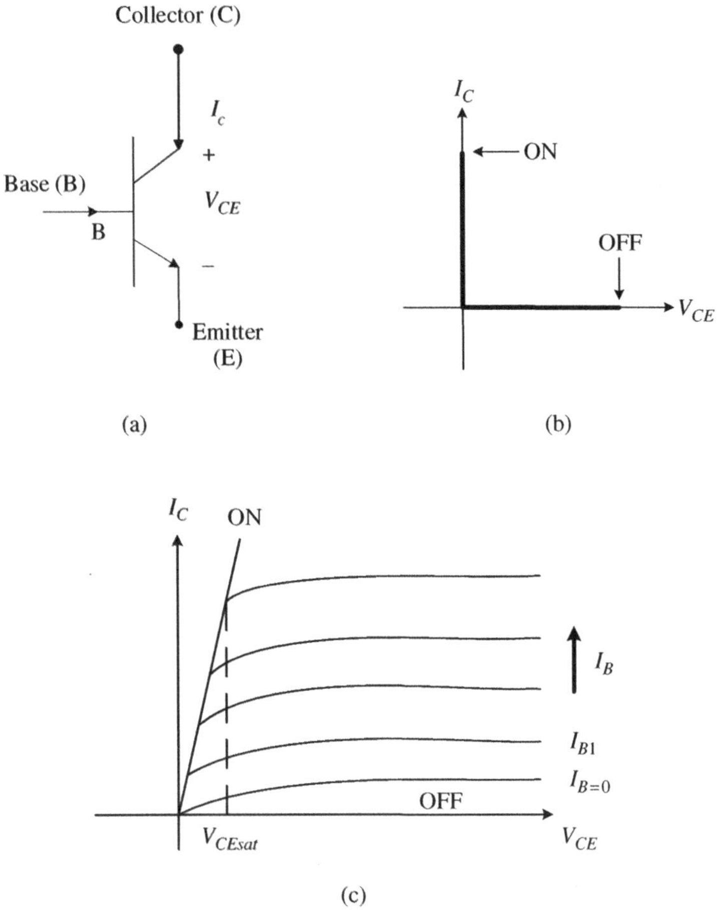

Figure 1.6 (a) BJT symbol representation, (b) ideal characteristics, and (c) *vi* behavior.

power connectors. The idealized attributes of the device as a switch are displayed in Figure 1.6 (b), ignoring the center area. The device has a positive collector current I_C with $v_{CE} = 0$ when it is in the ON state. The device supports positive v_{CE} with $I_C = 0$ when it is in the OFF-state. As a result, the BJT is a unidirectional voltage and current device. Figure 1.6 (c) displays the device's *vi* properties. The plot of the collector current against the collector to emitter voltage identifies the device, which operates in quadrant I. The device is divided into three areas, two of which function as switches, and the third of which is a linear amplifier. In the region where $v_{CE} < v_{CE(sat)}$, the device is ON; in the region below $i_B = 0$, it is OFF. Although the BJT has historical significance, other technologies such as the IGBT cover the majority of its functions.

1.2.3.2 Metal Oxide Semiconductor Field Effect Transistor (MOSFET)

The schematic representation for an n-channel, enhancement-type MOSFET is displayed in Figure 1.7 (a). Its power connectors are the drain (D) and source (S), and it features a control terminal called the gate (G), just like the BJT. The MOSFET functioning is controlled through a voltage between G and S terminals. Because of this, it is a voltage-controlled device as opposed to a current-controlled device like the BJT. Figure 1.7 (b) depicts the idealized properties of a MOSFET switch. In ON-state, the drain-to-source voltage is zero, and drain current flowing through it is positive. Similarly, when it is in OFF state, the device upholds a positive drain-to-source voltage, and drain current is zero. Figure 1.7 (c) displays the practical *vi* characteristics of the device. The MOSFET functions within three operating areas, just like the BJT. When the device is employed as a switch, two of its areas are utilized, and when it is used as an amplifier, the third region is utilized. In order to have the MOSFET in the off-state, gate-to-source voltage needs to be lower than a threshold voltage (V_T), which is the area beneath the OFF line.

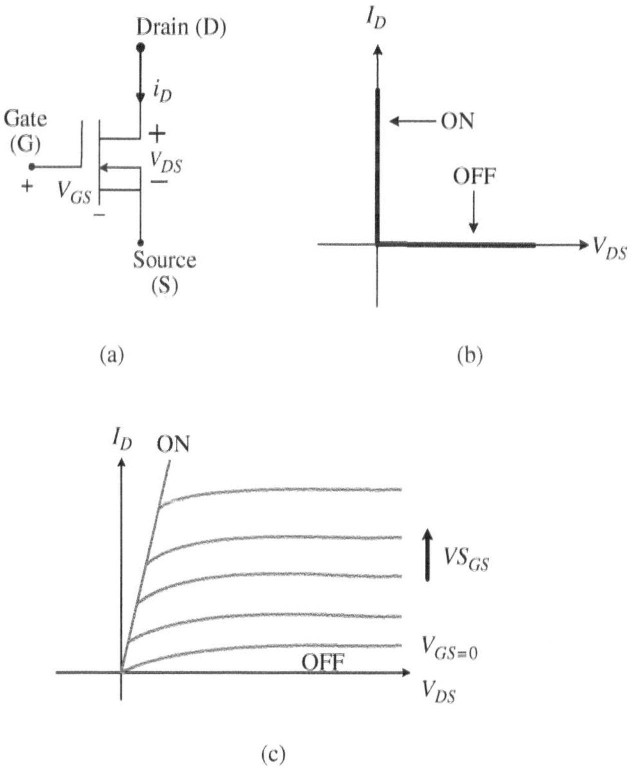

Figure 1.7 (a) The MOSFET symbol, (b) ideal characteristics, and (c) *vi* behavior.

Additionally, the slope of the ON line specifies the resistance of the device when it is turned on.

1.2.3.3 The Insulated Gate Bipolar Transistor (IGBT)

The IGBT is a hybrid device, sometimes referred to as a two-fold mechanism. It has a power port that looks like a BJT and a control port that looks like a MOSFET. Consequently, an IGBT integrates a BJT's less power conduction loss with an MOSFET's rapid switching. The circuit representation for an IGBT, which differs somewhat from that of a MOSFET with comparable terminal labels, is displayed in Figure 1.8 (a). The power connections are designated as collector (C) and emitter (E), whereas the control terminal is designated as gate (G). Figure 1.8 (b) displays the device's ideal characteristics. This device has the ability to conduct unidirectional current and block bidirectional voltage. Figure 1.8 (c) displays the

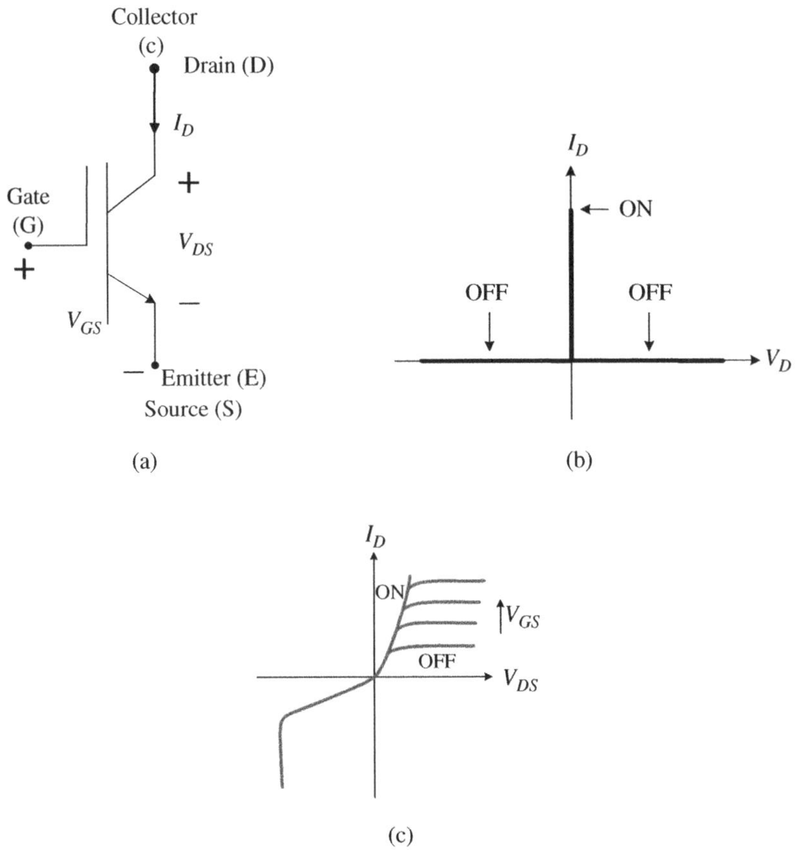

Figure 1.8 (a) IGBT symbol, (b) ideal characteristics of IGBT, and (c) *vi* behavior.

vi properties of an actual IGBT, indicating that the device functions in the first and third quadrants. An IGBT is less rapid than a MOSFET device, but it may switch to the ON state very quickly. Control of the IGBT is fully restored to the OFF-state by discharging the gate capacitance. IGBTs are commonly utilized for medium power switching applications like PV, electric vehicles (EVs), and wind, as well as high power applications like motor controllers.

1.2.3.4 Gate Turn-Off (GTO) Thyristor

The GTO functions analogous to a typical thyristor, with the exception that its manufacturing, design, and physical properties, a bipolar transistor may be used to switch it off with a negative gate current. The schematic representation for GTO is illustrated in Figure 1.9 (a). The *vi* behavior and ideal characteristics are also displayed in Figure 1.9 (b), and (c), respectively. GTOs, though less common now, are still applied in specific high-power circuits.

1.2.3.5 MOS-Controlled Thyristor (MCT)

The MCT, like the IGBT, is a hybrid or dual mechanism semiconductor that was made to have a thyristor power port and a MOSFET control port. Figure 1.10

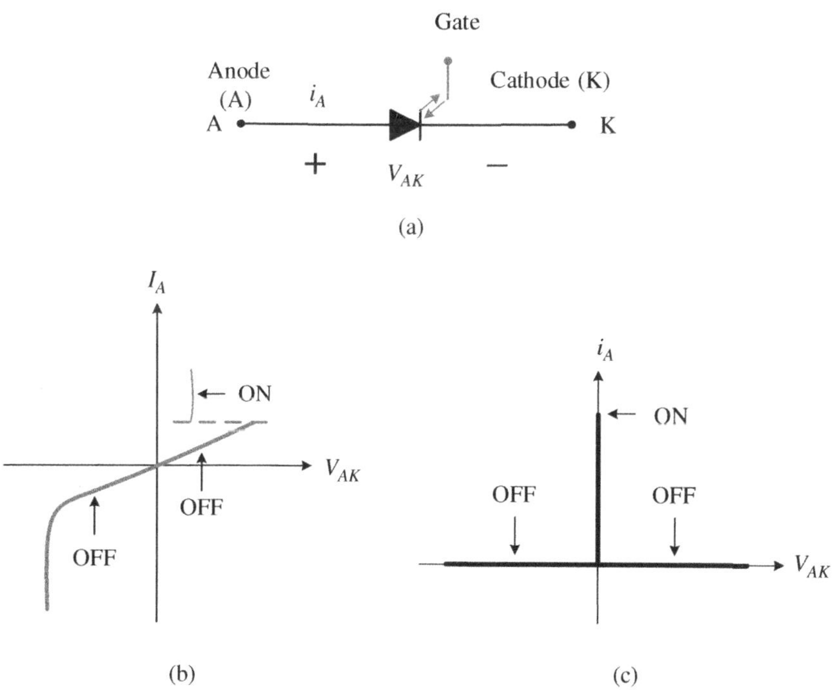

Figure 1.9 (a) GTO Symbol, (b) *vi* characteristics for GTO, and (c) ideal GTO behavior.

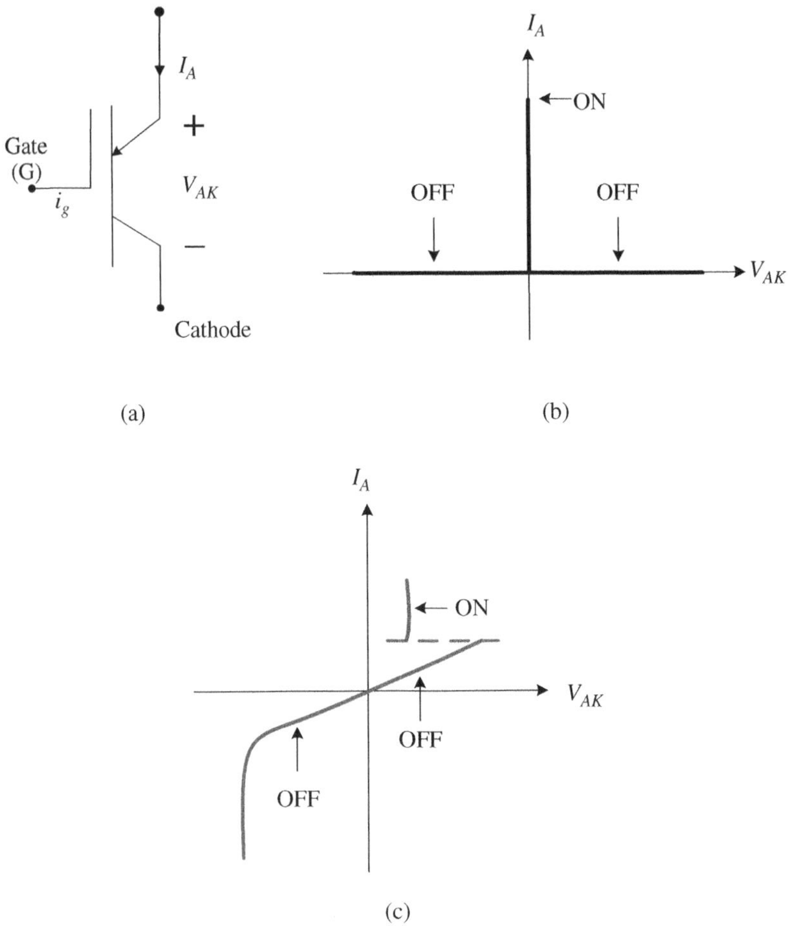

Figure 1.10 (a) MCT Symbol representation, (b) ideal MCT behavior, and (c) *vi* characteristics.

(a) displays the device's circuit symbol. Figure 1.10 (b) and (c) display both the idealized and practical device properties. With the exception of the MCT's simpler gate drive circuitry, the features are comparable to those of the GTO since MCT control circuit employs MOSFET, rather than a transistor. Therefore, MCTs are expected to operate at higher switching frequencies. MCTs, with their fast switching and bidirectional capability, are primarily used in niche applications but have been largely overshadowed by newer technologies like SiC (Silicon Carbide) and GaN (Gallium Nitride) devices, which demonstrate better performance and efficiency.

1.3 Fundamental Principles Related with Power Electronic Converters

In reality, switches have limitations compared to ideal switches. For instance, when a switch is on, it experiences a voltage drop (on-voltage) and carries a finite current. In the off state, a small leakage current may flow while supporting a finite voltage. Additionally, switching devices take time to change states, as seen by their rise, fall, storage, and delay periods; thus, the transition between ON and OFF states is not immediate. These non-idealities result in continual current and voltage across the switch, which causes two kinds of losses: switching loss during state transitions and conduction loss throughout on/off states. Power electronic converters are switch-mode circuits that use semiconductor-based devices to transfer power between two electrical circuits. There are four primary types of converters based on whether these systems are DC or AC: AC/DC, DC/DC, DC/AC, and AC/AC.

1.3.1 AC/DC Converters or Rectifiers

Rectifiers convert an alternating current (AC) voltage, typically from power distribution systems, into DC voltage. This DC voltage can either power a load or be further converted through additional stages, such as DC/DC or DC/AC converters [11]. Rectifiers are widely used in power systems but may introduce harmonics and low power factors if not properly implemented. Rectifier circuits, also known as AC/DC converters, can be generally categorized according to the kind of AC source and semiconductor switch utilized, as seen in Figure 1.11. Although poly-phase topologies are feasible for high power systems, rectifiers are typically

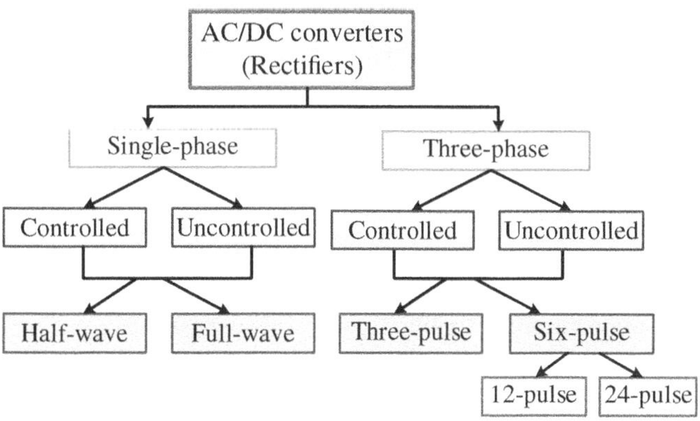

Figure 1.11 Types of AC/DC converters (rectifiers).

utilized in single-phase or three-phase systems. The switches employed in rectifier circuits may be either controlled (thyristor) or uncontrolled (diode). Single-phase rectifiers fall into one of two categories: full-wave or half-wave circuits. Depending on how many pulses the rectified output voltage contains, three-phase rectifiers are categorized as 3-pulse rectifiers, 6-pulse rectifiers, 12-pulse rectifiers, etc. The ratio of the rectifier output's fundamental frequency to the AC input determines the number of pulses.

Figure 1.12 (a) shows a generic schematic for a single-phase rectifier. The P- and N-switches are adjusted to either eternally close or open to complete the elementary single-phase rectifiers. A halfwave rectifier circuit would arise, for instance, if the bottom P-switch was always closed and all of the N-switches were continually open. The circuit that results is a controlled half-wave rectifier if the residual P-switch is a thyristor. Figure 1.12 (b) illustrates this structure, and Figure 1.12 (c) displays the waveforms that produced it. The angle α in this situation is the thyristor turn-on delay angle, and it is measured beginning at the input voltage

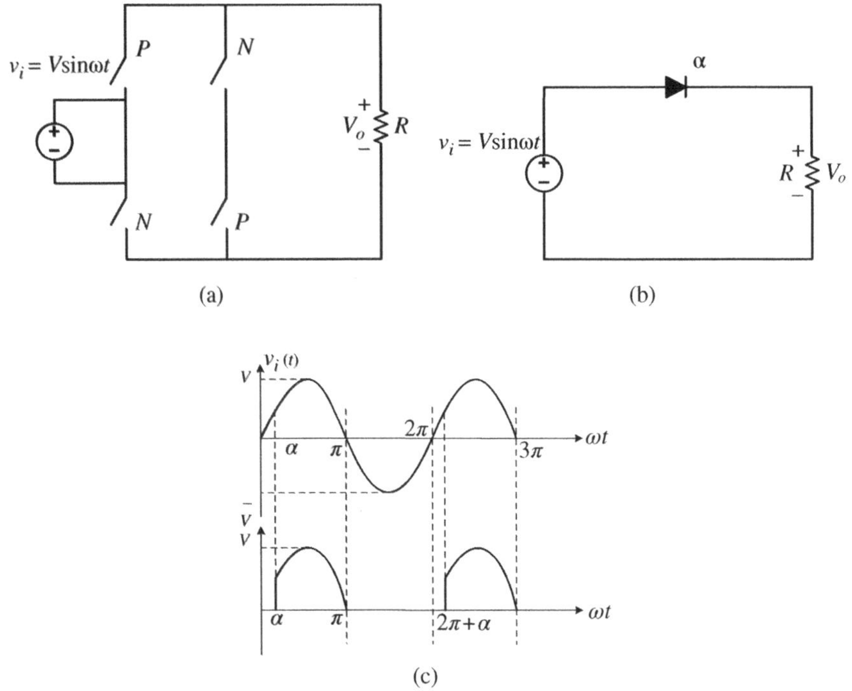

Figure 1.12 (a) General rectifier circuit (1-phase), (b) controlled rectifier, and (c) input and output voltage waveforms.

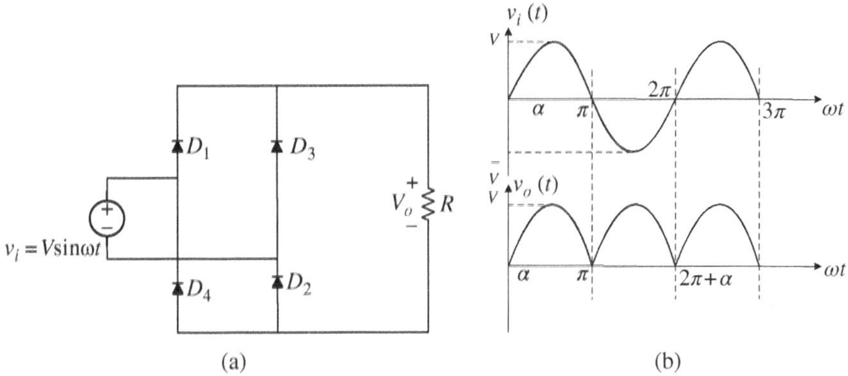

Figure 1.13 (a) Full bridge uncontrolled rectifier and (b) input, and output waveforms.

source's zero crossings. There is an average value (the necessary DC component) in the output waveform expressed as

$$V_{avg} = \frac{1}{T}\int_0^T v(t)dt = \frac{1}{2\pi}\int_\alpha^\pi V\sin\omega t \quad d(\omega t) = \frac{V}{2\pi}(1+\cos\alpha) \tag{1.1}$$

Therefore, the DC voltage value can be varied with the variation in α. When a diode is used in place of a thyristor to implement the remaining P-switch, a half-wave uncontrolled rectifier is achieved. When α is set to zero, the diode will activate when the supply voltage crosses zero. A full-bridge rectifier circuit with uncontrolled devices is depicted in Figure 1.13 (a), along with the respective waveforms in Figure 1.13 (b). For half-wave or full-wave types, there exist circuit topologies like one-, two-, and four-quadrant. Based on the arrangement of diodes and thyristors, some of them are uncontrolled, while others are completely controlled; they are typically categorized as hybrid or semi-controlled rectifiers. To provide further rectification capabilities, certain rectifiers utilize transformers for isolation.

1.3.2 DC/DC Converters

They change a DC voltage to another DC level, either for powering DC loads or for use in subsequent power conversions, such as DC/AC conversion. These converters, along with AC/DC converters, play a key role in high-voltage DC (HVDC) transmission systems. Figure 1.14 shows a basic schematic for a DC/DC converter, including every switch that might be used to link the input to the output. The circuit that results from using one N-switch, and another P-switch is seen in Figure 1.15 (a). Within a given time T, the switches are regulated to switch ON and OFF. When the N-switch is activated, the output voltage drops to zero; when the P-switch is activated, the output voltage is equal to the input voltage.

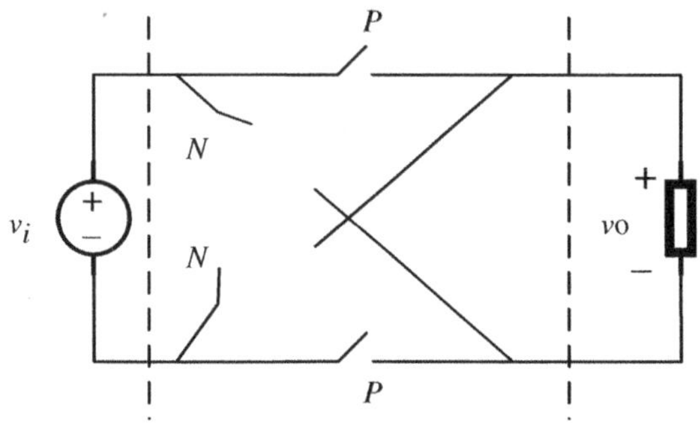

Figure 1.14 Generic DC/DC converter schematic.

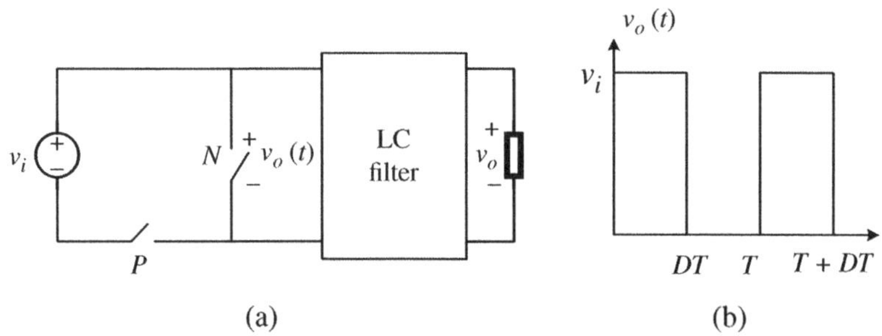

Figure 1.15 (a) Step-down isolated DC/DC converter and (b) corresponding waveforms.

The ON-time of switch P divided by the period T is known as the duty ratio or duty cycle. Figure 1.15 (b) displays the output voltage waveform. The voltage is filtered to a DC using an LC circuit since output voltage is a pulsating waveform. The output voltage (average) is computed as follows

$$V_{avg} = \frac{1}{T}\int_0^{DT} V_i dt = V_i D \tag{1.2}$$

As $D < 1$, the output voltage for this circuit is continually smaller than the input voltage. Figure 1.16 (a) displays the final circuit when every generalized converter switch is utilized. The output voltage equals the input voltage when the P-switches are turned on, and then it becomes negative input voltage when the N-switches are turned on. The waveform that results is seen in Figure 1.16 (b). Equation (1.3) provides the output voltage average or the DC component.

$$V_{avg} = \frac{1}{T}\left[\int_0^{DT} V_i dt + \int_{DT}^{T} -V_i dt\right] = V_i(2D-1) \tag{1.3}$$

With a different polarity, the output voltage is lower than the input voltage, according to Eq. (1.3). The output has a positive value for duty-cycle $D > 0.5$ and a

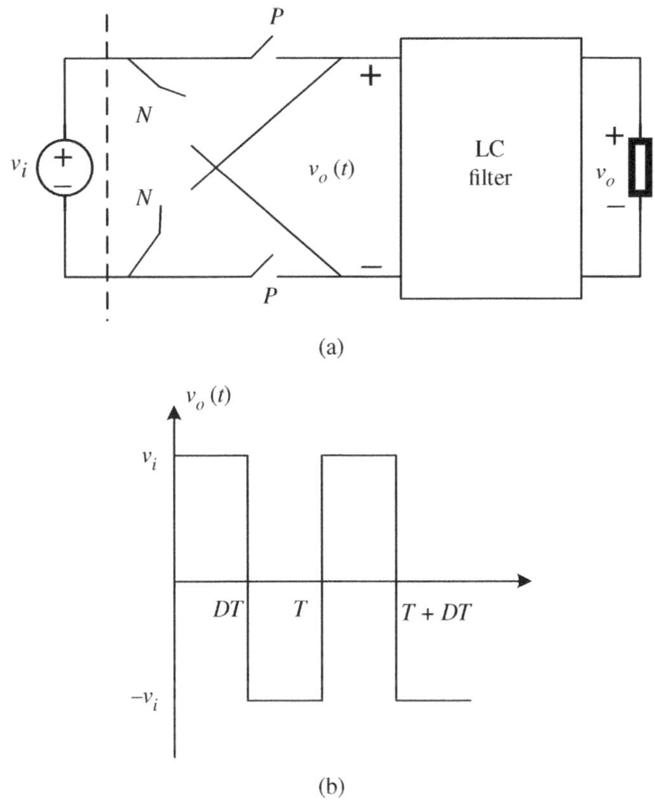

Figure 1.16 (a) Full-bridge non-isolated DC/DC converter schematic (buck) and (b) corresponding waveforms.

negative value for duty-cycle $D < 0.5$. The output voltage's harmonic contents are filtered by the LC circuit so that load voltage ripples are minimized. The output DC voltage is less than the input voltage, according to both of Eqs. (1.2) and (1.3). As a result, the converters are called buck or step-down converters. The boost and buck/boost converters are two further fundamental DC/DC converters [12]. When the DC output voltage exceeds the input voltage, the converter is said to be a boost converter. Another name for this design is a step-up converter, and Figure 1.17 illustrates a common design. The switches in this circuit are placed in between the capacitor and the inductor. Equation (1.4) provides the output-to-input DC voltage ratio if the converter is lossless. The output voltage obtained is permanently higher than the input voltage since $D < 1$.

$$V_{avg} = \frac{V_i}{D} \tag{1.4}$$

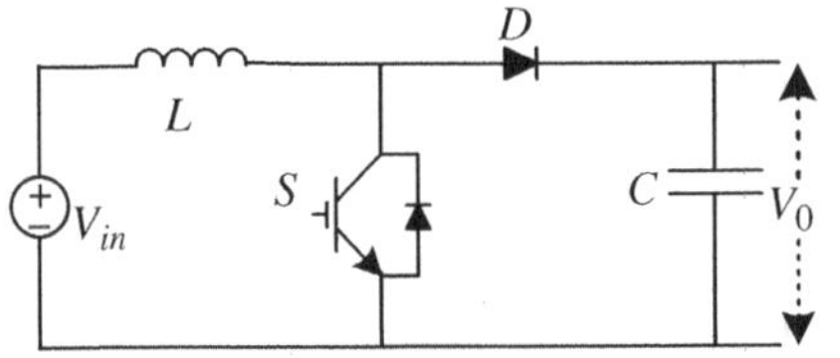

Figure 1.17 Boost DC/DC converter schematic.

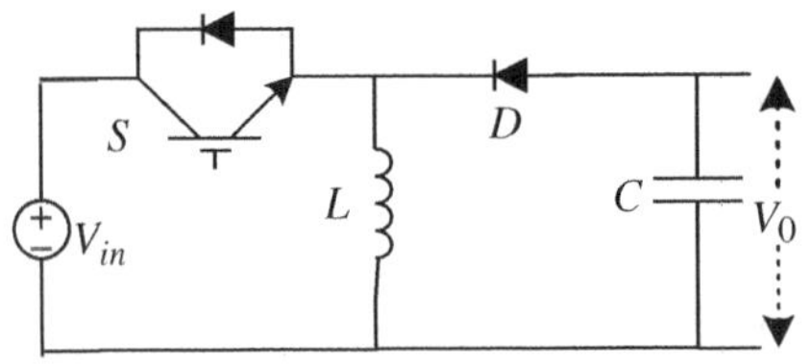

Figure 1.18 Buck/boost DC/DC converter schematic.

In Figure 1.18, a step up/down or buck/boost converter is displayed. A DC output voltage that is either greater or lower than the input DC voltage can be produced by this converter. Equation (1.5) provides the input/output conversion ratio.

$$\frac{V_o}{V_i} = \frac{D}{1 - D} \tag{1.5}$$

The converter functions as a buck converter when the duty ratio D is less than 0.5, and as boost converter when D is more than 0.5. An uncontrolled switch (diode) and a controlled switch (transistor) implement the switch for all of the fundamental converters that have been covered this far. When the transistor and diode are switched alternately, the continuous conduction mode (CCM) is the operating mode. In this mode, the duty ratio is the only factor influencing the output voltage-to-input voltage conversion ratio; the inductor current never falls to zero. If the inductor current is zero for a portion of the cycle, the mode of operation is known as the discontinuous conduction mode (DCM). When the switching frequency is lowered, the load resistance is raised, or circuit characteristics like inductance are reduced, the converter will function at DCM. As a result, the duty ratio and a number of those characteristics determine the output–input conversion ratio. Non-isolated converter topologies are the three fundamental DC/DC converters that have been covered this far [13]. Other non-isolated converters include Single-Ended Primary-Inductor Converter (SEPIC) converter, Cuk converter, and Zeta converter. Compared to non-isolated converters, isolated converters often feature more switches and filter components and use a transformer that is positioned directly between the input and output. These isolated DC/DC converters consist of Flyback converter, Forward converter,

Push-pull converter, Bridge converters, as well as advanced converters like dual active bridge (DAB), Resonant converter, etc.

1.3.3 DC/AC Converters

Essential power electronic devices known as DC/AC converters or inverters transform direct current (DC) from sources including batteries, fuel cells, solar panels, or rectified AC power into alternating current (AC). Applications such as like generating stand-alone AC outputs, powering motor drives, or incorporating RENs into the AC grid depend on this conversion. Inverters are often categorized based on their output configuration, either as single-phase or three-phase systems, and can further be distinguished as half-bridge or full-bridge converters depending on the circuit design and the desired output characteristics.

1.3.3.1 Single-Phase Inverter

Figure 1.19 (a) illustrates a general single-phase DC/AC inverter, while Figure 1.19 (b) displays the corresponding AC waveform when the P- and N-switches operate with a duty cycle (D) of 0.5, producing a square wave with no DC component, verifiable by substituting $D = 0.5$ in Eq. (1.3). A half-bridge converter, as shown in Figure 1.20 (a), uses two equal DC sources, one P-switch, and one N-switch, to generate a square wave output with period T, aligning with an AC fundamental frequency $f = 1/T$. The waveforms can be depicted in Figure 1.20 (b). By applying Fourier series to this square wave (Eq. (1.6)), the amplitude of each harmonic component is determined, and filtering the harmonics can be achieved with an LC resonant filter to ensure only the fundamental voltage appears across the load.

$$v_{on} = \frac{2}{T}\int_0^T v_o(t)\sin(n\omega_0 t)dt = \frac{4V_i}{\pi n}, \qquad n = 1, 3, 5, \ldots \tag{1.6}$$

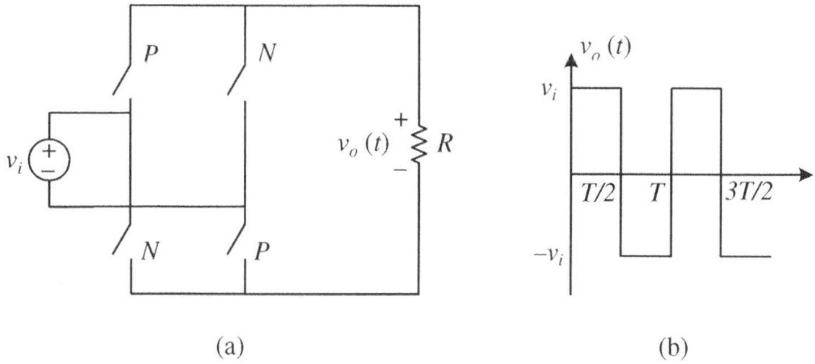

Figure 1.19 (a) DC/AC converter schematic (Full-bridge) and (b) waveforms.

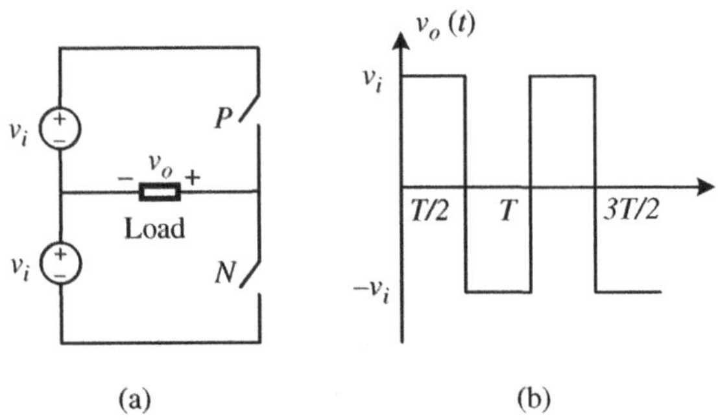

Figure 1.20 (a) DC/AC converter schematic (Half-bridge) and (b) waveforms.

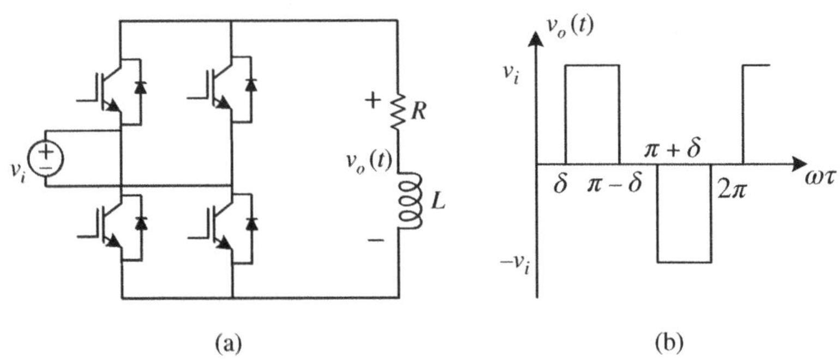

Figure 1.21 (a) DC/AC converter schematic (Full bridge) with variable AC voltage and (b) corresponding waveforms.

While half-bridge converters produce a fixed output voltage amplitude, full-bridge converters (Figure 1.21 (a)) can vary this amplitude. Additionally, a third level, or zero voltage level, can be added to create a three-level inverter, as shown in Figure 1.21 (b). By introducing a phase shift angle (δ) between the two transistor legs, the nth harmonic amplitude of the output voltage can be modulated, as derived through the Fourier series of v_o.

$$v_{on} = \frac{4}{T}\int_0^T v_0(t)\sin(n\omega_0 t)d(\omega t) = \frac{4V_i \cos n\delta}{\pi n}, \quad n = 1, 3, 5, \ldots \tag{1.7}$$

Several control strategies for harmonic cancellation and elimination have been put forth in the literature. It is important to discuss the basic modulation strategy known as the pulse-width modulation (PWM) technique [14]. By applying a zero voltage to the output voltage, the δ-control approach can lower the AC voltage's harmonic content by adding more zero-voltage segments. For instance, it can be demonstrated that both the third and fifth harmonic voltages can be removed by introducing two segments with $\delta = 30$, $\gamma = 12$, as seen in Figure 1.22 (a).

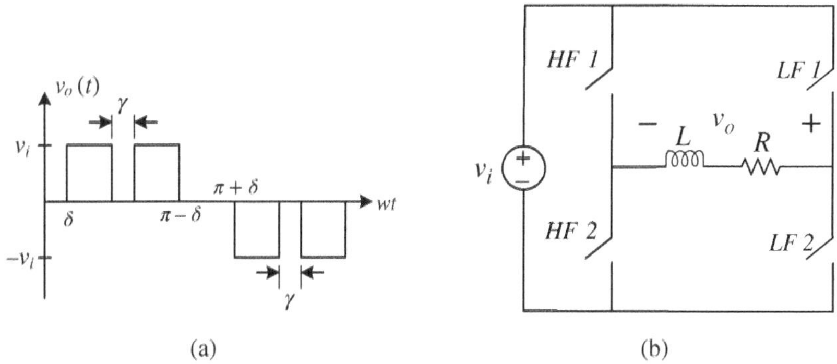

Figure 1.22 (a) Multiple harmonic elimination (PWM) and (b) PWM schematic.

Furthermore, if these zero-voltage zones can be introduced at high frequency (HF) with a variable width that is sinusoidally distributed at the fundamental frequency of the AC voltage, the resulting AC voltage will only comprise the fundamental AC component and components at the HF of the δ-control. Figure 1.22 (b) depicts a converter with switches controlled at either HF or low frequency (LF), which resembles to the full-bridge topology. The frequency at which the HF switches operate is the carrier frequency, which is much higher than the fundamental AC frequency. Whereas HF switches alternate at a set carrier frequency with various duty cycles, LF switches operate at the fundamental or modulating frequency. A sine wave signal at the modulating frequency and a triangle wave signal at the carrier frequency can be compared to determine the duty ratio. The outcome is a waveform with a consistent frequency and variable pulse-width.

1.3.3.2 Three-Phase Inverters

In power electronics, three-phase inverters are essential parts that transform DC electricity into three-phase AC power, usually for motor drive and industrial applications. They are formed through single-phase inverters (one leg branch). A common DC source is incorporated with three single-phase inverters. Voltage Source Inverters (VSIs) and Current Source Inverters (CSIs) are the two main types of inverters that allow for effective and regulated power conversion. VSIs are perfect for applications requiring precise voltage control, such motor drives and renewable energy systems, because they use a steady DC voltage input to produce a stable AC voltage output [15]. In contrast, CSIs provide a controlled AC current output from a continuous DC current input, which is useful for applications like inductive loads that demand current regulation. Because they offer reliable as well as flexible AC power regulation, both VSI and CSI kinds are crucial to existing energy systems.

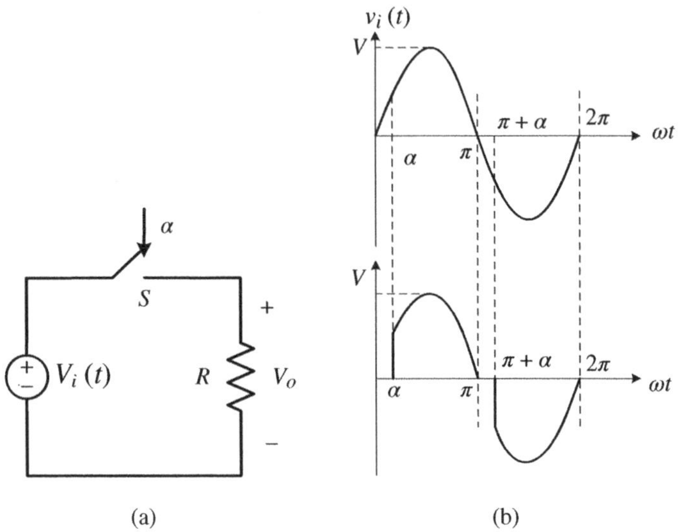

Figure 1.23 (a) VVFF AC/AC converter and (b) corresponding waveforms.

1.3.4 AC/AC Converters

AC/AC converters facilitate the connection between different AC sources, such as converting from single-phase to three-phase or vice versa. These converters are commonly used in applications that require adjustable AC outputs, like AC motor drives, and light dimmers, by transforming a fixed-voltage, fixed-frequency (FVFF) input to a single- or three-phase output with variable voltage and frequency (VVFF or VVVF). An example of a simple AC/AC converter that adjusts AC power for resistive loads is depicted in Figure 1.23 (a), with its corresponding waveform in Figure 1.23 (b). Known as an AC controller, this circuit varies the output voltage by adjusting the switch's turn-on delay angle (α), making it ideal for use in single-phase motor controls and lighting dimmers in household appliances. The input and output maintain the same fundamental frequency, but the output voltage's amplitude is regulated to meet specific application needs.

1.4 Case Study

Here are a few numerical examples to illustrate the fundamental concepts in power electronics, including AC–DC conversion, DC–AC conversion, DC–DC conversion, and thermal management.

1.4.1 AC/DC Converter (Rectifier)

Consider a single-phase full-wave rectifier, the AC input to which is 220 V (RMS), and the load resistance is 20 Ω. Further, calculation steps are as follows:

$$\text{Peak AC input voltage} = \sqrt{2} \times V_{rms} = 1.414 \times 220 = 311.08\ \text{V}$$

$$\text{Average DC output voltage} = \frac{2 \times V_{peak}}{\pi} = \frac{2 \times 311.08}{3.14} = 198.14\ \text{V}$$

$$\text{Output current} = \frac{V_{avg}}{R} = \frac{198.14}{20} = 9.9\ \text{A}$$

1.4.2 DC/DC Converter (Buck Converter)

A buck converter steps down a 100 V input DC to 24 V output DC with a load of 15 Ω. Assuming 95% efficiency, calculation steps are as follows

$$\text{Duty cycle } (D) = \frac{V_o}{V_{in}} = \frac{24}{100} = 0.24$$

$$\text{Output current} = \frac{V_o}{R} = \frac{24}{15} = 1.6\ \text{A}$$

$$\text{Output power} = V_o \times I_o = 24 \times 1.6 = 38.4\ \text{W}$$

$$\text{Input power} = \frac{\textit{Output Power}}{\eta} = \frac{38.4}{0.95} = 42.6\ \text{W}$$

$$\text{Input current} = \frac{\textit{Input Power}}{V_{in}} = \frac{42.6}{100} = 0.426\ \text{A}$$

1.4.3 DC/AC Inverter

A solar panel generates 36 V DC, which needs to be converted to 220 V AC RMS for household use. The inverter supplies a load demand of 200 W. The inverter uses PWM with an efficiency of 96%. Calculation steps can be as follows

$$\text{Peak AC voltage} = \sqrt{2} \times V_{rms} = 1.414 \times 220 = 311.08\ \text{V}$$

$$\text{Input power} = \frac{\textit{Output Power}}{\eta} = \frac{200}{0.96} = 208.34\ \text{W}$$

$$\text{Input current} = \frac{\textit{Input Power}}{V_{in}} = \frac{208.34}{36} = 5.78\ \text{A}$$

1.5 Challenges in Power Electronics

Modernizing the power grid presents significant challenges for ensuring the efficient and reliable operation of today's complex electric networks. The discussed challenges, detailed below, include issues brought about by renewable energy integration and the enhanced use of power electronics in the grid.

1.5.1 Renewable Energy Integration Challenges

The rapid growth of solar and wind energy has increased the penetration of renewable sources in power systems worldwide. While renewables offer the benefit of reducing the carbon footprint, their integration introduces technical challenges, particularly when relying on high levels of renewable generation [16]. Two critical challenges include climate-dependent variability and low inertia. Variability in renewable power sources can lead to an imbalance between generation and load demands, as these energy sources fluctuate based on weather and climate conditions. Addressing this imbalance requires robust system design that incorporates renewable generation, energy storage units, protection measures, and control systems to ensure system adequacy. This is achieved through the deployment of storage solutions, interconnections with neighboring power grids, hydroelectric backup, and operational de-rating of renewable power systems.

- **Large-Scale Energy Storage**: Implementing energy storage systems is a reliable approach to handle high levels of renewable generation. However, significant costs, the large physical size, and strategic placement of storage systems with varying technologies pose challenges. Ensuring that these systems are properly designed and sited is crucial for maintaining system reliability and resilience under conditions that could otherwise result in grid partitioning during uncertain events.
- **Cross-Border HVDC Interconnections**: Connecting with power systems in neighboring countries using HVDC systems is an effective solution to support high levels of renewable integration, as HVDC allows enhanced grid services. Smart HVDC control systems can contribute to power system adequacy in renewable-dominant networks. However, such cross-border support may introduce market-related challenges, particularly at high levels of renewables integration. HVDC links can also accommodate different types of renewable generation to address load-generation imbalances.

To maintain frequency stability, renewable power plants require frequency support mechanisms to manage their limited inertia. For example, synchronous generators offer inherent frequency support via stored rotational energy, helping balance demand and generation. In contrast, power-electronics-based renewable systems often lack this inertia, which increases the rate of change of frequency (ROCOF) and can trigger protective devices [17]. Without proper frequency support, cascading outages could occur, particularly in power systems dominated by power electronics. Introducing frequency-supportive capabilities, such as droop control within renewable converter control systems, allows these units to mimic synchronous generator behavior. However, this requires renewable systems to operate in a de-rated mode to manage frequency deviations, a process that can balance system reliability with market economics.

1.5.2 Challenges Arising from Power Electronics Integration

Power electronics play an increasingly essential role in modern energy systems, especially in microgrid applications and renewable energy integration [18]. Although they enhance controllability, flexibility, and operational efficiency, power electronics also introduce technical issues across various aspects of the system. A primary challenge of high power electronics integration is stability, due to interactions between converters and other components, such as passive elements, mechanical systems, and additional converters.

Power electronic converters incorporate several control mechanisms, including inner current control, grid synchronization, AC/DC voltage regulation, active/reactive power control, and frequency regulation. Each control system operates across a wide range of frequencies, from a few Hz to several kHz, and interacts dynamically with other power system components and control mechanisms. These interactions, if not properly managed, may cause issues such as sub-synchronous and super-synchronous oscillations, which can result in equipment fatigue over time. For instance, interactions between series capacitive compensation and synchronous generator torsional modes can cause sub-synchronous resonance (SSR), and similar oscillations can occur between power electronic converters and series-compensated systems.

The reliability of converter components is another significant concern, as these devices are prone to aging and failure, particularly under variable operating and environmental conditions [19]. Capacitors and semiconductors, which are critical yet vulnerable components within converters, face accelerated wear and failure due to power loading, vibration, temperature, and humidity. The reliability of these converters is thus time-dependent and affected by climate and operational conditions, making mission profile analysis necessary for accurate reliability assessments [20]. With mission profile data, the lifespan of converter components can be predicted using specific lifetime models that account for failure mechanisms. Monte Carlo simulations or theoretical analyses can calculate the probability of device failure, although these approaches are computationally demanding due to the detailed electro-thermal models required

[21]. For broad power system applications, converter reliability assessments traditionally rely on historical failure rates, which may lead to inaccuracies in reliability predictions.

Accurate prediction of power system reliability requires integrating condition-based and time-varying failure rates for converters. For system-wide assessments, including these detailed reliability models is essential for making sound planning and operational decisions. Considering the limited lifespan of converters during facility planning can optimize replacement strategies and maintenance practices. Similarly, integrating these reliability considerations in operational planning allows for condition-based decisions that affect power system reliability and performance.

1.6 Future Trends in Power Electronics

1.6.1 Wide-Bandgap (WBG) Semiconductors

WBG devices like Silicon Carbide (SiC) and Gallium Nitride (GaN) are set to transform power electronics, driven by their superior electrical properties over traditional silicon-based devices. These materials enable devices to operate at higher voltages, frequencies, and temperatures with greater efficiency and smaller form factors.

- **Silicon Carbide (SiC)**: SiC semiconductors are excellent for high-power applications such as industrial motor drives, converters in renewable energy based systems, and electric vehicle applications because of their high breakdown voltage and exceptional thermal conductivity. Because SiC devices have reduced switching losses, systems may be smaller and use less energy. SiC devices may also function in harsh conditions due to this feature, which eliminates the need for sophisticated cooling systems.
- **Gallium Nitride (GaN)**: GaN devices are known for their high electron mobility, which translates to faster switching speeds and lower conduction losses [22]. GaN is well suited for low-to-medium power applications, such as power supplies, adapters, and consumer electronics. GaN's ability to function effectively at higher switching frequencies allows designers to reduce the size of inductors and capacitors, creating compact, lightweight systems ideal for modern, and advanced power electronics applications. Figure 1.24 illustrates the material properties of GaN, as well as the advantageous outcomes from those properties.

The integration of WBG semiconductors is expected to improve overall energy efficiency, reduce system size and weight, and open new possibilities for compact, high-performance power systems. As manufacturing processes advance

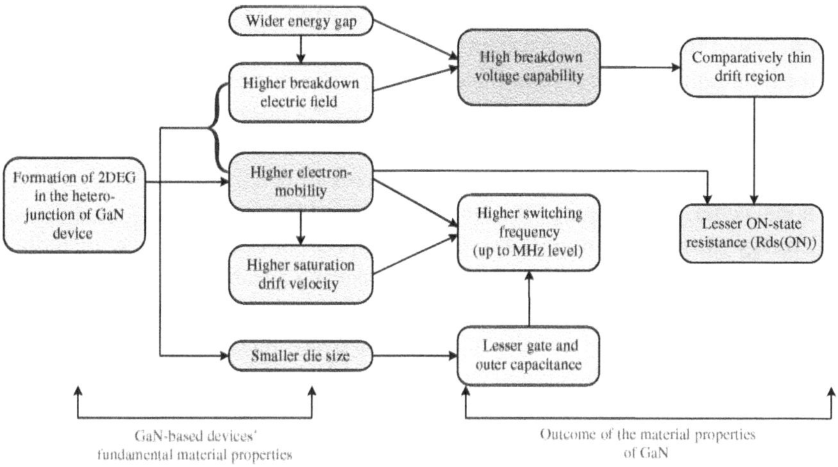

Figure 1.24 Material properties of GaN based devices.

and costs decrease, WBG materials will become more accessible for a wider range of applications, transforming sectors from consumer electronics to heavy industry.

1.6.2 Trends in Materials and Packaging for Better Performance

The transition to high-performance power electronics places a strong emphasis on the application of cutting-edge materials and innovative packaging to improve power density, reliability, and thermal management under difficult conditions. Heat dissipation is improved by methods like double-sided cooling, ceramic substrates, and 3D packaging, but thermal resistance is decreased by high-conductivity materials like diamond composites and aluminum nitride. Devices can function at greater temperatures owing to improved materials, which expands their application in renewable energy, automotive, and aerospace. These advancements promise more resilient, efficient, and durable power systems suitable for high-stress applications.

1.6.3 Role of Power Electronics in Sustainable Energy Systems

Since power electronics makes it possible to convert and regulate electrical power from a variety of renewable sources efficiently, it is essential to the development of sustainable energy systems [23]. Power electronics will be crucial in boosting grid stability, increasing energy efficiency, and supporting energy storage alternatives as the worldwide transition to renewable energy sources accelerates.

- **Grid Integration of Renewables**: RENs like wind and solar can be seamlessly integrated into the power system with the help of power electronics. Despite the fluctuating nature of renewable energy, power electronics contribute to grid reliability and stability through innovative inverter designs and grid-interactive converters. Droop control and grid-forming inverters are two examples of technologies that enable renewable-based systems to behave similarly to conventional generators, helping to regulate voltage and frequency [24].
- **Energy Storage Management**: The integration of energy storage, including batteries and supercapacitors, is vital for ensuring a stable energy supply from intermittent renewables. Power electronics make it possible to regulate storage systems' charging and discharging effectively, increasing battery life and energy efficiency. Bidirectional converters, which offer peak shaving, load balancing, and frequency regulation, are being used more and more to enable flexible energy flows between storage, renewable energy sources, and the grid.
- **Electrification of Transportation**: From electric vehicles (EVs) to electric trains and airplanes, power electronics are indispensable to the electrification of transportation networks. Converters are essential parts of EV drive systems because they allow batteries and motors to handle power effectively. Bidirectional onboard chargers also facilitate vehicle-to-grid (V2G) technology, which enables EVs to function as mobile energy storage devices that can support the grid or supply backup power during periods of high demand.
- **Smart Grids and Decentralized Energy Systems**: The advancement in smart grid technology, which optimizes energy distribution through real-time monitoring and control, depends heavily on power electronics. Advanced power converters and controllers are essential for managing energy flows in distributed generating systems, microgrids, and virtual power plants in order to ensure resilience, flexibility, and reliability. Additionally, prosumers can now actively participate in decentralized energy markets owing to power electronics, which simplify energy trading.

Power electronics will be essential for developing robust, adaptable, and efficient energy systems as the energy landscape moves toward renewable and decentralized models. Intelligent, sustainable energy networks will be made possible by the continuous developments in power electronics technology, which will also help integrate renewable energy sources and encourage the expansion of electric vehicles.

1.7 Conclusion

In conclusion, power electronics serves as the foundation for modern electrical systems by providing effective means of converting power through passive components and semiconductors. We may grasp the revolutionary influence of power electronics across a variety of applications by comprehending the basic concepts of AC–DC, DC–AC, AC–AC, and DC–DC conversions as well as the crucial significance of switching operations. Despite these developments, power electronic converters still have issues that require constant attention, namely with regard to operating efficiency and dependability. Wide-bandgap materials like SiC and GaN are expected to improve performance in the future, pushing power electronics' limits in the direction of higher efficiency, reliability, and adaptability. In the upcoming years, these developments will be crucial for meeting the energy demands of high-performance and sustainable systems.

Exercises

1.1 What are the primary functions of power electronics in the field of electrical engineering?

1.2 In what ways do power electronics contribute to renewable energy systems like solar PV and wind?

1.3 Elaborate the key components of power electronics systems.

1.4 What challenges arise from integrating power electronics with renewable energy sources?

1.5 Explain the concept of an "ideal switch" and its significance in power electronics.

1.6 What differentiates uncontrolled switches from semi-controlled switches in power electronics?

1.7 How does a thyristor operate differently from a diode, and what makes it a semi-controlled switch?

1.8 Why is a MOSFET considered a voltage-controlled device in power electronics?

1.9 How does an IGBT combine the properties of BJT and MOSFET?

1.10 What makes a Gate Turn-off (GTO) thyristor different from a standard thyristor?

1.11 What are the functions of AC/DC converters (rectifiers) function in power electronics?

1.12 What are the main types of DC/DC converters, and what is their purpose in power electronic systems?

1.13 How do DC/AC inverters facilitate the use of renewable energy in AC grids?

1.14 How do advanced materials and packaging trends impact the performance of converters in harsh environments?

References

1 M. H. Rahid, *Power Electronics (Academic Press Series in Engineering)*. Academic Press, 2001.

2 A. Elrayyah, Y. Sozer, and M. Elbuluk, "Microgrid-connected PV-based sources: A novel autonomous control method for maintaining maximum power," *IEEE Ind. Appl. Mag.*, vol. 21, no. 2, pp. 19–29, Mar. 2015. doi: 10.1109/MIAS.2014.2345822.

3 O. S. Chaudhary, M. Denaï, S. S. Refaat, and G. Pissanidis, "Technology and applications of wide bandgap semiconductor materials: Current state and future trends," *Energies*, vol. 16, no. 18, p. 6689, Sep. 2023. doi: 10.3390/en16186689.

4 A. Haque and A. Malik, *Machine Learning in Renewable Energy Systems for Smart Cities*. CRC Press, 2024, p. 28. [Online]. Available: https://www.taylorfrancis.com/chapters/edit/10.1201/9781032669809-13/machine-learning-renewable-energy-systems-smart-cities-ahteshamul-haque-azra-malik.

5 IRENA, *Renewable Energy Statistics 2024*. IRENA, 2024.

6 M. Islam, S. Mekhilef, and M. Hasan, "Single phase transformerless inverter topologies for grid-tied photovoltaic system: A review," *Renew. Sust. Energ. Rev.*, vol. 45, pp. 69–86, 2015. doi: 10.1016/j.rser.2015.01.009.

7 S. Mateen, A. Haque, F. S. Zehgeer, and F. I. Bakhsh, "Current stress minimization in dual active bridge converter for ultra-fast charging of electric vehicles," in *2023 11th National Power Electronics Conference (NPEC)*, Dec. 2023, pp. 1–6. doi: 10.1109/NPEC57805.2023.10385015.

8 A. Malik, A. Haque, and V. S. B. Kurukuru, "IoT-based monitoring and management for photovoltaic system," in *Fault Analysis and its Impact on Grid-Connected Photovoltaic Systems Performance*. Wiley, 2022, pp. 291–318.

9 H. Meshael, A. Elkhateb, and R. Best, "Topologies and design characteristics of isolated high step-up DC–DC converters for photovoltaic systems," *Electronics*, vol. 12, no. 18, p. 3913, 2023. doi: 10.3390/electronics12183913.

10 E. R. C. S. da Silva and M. E. Elbuluk, *Fundamentals of Power Electronics*, vol. 59. Springer, 2013.

11 J. G. Hwang, P. W. Lehn, and M. Winkelnkemper, "A generalized class of stationary frame-current controllers for grid-connected ACDC converters," *IEEE Trans. Power Deliv.*, vol. 25, no. 4, pp. 2742–2751, 2010. doi: 10.1109/TPWRD.2010.2045136.

12 M. Z. Hossain and N. A. Rahim, "Recent progress and development on power DC-DC converter topology, control, design and applications: A review," *Renew. Sust. Energ. Rev.*, vol. 81, pp. 205–230, Jan. 2018. doi: 10.1016/j.rser.2017.07.017.

13 T. Sutikno, R. A. Aprilianto, and H. S. Purnama, "Application of non-isolated bidirectional DC-DC converters for renewable and sustainable energy systems: A review," *Clean Energy*, vol. 7, no. 2, pp. 293–311, 2023. doi: 10.1093/ce/zkac070.

14 F. Blaabjerg, Y. Yang, K. Ma, and X. Wang, "Power electronics – the key technology for renewable energy system integration," in *2015 International Conference on Renewable Energy Research and Applications (ICRERA)*, Nov. 2015, pp. 1618–1626. doi: 10.1109/ICRERA.2015.7418680.

15 A. Malik, A. Haque, V. S. B. Kurukuru, M. A. Khan, and F. Blaabjerg, "Overview of fault detection approaches for grid connected photovoltaic inverters," *e-Prime Adv. Electr. Eng. Electron. Energy*, vol. 2, p. 100035, Jan. 2022. doi: 10.1016/J.PRIME.2022.100035.

16 M. Bahrami and Z. Khashroum, "Review of machine learning techniques for power electronics control and optimization," *Comput. Res. Prog. Appl. Sci. Eng.*, vol. 9, no. 3, pp. 1–8, 2023. doi: 10.61186/crpase.9.3.2860.

17 K. Beyer, R. Beckmann, S. Geißendörfer, K. von Maydell, and C. Agert, "Adaptive online-learning volt-var control for smart inverters using deep reinforcement learning," *Energies*, vol. 14, no. 7, 2021. doi: 10.3390/en14071991.

18 E. A. Jones, F. F. Wang, and D. Costinett, "Review of commercial GaN power devices and GaN-based converter design challenges," *IEEE J. Emerg. Sel. Top. Power Electron.*, vol. 4, no. 3, pp. 707–719, Sep. 2016. doi: 10.1109/JESTPE.2016.2582685.

19 A. Sangwongwanich and F. Blaabjerg, "Monte Carlo simulation with incremental damage for reliability assessment of power electronics," *IEEE Trans. Power Electron.*, vol. 36, no. 7, pp. 7366–7371, 2021. doi: 10.1109/TPEL.2020.3044438.

20 A. Schiffmacher *et al.*, "New lifetime model for advanced power semiconductor interconnects," in *Proceedings of the Electronic Components and Technology Conference, May*, 2022, pp. 473–477. doi: 10.1109/ECTC51906.2022.00081.

21 K. Hu, Z. Liu, Y. Yang, F. Iannuzzo, and F. Blaabjerg, "Ensuring a reliable operation of two-level IGBT-based power converters: A review of monitoring and fault-tolerant approaches," *IEEE Access*, vol. 8, pp. 89988–90022, 2020. doi: 10.1109/ACCESS.2020.2994368.

22 S. Folmer, M. Kosakowski, and R. Stala, "48-V input DC-DC high step-down converter in GaN-based design," *IEEE Access*, vol. 10, pp. 115958–115973, 2022. doi: 10.1109/ACCESS.2022.3218309.

23 F. Obeidat, "A comprehensive review of future photovoltaic systems," *Sol. Energy*, vol. 163, pp. 545–551, Mar. 2018. doi: 10.1016/j.solener.2018.01.050.

24 Y. Jiang, Y. Yang, S.-C. Tan, and S.-Y. R. Hui, "Distribution power loss mitigation of parallel-connected distributed energy resources in low-voltage DC microgrids using a Lagrange multiplier-based adaptive droop control," *IEEE Trans. Power Electron.*, vol. 36, no. 8, pp. 9105–9118, Aug. 2021. doi: 10.1109/TPEL.2021.3050506.

2

Introduction of AI and Utility for Power Electronics Applications

Suwaiba Mateen and Ahteshamul Haque

Advance Power Electronics and Research Laboratory, Jamia Millia Islamia (A Central University), New Delhi, India

2.1 Introduction

In the constantly changing field of power electronics, Artificial Intelligence (AI) plays a vital role for enhancing the reliability and sustainability. AI refers to the creation of computer systems capable of performing tasks that usually require human intelligence. These tasks involve learning, comprehending, thinking, and problem-solving. AI is the umbrella term for a range of methods and tools that allow machines to mimic cognitive processes [1]. It has a broad scope of applications in several areas, such as power electronics, power systems, banking, and automobiles. By integrating AI, these sectors can achieve significant advancements in efficiency, accuracy, and reliability.

The formal study of AI began in the mid-20th century, with Alan Turing's introduction of the Turing Test in 1950 to determine a machine's ability to exhibit intelligent behavior. The 1956 Dartmouth Conference is often considered the birthplace of AI as an academic discipline. The 1960s and 1970s saw the development of early AI programs, but progress was hampered by limited computational power and high expectations [2]. Renewed interest in the 1980s brought about the development of expert systems which paved the way for modern AI applications. The convergence of big data, enhanced computational power, and advanced algorithms in recent years has led to significant breakthroughs in AI.

AI's applications are vast, impacting healthcare, finance, automotive, and power electronics. In power electronics, AI optimizes the design, operation, and maintenance of the systems. AI algorithms can automate the tuning of power converters, improve efficiency, and reduce energy loss. For instance, AI-driven predictive maintenance in industrial power systems helps anticipate failures

Artificial Intelligence for Power Electronics, First Edition.
Edited by Ahteshamul Haque, Saad Mekhilef, and Azra Malik.

and schedule maintenance, thereby minimizing downtime and enhancing productivity and reliability. In grid management, AI algorithms predict power demand and optimize energy distribution, ensuring a reliable and efficient power supply. These systems help utilities make informed decisions about energy production and distribution, balancing load and preventing outages. AI significantly enhances predictive maintenance in power electronics. Machine learning models analyze operational data from power systems to detect anomalies and predict component failures before they occur. This proactive approach reduces maintenance costs and prevents unexpected equipment failures, ensuring continuous and reliable operation [3]. It plays a crucial role in integrating renewable energy sources into power grids. AI algorithms forecast solar and wind-energy production by analyzing weather data, enabling better planning and utilization of renewable resources. This helps in stabilizing the grid and maximizing the use of clean energy. Advanced control algorithms, such as those based on reinforcement learning, optimize the performance of power converters and inverters. This leads to better regulation of voltage and frequency, improving the overall stability and efficiency of power systems [4]. Smart grids use AI to monitor and control various components, such as substations, transformers, and distribution lines. AI contributes to the development of efficient power electronics systems for electric vehicles (EVs). The algorithms optimize battery management systems, extending the life of batteries and improving charging efficiency. Additionally, AI helps in the design and control of power electronics for EV drivetrains, enhancing performance and energy efficiency [5]. Furthermore, AI facilitates the exploration of new materials and topologies, pushing the boundaries of what is possible in power electronics. It contributes to sustainability in power electronics by optimizing energy usage and reducing waste. In renewable energy systems, AI improves the prediction and integration of variable energy sources, promoting the use of sustainable energy.

AI techniques for power electronics can be classified into four main areas: Expert Systems (ES), Fuzzy Logic (FL), metaheuristic methods, and Machine Learning (ML) [6]. These techniques have evolved significantly, offering powerful tools for optimizing power electronics. ES uses rules and a knowledge base to emulate human decision-making, providing precise computing solutions for diagnostic and control applications. Developed by Lotfy Zadeh, FL handles approximate reasoning, making it ideal for control systems in varying and unpredictable environments [7]. Genetic Algorithms (GA) optimize solutions through mechanisms like mutation and selection, useful for design parameter optimization and complex scheduling problems. ML involves algorithms that improve automatically through experience. ML has revolutionized power electronics by enabling systems to learn, adapt, and enhance performance over time.

This chapter deals with the integration of AI techniques in power electronics, highlighting their transformative impact on the field. The primary aim of this chapter is to understand the pivotal role of AI in advancing power electronics. It aims to give readers a thorough knowledge of AI techniques and their applications in the field of power electronics. By the end of the chapter, readers should be able to:

- Understand the fundamental concepts and classifications of AI relevant to power electronics.
- Identify the benefits and different techniques utilized in this field.
- Understand the applications of AI with respect to design, control, and maintenance in power electronic systems.
- Recognize the practical applications and real-world examples of AI in power electronics.

The chapter has a total of seven sections. Section 2.2 discusses the intersection of AI and power electronics. Different AI techniques in power electronics such as expert systems, fuzzy logic, metaheuristic methods, and machine learning are discussed in Section 2.3. Section 2.4 analyzes the applications of AI with respect to design, control, and maintenance in AI, and its real-time examples and implementation are shown in Section 2.5. Challenges and limitations for AI in power electronics are discussed in Section 2.6, and chapter is concluded in Section 2.7.

2.2 Intersection of Artificial Intelligence and Power Electronics

AI is a transformative technology. It has the power to bring more intelligence, adaptability, and efficiency in Power Electronics systems. These AI techniques can create better decision-making, improve performance, and offer predictive understanding that are considered necessary for the currently used power electronics. Using AI technologies, power electronic systems can become more efficient and reliable while increasing their adaptation to the flexible operation of modern energy applications as shown in Figure 2.1.

AI helps to improve decision-making in power electronics by learning from massive data sets and using that information to both provide insights as well optimize system performance. Complex datasets can be analyzed using techniques like ML and GA, which can identify patterns and connections that conventional methods might miss. This leads to more informed and efficient decisions in design, control, and maintenance [8]. AI also helps in improving the system efficiency. For example, the AI algorithms can optimize the switching patterns in power converters, leading to reduced power loss and enhanced overall system performance. This

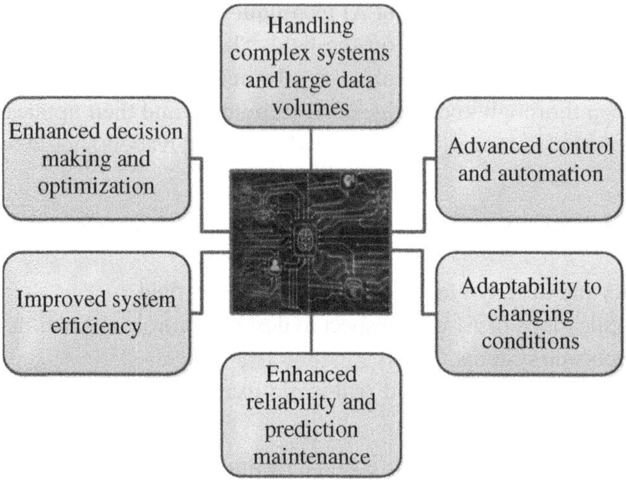

Figure 2.1 Benefits of AI in power electronics.

is particularly crucial in applications like renewable energy systems, where efficiency directly impacts energy yield and sustainability. Some AI techniques such as artificial neural network (ANN) and ES can analyze operational data to predict potential failures and diagnose issues before they lead to significant downtime. This approach reduces maintenance costs and extends the lifespan of power electronic components [9].

AI provides adaptive control mechanisms that respond dynamically to changing operating conditions. Fuzzy Logic (FL) and Reinforcement Learning (RL) systems can adjust control parameters in real-time, ensuring optimal performance even under varying loads and environmental conditions. This adaptability is critical in applications like EVs and smart grids, where operational conditions can fluctuate widely [10]. AI enhances control strategies in power electronics, making systems more autonomous and reducing the need for human intervention. For example, AI-driven control systems in smart grids can balance loads, integrate renewable energy sources, and manage energy storage systems more efficiently than traditional control methods.

Power electronic systems are becoming increasingly complex, with numerous interconnected components and vast amounts of data generated. AI is well suited to handle this complexity, providing tools to manage and analyze large datasets efficiently. This capability is essential for modern power systems, which require sophisticated monitoring and control to maintain stability and performance [11]. AI's capacity to process vast volumes of data efficiently makes it indispensable for everything from design to operation and maintenance, as well as for prolonging lifespan performance. Some of the advantages of intersection of AI and power electronics, discussed in [12], are:

- Achieving optimal maintenance under real-world situations can increase product lifetime by as much as 30%.
- Up to 30% less is spent on maintenance when condition-state analysis and routine monitoring are performed. As a result, downtimes can be decreased and factory output can increase by 10%.
- Digital motors benefit from AI and data analytics by being able to identify crucial operations that might otherwise result in inefficiencies.
- By gaining access to the most recent condition states, service providers may react swiftly and remotely, minimizing the likelihood of interruptions.
- Defects and faults can be reported right away so that fixes can be implemented quickly and safety concerns and damages can be avoided. In industrial settings, this proactive approach to maintenance improves operational reliability and safety.

2.3 AI Techniques in Power Electronics

The AI techniques can be broadly divided into expert systems, fuzzy logic, meta-heuristic approaches, and machine learning. According to [13], their respective usage percentages are 0.9%, 21.3%, 32.0%, and 45.8%.

2.3.1 Expert Systems (ES)

Expert systems are among the earliest AI methods effectively implemented in industrial applications. They function as databases that integrate expert knowledge into a Boolean logic framework, simulating human reasoning through IF–THEN rules. These systems can answer complex queries based on their databases, simulation data, facts, and statements. Technical details on expert systems can be found in [14], with exemplary applications discussed in [15]. An ES block diagram showing different elements and their interconnection is shown in Figure 2.2. The basic elements of ES include knowledge base, inference engine, user interface, and explanation subsystem for user education. The knowledge base, which is made up of expert knowledge backed by databases, is the central component of the ES.

Despite their early prominence, expert systems account for only 0.9% of AI applications, as analyzed by the Sankey diagram in [13]. This limited use is due to their reliance on specific system principles and rules, making them suitable only for well-defined domains with solid expert rules. Additionally, advancements in

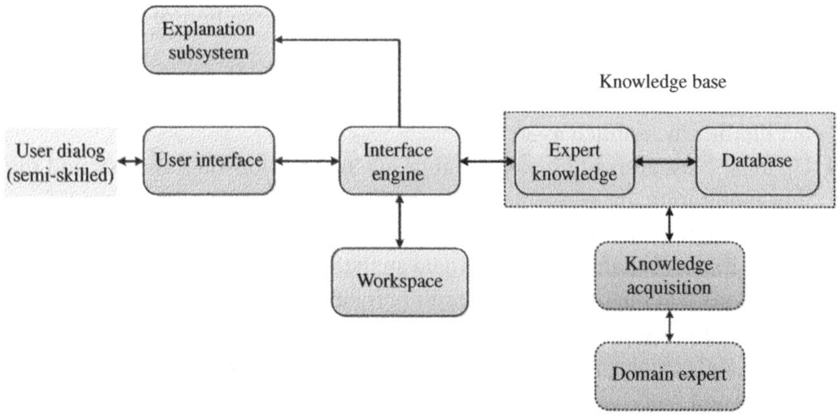

Figure 2.2 Block diagram of an expert system.

computational platforms have enabled other AI methods, such as fuzzy logic and ML, to surpass expert systems in inference and approximation capabilities.

2.3.2 Fuzzy Logic (FL)

Fuzzy logic, like ES, is a rule-based method, that expands on Boolean logic to create a multivalued framework. This makes it an ideal tool for handling system uncertainties and noisy measurements. Fuzzy logic initially performs fuzzification, converting inputs into fuzzy sets represented by membership functions that range from 0 to 1. Following the inference step's processing of these fuzzy inputs using fuzzy rules, the outcomes are de-fuzzified to produce a crisp output value. This process allows for nonlinear mapping between input and output.

Fuzzification, rule inference, a knowledge base, and defuzzification are the four main parts of a fuzzy logic system [16]. Fuzzification is the process of transforming input language variables into membership functions, which can take on various shapes and sizes, such as bell, trapezoidal, Gaussian, triangular, singleton, and custom shapes. Then, using IF–THEN rules that are stored in the knowledge base, the inference module combines these signals. After inference, defuzzification converts the fuzzy results into a crisp output.

For example, a fuzzy rule might be:

Antecedent: IF Temperature is High AND Humidity is Low,
Consequent: THEN Fan Speed is High

Both the antecedent and consequent degrees of fulfilment are determined by their membership functions. Fuzzy inference schemes can be categorized as Mamdani-type or Takagi–Sugeno–Kang-type (TSK-type). In the Mamdani-type scheme, both antecedent and consequent are shape-based functions [17]. In the TSK-type, while the antecedent remains similar to the Mamdani-type, the consequent is represented by singleton values [18]. Typically, the Mamdani-type requires more fuzzy sets than the TSK-type for the same task. The TSK-type's consequent membership functions can be linear or constant, providing more power and accuracy in non-linear approximations.

Expert experience is crucial in designing membership functions and fuzzy rules. This expertise makes fuzzy logic particularly suited to applications where expert knowledge is available. Furthermore, fuzzy logic can be combined with other AI techniques to create hybrid methods, leveraging prior information and expert experience effectively [19].

2.3.3 Metaheuristic Methods

Either deterministic programming techniques (such linear or quadratic programming) or non-deterministic programming techniques (referred to as metaheuristic techniques) can be used to find solutions when optimizing tasks for particular applications. Deterministic methods require calculating gradients and Hessian matrices, which can be challenging due to the complexity of many optimization tasks in power electronics. However, for a wide range of optimization issues, metaheuristic techniques provide a versatile, all-encompassing tool that is effective, scalable, and less reliant on specific knowledge.

Natural processes and biological evolution serve as inspiration for metaheuristic approaches. As an illustration, particle swarm optimization (PSO) is based on the social behavior of flocking birds, whereas GA mimic natural selection. These methods explore optimal solutions through trial and error and can be divided into trajectory-based and population-based methods.

2.3.3.1 Trajectory-Based Methods

Trajectory-based methods, such as tabu search and simulated annealing, involve a single candidate solution that evolves based on specific rules [20]. The effectiveness of these methods depends on the quality and efficiency of the rules applied. Although simple, these methods tend to have slower convergence speeds and are more likely to find local rather than global solutions, especially in non-convex optimization tasks.

2.3.3.2 Population-Based Methods

Population-based methods, including GA, PSO, Ant Colony Optimization (ACO), differential evolution, and immune algorithms, start with multiple candidate solutions. In each iteration, these solutions are diversified or combined to improve the overall population quality. These methods excel in convergence speed and global searching capability, making them ideal for large-scale optimization tasks [21]. However, they also require more computational resources, which can be a limitation for online applications where efficiency and speed are critical.

Table 2.1 summarizes various metaheuristic methods used in power electronics, highlighting their advantages and limitations. Most optimization tasks in power electronics are solved using population-based methods due to their numerous advantages. Besides the widely used methods, several emerging approaches are also applied on a limited scale, such as Ant Lion Optimization (ALO), colonial competitive algorithm, and teaching-learning-based optimization. Choosing the best method depends on the specific application. GA and PSO are the widely used metaheuristic methods in power electronics, representing evolutionary algorithms and intelligence algorithms, respectively. These methods serve as the foundation for many variants. It's important to note that while metaheuristic methods do not guarantee finding a global optimum, they typically provide satisfactory and solutions for most real-world applications.

2.3.4 Machine Learning

Machine learning aims to automatically identify patterns and regularities through experience, utilizing collected data or trial-and-error interactions [30]. In power electronics, machine learning applications are divided into supervised learning, unsupervised learning, and reinforcement learning (RL) [30].

2.3.4.1 Supervised Learning

Supervised learning in power electronics aims to establish functional relationships between input and output pairs in a training dataset. This method is particularly useful in scenarios where system models are challenging to formulate. Supervised learning tasks are typically categorized into classification and regression. For classification tasks, the objective is to predict discrete categories for input data. A common example in power electronics is the classification of power quality events, where the model identifies and categorizes different types of disturbances, such as voltage sags, swells, and harmonics, based on the input signal characteristics. This helps in monitoring and maintaining the quality of power supply in electrical grids. In contrast, regression tasks involve predicting continuous variables, such as forecasting the load demand on an electrical grid. Once trained, a supervised learning model should generalize well to new data not present in the training set. This generalization capability is crucial for the model's performance, as it determines how effectively the model can handle new and unseen data points.

There are several types of supervised learning methods, each with its unique mechanisms and applications. Connectionism-based methods, such as neural networks (NN), store knowledge in the weights and structure of the network.

Table 2.1 Metaheuristic Methods in Power Electronics.

Type	Algorithm	Advantages	Limitations	Applications
Trajectory-based methods	Tabu search method	• Simple implementation • Flexibility • Fast convergence speed	• Parameter sensitivity • No guarantee of global optimum • Memory requirements	Design of brushless DC (BLDC) motor [22]
	Simulated annealing method	• Simple implementation • Versatility	• Parameter sensitivity • No guarantee of optimal solution	Control of an automatic voltage regulator [23]
Population based methods	Genetic algorithm (GA)	• Global search capability • Flexible and robust • Parallel capability	• Parameter sensitivity • Slow convergence speed • Premature convergence	Design of inductor-integrated high-power medium-frequency transformer [24]
	Particle swarm optimization (PSO)	• Global convergence • Parallel capability	• Parameter sensitivity • Stagnation	Estimation of solar cell parameters [25]
	Ant colony optimization (ACO)	• Effective for combinatorial problems • Parallel capability	• Parameter sensitivity • Scalability issues	Control of a three-phase inverter by frequency-optimized random switching frequency SVPWM [26]
	Differential evolution (DE)	• Global convergence • Parallel capability • Few parameters	• Sensitivity to initial population • Poor performance on discrete problems • Computationally expensive for large problems	Control of a six-phase permanent magnet synchronous motor (PMSM) for an electric power steering system [27]
	Immune algorithm (IA)	• Adaptive learning • Versatile	• Computational complexity • Implementation complexity	Control of sinusoidal pulse width modulation (SPWM) inverter [28]
	Biogeography-based optimization (BBO)	• Versatile • Realistic problem modeling	• Parameter sensitivity • Complex implementation	Design of a robust controller for a nonlinear system [29]

These methods have seen significant improvements in robustness and dynamic performance. Additionally, using recurrent units, or convolutional structures, enhances performance on time-series data. These advancements make NNs suitable for applications like intelligent controllers [31].

Probabilistic graphical methods are another group of supervised learning. These methods use diagrams to show how variables depend on each other. This approach helps to understand the model better and handles the uncertainty effectively. An example of these methods utilized in power electronic systems is Bayesian networks. Memory-based methods, also called kernel methods, keep the training dataset and identify the key data points, such as support vectors in Support Vector Machines (SVM). These methods work well for small datasets but need more memory and take longer to compute because they must retain the training dataset during testing. Table 2.2 gives a detailed analysis of advantages and limitations and uses of different supervised learning methods in power electronics.

2.3.4.2 Unsupervised Learning

Unlike supervised learning, which uses input–output data pairs, unsupervised learning works without labeled output data. In power electronics, unsupervised learning can be divided into two types: data clustering and data compression. Data clustering aims to spot patterns and put similar data points into clusters. A typical example is to figure out health states from continuous degradation data when monitoring power electronic converters. By grouping the data, similar characteristics end up together, which helps to identify different health states. Data compression, on the other hand, tries to cut down the size of the dataset by getting rid of extra information. Unsupervised learning algorithms are often used for data pre-processing before further analysis, such as fault diagnosis. While this step is optional, it can significantly reduce computational load and enhance the accuracy of subsequent analyses. Table 2.3 provides a summary of common unsupervised learning methods used in power electronics.

2.3.4.3 Reinforcement Learning

Reinforcement learning (RL) focuses on finding an optimal action strategy that maximizes rewards for a given task, essentially addressing a dynamic programming or optimization problem and does not require a pre-existing training dataset. Through trial-and-error interactions with systems or simulation models, this technique is built, enabling RL to gradually gain expertise and acquire tactics to accomplish particular objectives. Theoretically, RL operates as a Markov decision process [49]. The training process involves developing a Q-table that defines an action selection policy aimed at maximizing total expected rewards over time. This Q-table serves as a policy matrix that records the optimal action for given

Table 2.2 Supervised Learning Methods in Power Electronics.

Type	Method	Variant	Advantages	Limitations	Applications
Neural network (NN) methods	Conventional NN	Feed forward neural network (*FFNN*)	• Simplicity • Faster to train	• No memory • Performance degrades with large datasets	Channel estimation for V2X systems [32]
		Radial basis function network (RBFN)	• Fast training • Simple architecture	• Overfitting • May not scale well to very large datasets	Fault diagnosis of electronic current transformers [33]
	NN with fuzzy logic	Adaptive neuro-fuzzy inference system (ANFIS)	• Powerful learning capability • Transparency	• Computationally Intensive • More data requirements	Sliding Mode control for a DC servo system [34]
		Fuzzy neural network (FNN)	• Fast training • Improved interpretability	• Higher computational requirements • Complex design	Non-linear system modeling [35]
	NN with recurrent unit	Recurrent neural network (RNN)	• Better transient and dynamic capability • Better sensitivity	• Slow training speed • Difficult training for long sequences	Control of variable-speed wind-energy conversion systems [36]
		Long short-term memory (LSTM)	• Effective at capturing long term dependencies • Robust to vanishing gradient problem	• Computationally more intensive • Complex architecture	Detection of multiple open-circuit switch faults in wind turbine systems [37]
		Echo state network (ESN)	• Simplified training process • Good performance	• Careful tuning of the reservoir parameters • Less flexible	Fault diagnosis of voltage-source inverters for PMSM drives [38]

(Continued)

Table 2.2 (Continued)

Type	Method	Variant	Advantages	Limitations	Applications
	NN with convolutional structure	Convolutional neural network (CNN)	• Reduced number of parameters • Easier to train	• Require large amounts of labeled data • Computationally intensive	Fault detection and identification for multilevel modular converter (MMC) [39]
		Recurrent convolutional neural network (RCNN)	• Captures spatial features and temporal dependencies • Useful for time series prediction	• Higher computational and memory demands • Complex	Prediction of remaining useful life (RUL) of rotating machinery [40]
Probabilistic graphical methods	Directed acyclic graphs	Bayesian networks (BN)	• Better interpretability • Efficient computation of marginal probabilities and conditional independence	• Requires expert knowledge • Computationally intensive	Open-circuit fault detection in DC/AC inverter [41]
	Temporal probabilistic model with directed edges	Hidden Markov model (HMM)	• Efficient algorithms • Used for time-series analysis and sequence prediction tasks	• Assumes independence between observations • Limited in capturing long range dependencies	Detection of a multi-stage network attack (MSA) [42]
Kernel method	Conventional kernel method	Gaussian Processes	• Flexible • Effective for small datasets	• Computationally complex • Slow predictions for large dataset	RUL of discrete insulated gate bipolar transistor (IGBT) devices [43]
	Sparse kernel method	Support vector machine (SVM)	• Versatile • Effective in high dimensions	• Slow training • Difficult to interpret	Fault detection in grid connected single-phase PV inverter [44]
		Relevance vector machine (RVM)	• Flexible • Sparse than SVM	• Parameter sensitivity • Complex implementation	Fault diagnosis of cascaded H-bridge multilevel inverter [45]

Table 2.3 Unsupervised Learning Methods in Power Electronics.

Type	Method	Advantages	Limitations	Applications
Clustering	*k*-means	• Fast convergence • Simple and efficient	• Sensitive to initialization • Fixed number of clusters	Detection of open-circuit fault in inverter [41]
	Self-organizing maps (SOMS)	• Useful for visualizing • Preservation of topology	• Parameter sensitivity • Complex	Identification of the degradation state of IGBTs [46]
Data compression	Principle component analysis (PCA)	• Simple and efficient • Reduces noise	• Assumes linear relationships • Sensitive to scaling	Fault prognosis of SiC MOSFETS [47]
	t-Distributed stochastic neighbor embedding (t-SNE)	• Excellent for visualizing data in 2D or 3D • Better in clustering	• Computationally complex • Non-deterministic	Fault detection and classification for a photovoltaic array [48]

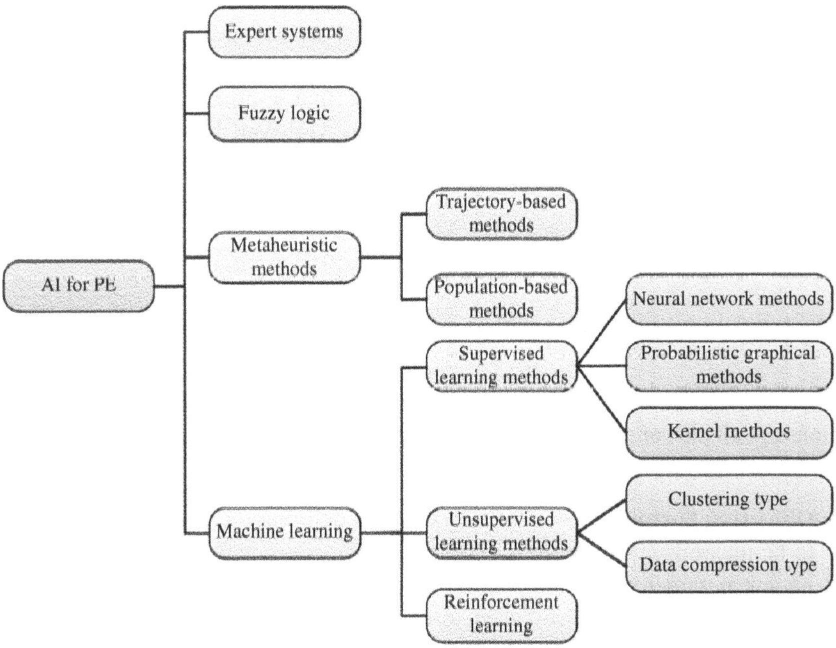

Figure 2.3 Different AI techniques in PE.

conditions. One notable application of RL in power electronics is Maximum Power Point Tracking (MPPT) [50]. RL's strength lies in its ability to learn from system interactions rather than relying on existing datasets, making it particularly advantageous for systems with limited prior knowledge or where modeling is complex.

The broader categories for different AI techniques utilized in power electronics are shown in Figure 2.3.

2.4 Applications of AI in Power Electronics

AI has a wide range of applications, significantly enhancing system performance, reliability, and efficiency. The boarder classifications in applications include design, control, and maintenance. In design, AI techniques such as GA and NN converter parameters and thermal management systems result in more efficient and compact power converters. For control, AI-based algorithms like RL and Adaptive Neuro-Fuzzy Inference System (ANFIS) improve MPPT in renewable energy systems and optimize converter control for better battery management and response times. Smart grid operations benefit from AI through enhanced load forecasting, demand response, and power quality improvement, using machine learning to predict consumption patterns and design advanced power filters. In maintenance, AI revolutionizes fault diagnosis, prognostics, and condition monitoring, enabling predictive maintenance that minimizes downtime and extends the life of power electronic devices. AI-driven battery management systems accurately estimate state-of-charge and predict battery health, enhancing the reliability and safety of electric vehicles and grid storage. Additionally, autonomous power systems utilize AI to detect and isolate faults in real-time, ensuring continuous and efficient operation.

2.4.1 Design Using AI

Topology selection, component sizing, circuit synthesis, and reliability considerations are all part of power electronics design. The typical procedure for designing includes four steps:

- **Objective Formulation**: This step involves defining the desired design goals to be maximized or minimized. Common objectives include optimizing component parameters, weight, volume, cost, heatsink patterns, area, and power loss [24]. These objectives need to be translated into explicit mathematical expressions, either as a single objective or multiple objectives. For multiple objectives, the optimization can be done by either maximizing a scalar function that weights the objectives or by directly optimizing an objective vector, using methods like the Pareto front to determine the optimal solution [5].

- **Constraint Space**: This defines the feasible space, boundaries, relationships, and limitations that the objective function must adhere to. Constraints can be linear or non-linear equalities and inequalities that are based on real-world design specifications such as volume, lifetime attributes, and cost.
- **Solution Exploration**: In this step, the optimization problem is solved by adjusting decision variables within the constraint space to maximize or minimize the objective functions. AI methods, particularly metaheuristic methods, are often employed in this step to explore potential solutions effectively.
- **Performance Evaluation**: Candidate solutions are tested against predefined objectives using simulation, hardware-in-the-loop testing, prototype experiments, etc. The results are used to refine and improve the design iteratively.

Rather than following a strict sequential strategy, the design task is often an iterative process of trial and error. Evaluations at each step may lead to reformulating the task, such as adjusting objectives, modifying constraint spaces, or reconfiguring programming methods. Traditional design approaches in power electronics can be time consuming and may require multiple iterations, relying heavily on expert experience and intuition. AI methods can mitigate these drawbacks by enhancing objective formulation and solution exploration, leading to faster convergence and improved design performance.

2.4.2 Control Using AI

Control applications using AI methods can be broadly categorized into optimization and regression tasks. Similar to the optimization processes used in the design phase, metaheuristic approaches are frequently employed in control application optimization. These methods are employed to address complex control challenges, such as tuning control parameters, minimizing power loss, and improving system stability and efficiency. Optimization tasks aim to find the best possible control strategy that maximizes or minimizes a specific objective function. This could involve adjusting control parameters to achieve optimal performance under varying operating conditions. Metaheuristic methods, such as GA, PSO, and ACO [26], are particularly effective in exploring a wide range of potential solutions and converging on an optimal control strategy.

On the other hand, regression tasks in control applications focus on predicting continuous output variables based on input data. This is essential for applications like model predictive control, where accurate predictions of future system states are crucial for making informed control decisions. AI techniques, such as NN and SVM, are commonly used for regression tasks due to their ability to learn complex input–output relationships from historical data [36]. The regression-related task is generally organized in terms of FL-, NN-, and RL-based controllers.

2.4.3 Maintenance Using AI

Because of their complex and challenging operating settings, power electronic systems have a high risk of catastrophic failures even with thorough reliability features incorporated into design and control [51]. For field applications, it is essential to guarantee the dependability and reliability of power electronic components and systems. To make sure that these systems operate as intended, preventive maintenance tasks including condition monitoring, anomaly detection, fault diagnosis, and remaining usable life (RUL) prediction are crucial. These endeavors conform to the Prognostics and Health Management (PHM) standard framework for electronic systems established by IEEE [52]. For power electronics systems the maintenance activities consist of three sub-sections [13]:

- **Offline Training and Knowledge Learning**: This phase incorporates a variety knowledge sources, such as failure mode and effects analyses, simulation data, historical monitoring data, and test results from accelerated aging. Ensemble methods and fusion techniques are often employed to enhance performances. As a result, population data is used to properly describe offline physical system dynamics and behaviors.
- **Condition Monitoring and Health Assessment**: This phase involves assessing the health of the system through online condition monitoring in real-world applications. The offline model is customized to the specific unit in service by adjusting model parameters to reflect the operational environment and load conditions. Key functions such as non-invasive parameter identification, data preprocessing (like data cleaning), feature extraction, anomaly detection, fault diagnosis, and RUL prediction are done. This continuous monitoring provides valuable insights for decision-making.
- **Management and Decision-Making**: In this phase, health assessment information supports optimal decision-making. Feedback from health assessments allows for adjustments in control policies, such as power routing, to maximize system performance based on real-time health status. Additionally, this information facilitates the development of economical maintenance policies, enabling condition-based and predictive maintenance strategies.

A detailed systematic flowchart of maintenance activities is discussed by authors in [13] and is shown in Figure 2.4.

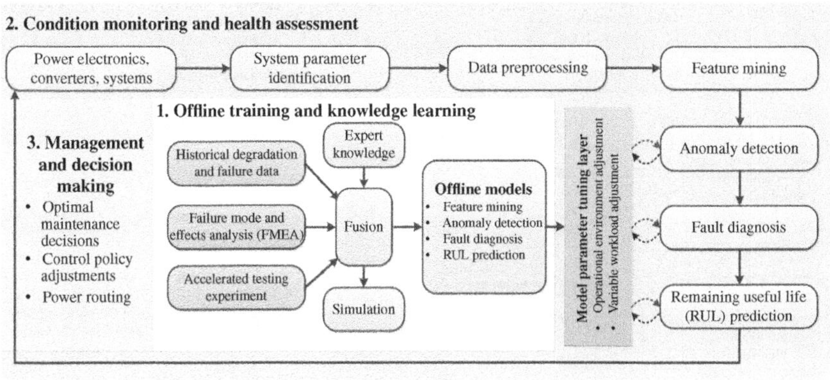

Figure 2.4 Maintenance flowchart in AI for PE.

2.5 Case Studies and Real-World Examples

Simple real-time examples of AI in designing, controlling, and maintenance of power electronic systems using different simulation platforms are discussed in this sub-section.

2.5.1 Modeling of a Non-linear System Using NN

In this example, future plant performance is predicted using a neural network model of a non-linear plant. The system is modeled with Simulink in MATLAB. The system identification process, or the determination of the NN plant model, is the initial phase. Next, the controller makes predictions about future performance based on the plant model. The block diagram of a non-linear plant coupled with an NN model is shown in Figure 2.5. The non-linear plant block contains the continuous stirred tank reactor (CSTR) plant model. The reference signal is a random signal, and the control signal is connected as the input of the plant model from the NN Predictive Controller block.

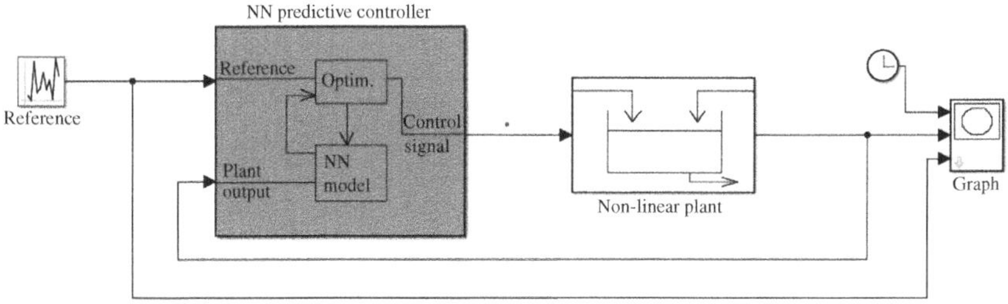

Figure 2.5 Non-linear plant combined with an NN model in MATLAB Simulink.

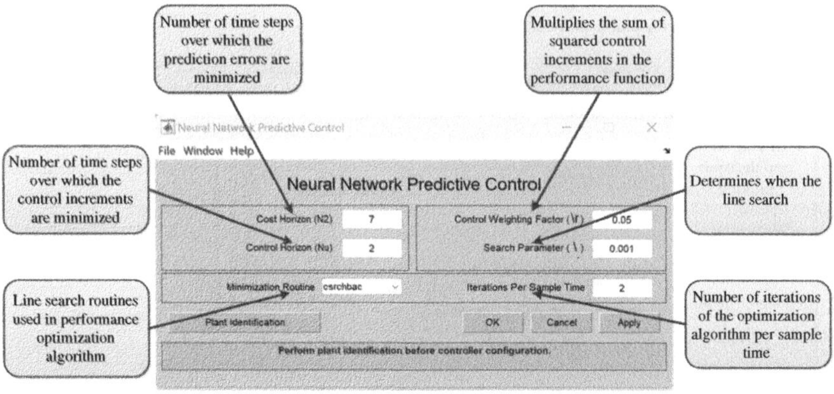

Figure 2.6 Neural network predictive control block in MATLAB Simulink.

The NN Predictive Controller block for designing is shown in Figure 2.6. Each parameter and its impact on the designing process are also discussed in the figure. Similarly, the plant identification block is shown in Figure 2.7. Before implementing the controller, it's necessary to develop the neural network plant model. This model forecasts future outputs of the plant. These predictions are then utilized by the optimization algorithm to determine the control inputs that will enhance

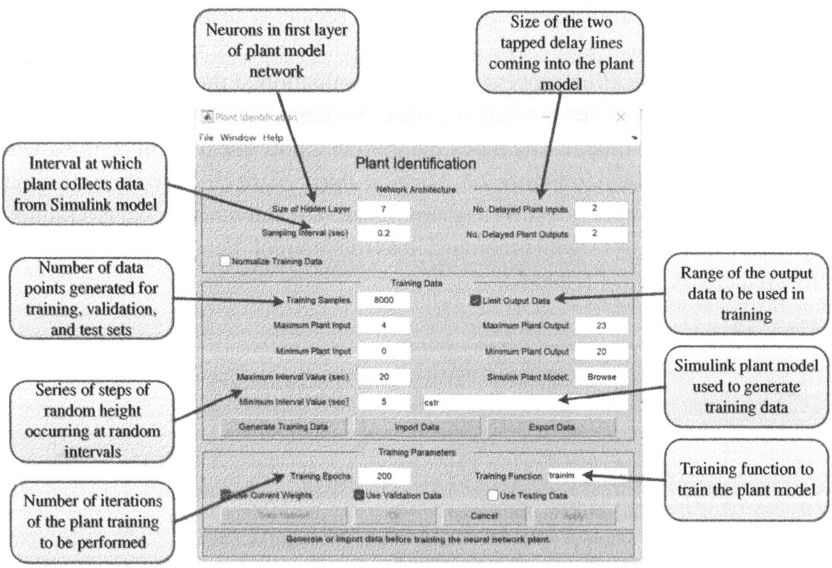

Figure 2.7 Plant identification block in MATLAB Simulink.

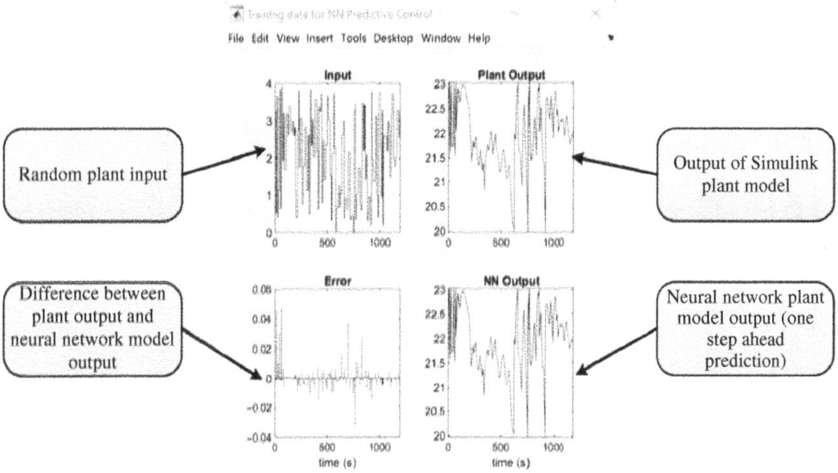

Figure 2.8 Training data for the example.

its performance. In the given example, the plant model neural network includes a single hidden layer. The size of this layer, time steps for which prediction errors and control increments are minimized, iterations per sample, and the training function can all be adjusted within this configuration window.

After generating the training data and training the network, plant model training begins. The chosen training algorithm is followed as the training progresses. Figure 2.8 illustrates how the reaction of the final plant model is exhibited following training.

After the simulation is run, the plant output and the reference signal are shown in the graph, as in Figure 2.9. light gray is the reference signal, while dark gray represents the plant output.

2.5.2 Control of a Boost Converter Using ANFIS

To simulate a closed-loop boost converter in MATLAB Simulink, model the circuit as shown in Figure 2.10. For this example, an input of 12 V is boosted to the value of 20 V with constant power of 100 W. The switching frequency is set at 25 kHz. Keeping the ripple in output voltage at 1%.

Using the "anfisedit" command in MATLAB command window, a Neuro Fuzzy Designer block opens. The data can be loaded from a file or from workspace. In the given example the data for training is uploaded from the workspace on the Neuro-Fuzzy Designer. The input and output data of a proportional integral (PI) controller is used to train the ANFIS block. The next step in the training process is to reduce the error between the network's actual and expected outputs.

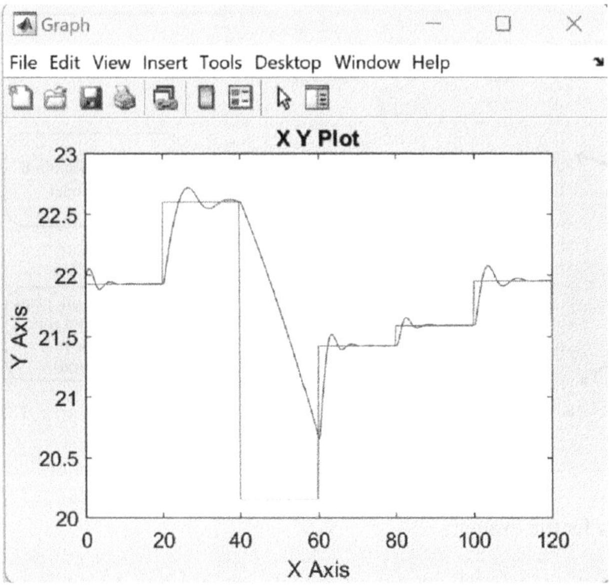

Figure 2.9 Waveform of the reference signal (light gray) and plant output (dark gray).

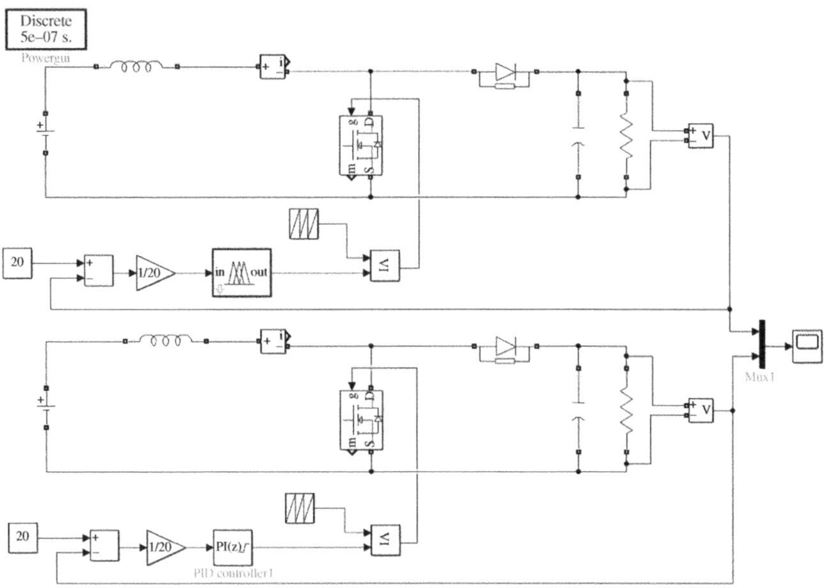

Figure 2.10 Schematic of closed-loop boost converter using ANFIS and proportional integral derivative (PID) controller in MATLAB Simulink.

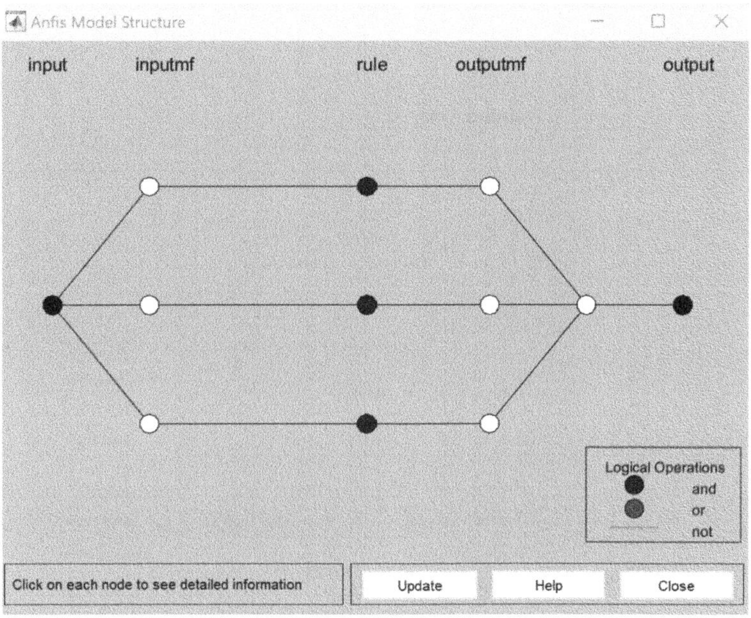

Figure 2.11 ANFIS model structure.

Usually, a hybrid training algorithm with forward pass and backward pass phase oversees this process. The node outputs are passed forward to layer 4 during the forward pass phase. In this stage, the least squares estimation algorithm is utilized to adjust the consequent parameters in layer 4. The goal of this approach is to reduce the system output's sum of squared errors. The layer one membership sets in the hybrid method are fine-tuned during the back pass phase. During this phase of training, the gradient descent technique is utilized to optimize the membership parameters, while the error signals are propagated backward from the output. Until the output error is minimum, this procedure is carried out in every training iteration. Figure 2.11 shows the ANFIS model structure obtained for the above example.

The ANFIS control in this charger is designed to have one input – the error output voltage. There are three products of triangular membership functions. The Neuro-Fuzzy plug-in in MATLAB automates the training process and tunes the ANFIS controller's performance using the hybrid learning technique. ANFIS is trained for 50 epochs and the error value is 0.018 as shown in Figure 2.12. Using this tool, it is very simple to evaluate the data after the training phase is over and import it into the Simulink model

The desired output voltage of 20 V is obtained using both a PI controller and an ANFIS control. In the Figure 2.13, light gray is the output voltage using ANFIS

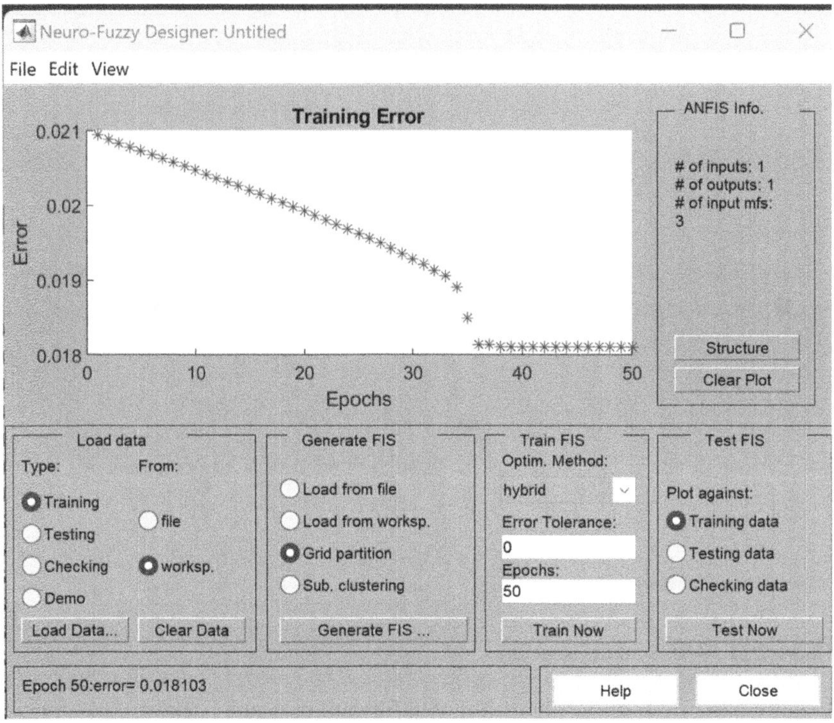

Figure 2.12 Neuro-fuzzy designer in MATLAB Simulink.

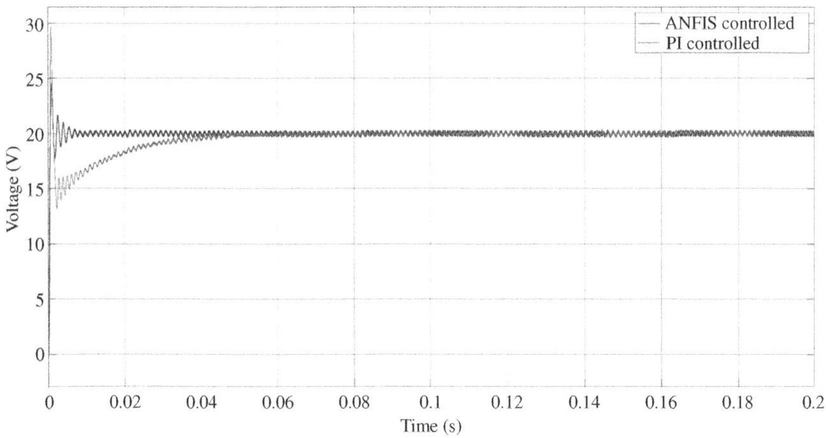

Figure 2.13 Output voltage waveform for PI-controlled (dark gray) and ANFIS-controlled (light gray) boost converter.

controller and dark gray for PI controller. It can be seen from the plot, the overshoot for PI-controlled circuit is almost 30 V whereas for ANFIS-controlled circuit it is around 24 V. Also, the AI-controlled circuit reaches the steady state early compared to the conventional PI-controlled boost converter.

2.5.3 Classification of Faults in Grid Tied PV System Using kNN

In this example a 3-ϕ grid connected photovoltaic (PV) system is simulated in Typhoon HIL. Different faults were injected in the system throughout its running, and the data points are noted. The generated data is then trained using the proposed machine classifier. The system has a power rating of 16 kW and reactive power rating of 5kVAR. The different faults injected in the system are Low-Voltage Ride Through (*LVRT*), High-Voltage Ride Through (*HVRT*), and Frequency Ride Through (*FRT*). Figure 2.14 shows the different faults and the corresponding waveforms of voltage and current.

The dataset is derived from simulated cases, where various conditions are categorized into High FRT (HFRT), HVRT, Low FRT (LFRT), LVRT, and No-fault

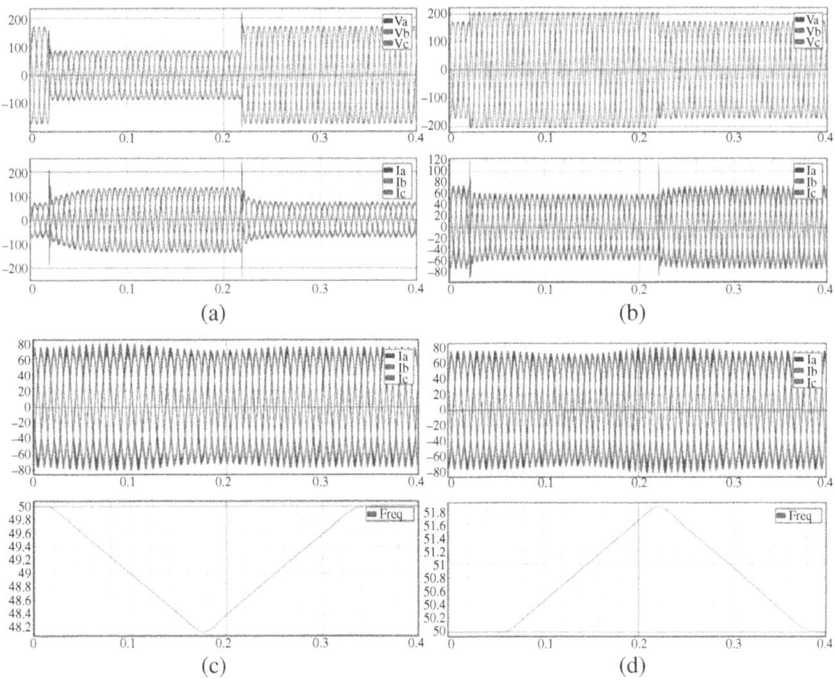

Figure 2.14 Faults injected in the 3-phase grid connected system. (a) LVRT, (b) HVRT, (c) LFRT, and (d) HFRT.

1	Time	Ia	Ib	Ic	Va	Vb	Vc	id	iq	P	Q	Idc_link	CLASS
2	0	64.10452	-43.7126	-20.3919	166.5536	-66.039	-100.515	59.57806	-19.6265	14991.9	4993.752	0	No fault
3	1.00E-05	60.59086	-40.9018	-19.689	168.4723	-66.2356	-102.236	59.57806	-19.6265	14991.9	4993.752	0	No fault
4	2.00E-05	61.71012	-40.4277	-21.2824	170.6449	-67.1607	-103.485	59.57806	-19.6265	14991.9	4993.752	0	No fault
5	3.00E-05	65.71187	-42.391	-23.3208	171.049	-67.5066	-103.542	59.57806	-19.6265	14991.9	4993.752	0	No fault
6	4.00E-05	68.3208	-44.6561	-23.6648	169.4051	-66.5909	-102.815	59.57806	-19.6265	14991.9	4993.752	63.71534	No fault
7	5.00E-05	67.12295	-45.2295	-21.8934	167.539	-64.9912	-102.547	59.57806	-19.6265	14991.9	4993.752	64.94059	No fault
8	6.00E-05	64.16396	-44.053	-20.111	167.5587	-63.9983	-103.561	59.57806	-19.6265	14991.9	4993.752	66.13616	No fault
9	7.00E-05	63.27986	-42.9051	-20.3748	169.2436	-63.9986	-105.244	59.57806	-19.6265	14991.9	4993.752	67.32772	No fault
10	8.00E-05	66.16715	-43.2813	-22.8858	170.8266	-64.2663	-106.561	59.57806	-19.6265	14991.9	4993.752	24.83149	No fault
11	9.00E-05	70.57836	-44.2631	-26.3153	170.5651	-63.5677	-106.997	59.57806	-19.6265	14991.9	4993.752	26.24136	No fault
12	1.00E-04	72.72046	-43.999	-28.7215	168.6119	-62.1951	-106.417	59.57806	-19.6265	14991.9	4993.752	0	No fault
13	0.00011	70.54805	-42.2768	-28.2712	166.4456	-61.2437	-105.201	59.57806	-19.6265	14991.9	4993.752	0	No fault
14	0.00012	65.70742	-40.5159	-25.1915	165.9234	-61.1253	-104.799	59.57806	-19.6265	14991.9	4993.752	0	No fault
15	0.00013	62.15007	-40.1019	-22.0482	167.3572	-61.4575	-105.899	59.78982	-19.9112	15000.2	4993.081	0	No fault
16	0.00014	62.48467	-41.1453	-21.3394	169.2858	-61.4745	-107.812	59.78982	-19.9112	15000.2	4993.081	0	No fault
17	0.00015	65.70297	-42.4469	-23.2561	169.876	-60.7272	-109.148	59.78982	-19.9112	15000.2	4993.081	0	No fault
18	0.00016	68.25996	-42.6348	-25.6252	168.5806	-59.476	-109.105	59.78982	-19.9112	15000.2	4993.081	0	No fault
19	0.00017	67.39085	-41.4124	-25.9784	166.5835	-58.399	-108.184	59.78982	-19.9112	15000.2	4993.081	24.41077	No fault
20	0.00018	63.77702	-39.4611	-24.3159	165.9049	-57.7244	-108.181	59.78982	-19.9112	15000.2	4993.081	25.8184	No fault
21	0.00019	61.03331	-37.6435	-23.3898	167.1177	-57.4786	-109.638	59.78982	-19.9112	15000.2	4993.081	27.2229	No fault

Figure 2.15 First 20 data points of "No fault" class.

classes. The "No fault" class represents the standard operational state of the grid-connected PV system. Each class comprises 40,001 samples, resulting in a matrix size of 40,001 × 11. Figure 2.15 shows the first 20 data points of characteristics including I_a, I_b, I_c, V_a, V_b, V_c, i_d, i_q, P, Q, and the dc link current. The points need to be labeled as shown.

Subsequently, the data is trained using a kNN classifier with a prediction speed of 1400 observations per second. The classifier considers 100 neighbors and employs the Euclidean distance metric, achieving an accuracy of 98.58% with a training duration of 110.69 seconds. The confusion matrix for the trained samples is illustrated in Figures 2.16 and 2.17.

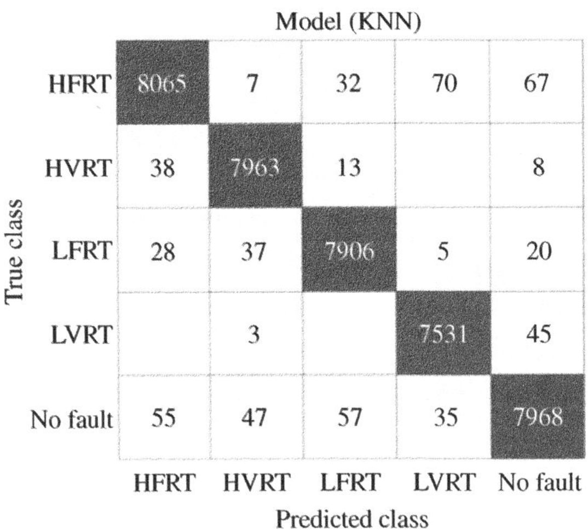

Figure 2.16 Confusion matrix for trained samples.

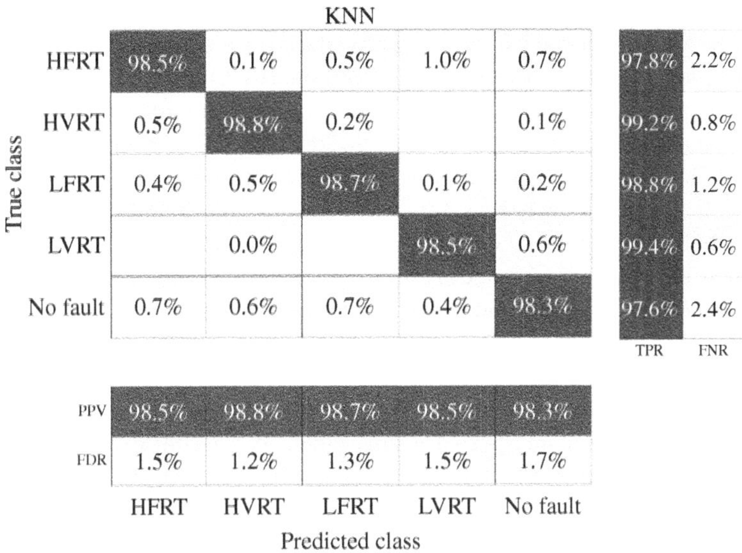

Figure 2.17 TPR, FNR, PPV, and FDR for each class.

Figure 2.15 displays the accurate and inaccurate classifications within the trained dataset. The diagonal dark gray values represent correct classifications, while the off-diagonal elements indicate misclassifications. Among all samples in the matrix, 567 are misclassified. Figure 2.16 calculates metrics, i.e. true-positive rate (*TPR*), false-negative rate (*FNR*), positive predictive values (*PPV*), and false discovery rates (*FDR*), for each class. These results show that the islanding classifiers mechanism was effectively trained with an efficiency of 98.58%.

2.6 Challenges and Limitations

Over many decades, a great deal of research has been conducted on the application of AI techniques in the field of power electronics. But the advantages of AI come with a few significant obstacles that stand in the way of using AI practically to solve problems [53]. The implementation of AI in power electronics presents several challenges and limitations that hinder its widespread adoption and effectiveness. One major issue with AI is data availability and its quality [54]. The models require large, high-quality datasets for training, but collecting relevant data from systems can be difficult due to operational variability and confidentiality issues. As a result, models trained on limited or synthetic data may not fully capture the complexities of real-world scenarios. Consider the case of predictive maintenance in power electronics, where AI models are used to predict the remaining useful life (RUL) of components like capacitors or switches. Ideally, this would require a vast amount of historical operational data, including data from both healthy and faulty components, to train the model. However, collecting this data from industrial environments is often difficult. Companies may have only a few instances of component failures, making it challenging to create a diverse dataset. Furthermore, the data that is available may not cover a wide range of operating conditions, limiting the model's ability to generalize. Similarly, in power electronics, a researcher might use software like MATLAB/Simulink to simulate the behavior of a DC–DC converter under different operating conditions. While the simulated data can be used to train an AI model, it may not capture real-world factors such as component aging, parasitic effects, or unexpected disturbances. As a result, the AI model may perform well in simulation but fail to deliver the same accuracy in a real deployment.

Another challenge lies in model generalization. Power electronics systems are often customized for specific applications, and AI models trained on one system may not perform well in others. This lack of generalization necessitates extensive retraining. For example, an AI model trained on data from an EV charging station located in a coastal region, where the climate is mild and humidity levels are high, might not generalize well to a similar charging station located in a desert region, where the climate is dry and temperatures can soar to extreme highs [55]. Differences in environmental conditions, such as the impact of high temperatures on the performance of power electronics components, may lead to reduced accuracy in the AI model's prediction.

Many AI models, particularly deep learning models, are computationally intensive and can introduce significant latency when processing data. In systems, where decisions must be made within microseconds or milliseconds, this latency can be unacceptable. For example, an AI-based control system for an active power filter (APF) that dynamically compensates for harmonics in an electrical grid. The APF needs to detect and mitigate harmonics in real-time to maintain power quality [56]. If the AI model used to identify harmonic patterns introduces latency due to its computational complexity, the system might not respond quickly enough, leading to harmonic distortion that could damage equipment. Lastly, energy efficiency poses a limitation. AI models often require more energy in comparison, which conflicts with the efficiency goals in power electronics. For example, in case of an EV, which uses an AI model to optimize switching patterns of on-board converters, if the model is complex and requires some processing power, the additional energy consumed by the control electronics might nullify some of the efficiency gains achieved by the optimization. While

AI holds great promise for enhancing power electronics through improved efficiency, fault detection, and real-time adaptability, addressing the associated challenges is essential for its successful implementation and integration into existing systems.

2.7 Conclusion

This chapter analyzes how power electronics and artificial intelligence (AI) interact. There is discussion of AI techniques such as expert systems, fuzzy logic, machine learning, and metaheuristic approaches. These techniques are essential for improving the dependability and efficiency of power electronic systems. Intelligent decision-making and adaptive control are made easier by AI approaches, and these are essential for maximizing system performance in a variety of operating environments. Power electrical system design, control, and maintenance have been completely transformed by AI particularly the machine learning. While unsupervised learning approaches are highly useful for data grouping and anomaly detection, supervised learning algorithms are superior in tasks like fault diagnosis and predictive maintenance. Through a process of trial and error, reinforcement learning maximizes the efficiency and reliability of control strategies.

AI has a wide range of applications in power electronics. It streamlines system architecture and component selection throughout the designing phase, cutting down on design time and improving performance. Regression and optimization algorithms powered by AI are useful for control applications because they guarantee optimal performance and adaptive responses to changing conditions. AI greatly increases system reliability and decreases downtime in maintenance by enabling predictive and condition-based maintenance techniques. The integration of AI and power electronics offers numerous advantages but also faces significant challenges. The chapter includes a few real examples of power electronic systems that use AI in MATLAB Simulink. These illustrations highlight the practical benefits of AI. These case studies show how AI can tackle challenging issues and provide observable gains in performance and efficiency, ranging from AI-optimized boost converters to sophisticated fault detection systems.

In summary, smarter and more dependable systems are possible by the intersection of AI and power electronics. AI technologies will find more applications in power electronics as they develop, spurring more breakthroughs and developments in this area. This chapter's observations and illustrations highlight how crucial it is to embrace AI in order to address the growing demands and challenges in power electronics.

Exercises

2.1 Elaborate on the applications of AI in the modern world.

2.2 How can AI-driven condition monitoring and predictive maintenance extend the lifespan of power electronic components?

2.3 How does AI's ability to process large datasets contribute to the stability and performance of complex power electronic systems?

2.4 What are the different AI techniques used in power electronics? Discuss each in brief with their usage percentages.

2.5 What distinguishes trajectory-based metaheuristic methods from population-based methods in terms of optimization strategies?

2.6 Give advantages and limitations of various algorithms of trajectory-based metaheuristic methods and population-based methods.

2.7 Name the types of supervised learning methods in power electronics and different methods that fall under each category.

2.8 Distinguish between connectionism-based methods and probabilistic graphical methods.

2.9 Name the types of unsupervised learning methods in power electronics and different methods that fall under each category.

2.10 Discuss the stepwise procedure of designing using AI in power electronics.

2.11 Why are AI techniques like Neural Networks (NN) and Support Vector Machines (SVM) effective for regression tasks in control applications?

2.12 What are the three different sub-sections in maintenance activities using AI? Draw and explain the flowchart in detail.

2.13 What is the significance of the NN Predictive Controller block and Plant Identification block in MATLAB Simulink, and how does it contribute to system design?

2.14 Define the following parameters from the NN Predictive Controller block and Plant Identification block:
(a) Cost Horizon (N2)
(b) Control Horizon (Nu)
(c) Control Weighing factor
(d) Training epochs
(e) Delayed plant inputs

2.15 How does the Adaptive Neuro-Fuzzy Inference System (ANFIS) differ from conventional controllers like PID in the context of boost converters?

2.16 What is the significance of the overshoot difference between the ANFIS- and PI-controlled boost converter circuits, as shown in the results? How does the steady-state response of the ANFIS-controlled circuit compare to the PI-control in terms of reaching steady state?

2.17 How does limited data availability affect the prediction accuracy of AI models used for estimating the Remaining Useful Life (RUL) of components in power electronics?

2.18 Why is energy consumption by AI models a concern in applications like optimizing the switching patterns of EV on-board converters?

References

1 J. Heaton, "Ian Goodfellow, Yoshua Bengio, and Aaron Courville: Deep learning," *Genet. Program. Evolvable Mach.*, vol. 19, no. 1–2, pp. 305–307, Jun. 2018. doi: 10.1007/s10710-017-9314-z.
2 B. Gonçalves, "Can machines think? The controversy that led to the Turing test," *AI Soc.*, vol. 38, no. 6, pp. 2499–2509, Dec. 2023. doi: 10.1007/s00146-021-01318-6.
3 I. Bandyopadhyay, P. Purkait, and C. Koley, "Performance of a classifier based on time-domain features for incipient fault detection in inverter drives," *IEEE Trans. Ind. Inform.*, vol. 15, no. 1, pp. 3–14, Jan. 2019. doi: 10.1109/TII.2018.2854885.
4 O. Zandi and J. Poshtan, "Voltage control of DC–DC converters through direct control of power switches using reinforcement learning," *Eng. Appl. Artif. Intell.*, vol. 120, p. 105833, Apr. 2023. doi: 10.1016/j.engappai.2023.105833.

5 M. D'Antonio, C. Shi, B. Wu, and A. Khaligh, "Design and optimization of a solar power conversion system for space applications," *IEEE Trans. Ind. Appl.*, vol. 55, no. 3, pp. 2310–2319, May 2019. doi: 10.1109/TIA.2019.2891228.

6 B. K. Bose, "Artificial intelligence techniques: How can it solve problems in power electronics?: An advancing frontier," *IEEE Power Electron. Mag.*, vol. 7, no. 4, pp. 19–27, Dec. 2020. doi: 10.1109/MPEL.2020.3033607.

7 P. Prokopowicz, J. Czerniak, D. Mikołajewski, Ł. Apiecionek, and D. Ślęzak, Eds., *Theory and Applications of Ordered Fuzzy Numbers*, vol. 356. Cham: Springer International Publishing, 2017. doi: 10.1007/978-3-319-59614-3.

8 J. Molina, *Artificial Intelligence in Power Electronics*. EE Power, Tech Insights, 2019.

9 P. Qashqai, H. Vahedi, and K. Al-Haddad, "Applications of artificial intelligence in power electronics," in *2019 IEEE 28th International Symposium on Industrial Electronics (ISIE)*, Jun. 2019, pp. 764–769. doi: 10.1109/ISIE.2019.8781216.

10 W. Wang *et al.*, "Secure-enhanced federated learning for AI-empowered electric vehicle energy prediction," *IEEE Consum. Electron. Mag.*, vol. 12, no. 2, pp. 27–34, Mar. 2023. doi: 10.1109/MCE.2021.3116917.

11 E&ICT Academy, IIT Kanpur, *Unleashing the Power: Exploring the Intersection of Artificial Intelligence and Internet of Things (IoT) for Smarter Solutions*. Newsletter, 2023.

12 R. Hellinger, "The impact of AI on the entire power electronics lifecycle," *pcim Europe*, 2024. https://www.power-and-beyond.com/the-impact-of-ai-on-the-entire-power-electronics-lifecycle-w-66c5e910cc7c8/?r=ext.

13 S. Zhao, F. Blaabjerg, and H. Wang, "An overview of artificial intelligence applications for power electronics," *IEEE Trans. Power Electron.*, vol. 36, no. 4, pp. 4633–4658, Apr. 2021. doi: 10.1109/TPEL.2020.3024914.

14 A. Mellit and S. A. Kalogirou, "Artificial intelligence techniques for photovoltaic applications: A review," *Prog. Energy Combust. Sci.*, vol. 34, no. 5, pp. 574–632, Oct. 2008. doi: 10.1016/j.pecs.2008.01.001.

15 B. K. Bose, "Artificial intelligence techniques in smart grid and renewable energy systems—Some example applications," *Proc. IEEE*, vol. 105, no. 11, pp. 2262–2273, Nov. 2017. doi: 10.1109/JPROC.2017.2756596.

16 M. G. Simoes, B. K. Bose, and R. J. Spiegel, "Design and performance evaluation of a fuzzy-logic-based variable-speed wind generation system," *IEEE Trans. Ind. Appl.*, vol. 33, no. 4, pp. 956–965, 1997. doi: 10.1109/28.605737.

17 V. S. Reddy and A. S. Kannan, "Optimized fuzzy type 1 Mamdani controller to enhance transient stability of AC-power system," *Wirel. Netw.*, vol. 30, pp. 7617–7632, Apr. 2024. doi: 10.1007/s11276-024-03738-w.

18 M. S. Chabani *et al.*, "Takagi–Sugeno fuzzy logic controller for DFIG operating in the stand-alone mode: Simulations and experimental investigation," *Arabian J. Sci. Eng.*, vol. 48, no. 11, pp. 14605–14620, Nov. 2023. doi: 10.1007/s13369-023-07704-0.

19 K. Chrysafiadi, "The role of fuzzy logic in artificial intelligence and smart applications," in *Fuzzy Logic-Based Software Systems. Learning and Analytics in Intelligent Systems*, vol. 34. Cham: Springer, 2023, pp. 25–29. doi: 10.1007/978-3-031-44457-9_2.

20 Q. Wang and S. Niu, "Design, modeling, and control of a novel hybrid-excited flux-bidirectional-modulated generator-based wind power generation system," *IEEE Trans. Power Electron.*, vol. 33, no. 4, pp. 3086–3096, Apr. 2018. doi: 10.1109/TPEL.2017.2704103.

21 J. Zhang, Y. Shi, and Z.-H. Zhan, "Power electronic circuits design: a particle swarm optimization approach," in X. Li *et al.*, Eds., *Simulated Evolution and Learning. SEAL 2008. Lecture Notes in Computer Science*, vol. 5361. Berlin, Heidelberg: Springer, 2008, pp. 605–614. doi: 10.1007/978-3-540-89694-4_61.

22 J. Pakdeeto, S. Wansungnoen, K. Areerak, and K. Areerak, "Optimal speed controller design of commercial BLDC motor by adaptive Tabu search algorithm," *IEEE Access*, vol. 11, pp. 79710–79720, 2023. doi: 10.1109/ACCESS.2023.3300233.

23 D. Izci, S. Ekinci, and H. L. Zeynelgil, "Controlling an automatic voltage regulator using a novel Harris hawks and simulated annealing optimization technique," *Adv. Control Appl.*, vol. 6, no. 2, Jun. 2024. doi: 10.1002/adc2.121.

24 B. Chen, X. Liang, and N. Wan, "Design methodology for inductor-integrated Litz-wired high-power medium-frequency transformer with the nanocrystalline core material for isolated DC-link stage of solid-state transformer," *IEEE Trans. Power Electron.*, vol. 35, no. 11, pp. 11557–11573, Nov. 2020. doi: 10.1109/TPEL.2020.2987944.

25 M. Issa, A. M. Helmi, and M. Ghetas, "Estimation of solar cell parameters through utilization of adaptive sine–cosine particle swarm optimization algorithm," *Neural Comput. Appl.*, vol. 36, no. 15, pp. 8757–8773, May 2024. doi: 10.1007/s00521-024-09534-z.

26 S. Zhang, X. Wang, K. Zhou, X. Shao, and J. Liu, "Chaotic ant colony algorithm-based frequency-optimized random switching frequency SVPWM control strategy," *J. Power Electron.*, vol. 23, no. 11, pp. 1688–1702, Nov. 2023. doi: 10.1007/s43236-023-00664-6.

27 Y.-C. Hung, F.-J. Lin, J.-C. Hwang, J.-K. Chang, and K.-C. Ruan, "Wavelet fuzzy neural network with asymmetric membership function controller for electric power steering system via improved differential evolution," *IEEE Trans. Power Electron.*, vol. 30, no. 4, pp. 2350–2362, Apr. 2015. doi: 10.1109/TPEL.2014.2327693.

28 J. Yuan, B. Chen, B. Rao, W. Wang, C. Tian, and X. Xu, "Possible analogy between the optimal digital pulse width modulation technology and the equivalent optimisation problem," *IET Power Electron.*, vol. 5, no. 7, pp. 1026–1033, Aug. 2012. doi: 10.1049/iet-pel.2011.0237.

29 M. I. Abdo and E. A. Elsheikh, "Optimization of a fractional-order interval type-2 fuzzy PID controller based on BBO for real-time applications," *Franklin Open*, vol. 8, p. 100121, Jun. 2024. doi: 10.1016/j.fraope.2024.100121.

30 A. Haque, K. V. S. Bharat, F. S. Zehgeer, and A. Malik, "Machine learning in power electronics for smart cities," in *Smart Cities: Power Electronics, Renewable Energy, and Internet of Things*. Boca Raton: CRC Press, 2023, pp. 253–286. doi: 10.1201/9781032669809-12.

31 W. Wang *et al.*, "Training neural-network-based controller on distributed machine learning platform for power electronics systems," in *2017 IEEE Energy Conversion Congress and Exposition (ECCE)*, Oct. 2017, pp. 3083–3089. doi: 10.1109/ECCE.2017.8096563.

32 P. T. Mehr, K. Koufos, K. El Haloui, and M. Dianati, "Low-complexity channel estimation for V2X systems using feed-forward neural networks," *IET Commun.*, vol. 18, no. 13, pp. 789–779, Jun. 2024. doi: 10.1049/cmu2.12788.

33 P. Yang, T. Wang, H. Yang, C. Meng, H. Zhang, and L. Cheng, "The performance of electronic current transformer fault diagnosis model: Using an improved whale optimization algorithm and RBF neural network," *Electronics*, vol. 12, no. 4, p. 1066, Feb. 2023. doi: 10.3390/electronics12041066.

34 J. George and G. Mani, "A portrayal of sliding mode control through adaptive neuro fuzzy inference system with optimization perspective," *IEEE Access*, vol. 12, pp. 3222–3239, 2024. doi: 10.1109/ACCESS.2023.3348836.

35 H. Han, J. Wang, Z. Liu, H. Yang, and J. Qiao, "Self-organizing robust fuzzy neural network for nonlinear system modeling," *IEEE Trans. Neural Netw. Learn. Syst.*, pp. 1–13, 2023. doi: 10.1109/TNNLS.2023.3334150.

36 W.-M. Lin and C.-M. Hong, "A new elman neural network-based control algorithm for adjustable-pitch variable-speed wind-energy conversion systems," *IEEE Trans. Power Electron.*, vol. 26, no. 2, pp. 473–481, Feb. 2011. doi: 10.1109/TPEL.2010.2085454.

37 Z. Y. Xue, K. S. Xiahou, M. S. Li, T. Y. Ji, and Q. H. Wu, "Diagnosis of Multiple Open-Circuit Switch Faults Based on Long Short-Term Memory Network for DFIG-Based Wind Turbine Systems," *IEEE J. Emerg. Sel. Top. Power Electron.*, vol. 8, no. 3, pp. 2600–2610, Sep. 2020. doi: 10.1109/JESTPE.2019.2908981.

38 Z. Huang, Z. Wang, X. Yao, and H. Zhang, "Multi-switches fault diagnosis based on small low-frequency data for voltage-source inverters of PMSM drives," *IEEE Trans. Power Electron.*, vol. 34, no. 7, pp. 6845–6857, Jul. 2019. doi: 10.1109/TPEL.2018.2875718.

39 S. Kiranyaz, A. Gastli, L. Ben-Brahim, N. Al-Emadi, and M. Gabbouj, "Real-time fault detection and identification for MMC using 1-D convolutional neural networks," *IEEE Trans. Ind. Electron.*, vol. 66, no. 11, pp. 8760–8771, Nov. 2019. doi: 10.1109/TIE.2018.2833045.

40 M. Ma and Z. Mao, "Deep recurrent convolutional neural network for remaining useful life prediction," in *2019 IEEE International Conference on Prognostics and Health Management (ICPHM)*, Jun. 2019, pp. 1–4. doi: 10.1109/ICPHM.2019.8819440.

41 S. S. Moosavi, A. Kazemi, and H. Akbari, "A comparison of various open-circuit fault detection methods in the IGBT-based DC/AC inverter used in electric vehicle," *Eng. Fail. Anal.*, vol. 96, pp. 223–235, Feb. 2019. doi: 10.1016/j.engfailanal.2018.09.020.

42 Q. Wang, W. Wang, Y. Wang, J. Ren, and B. Zhang, "Multi-stage network attack detection algorithm based on Gaussian mixture hidden Markov model and transfer learning," *IEEE Trans. Autom. Sci. Eng.*, pp. 1–15, 2024. doi: 10.1109/TASE.2024.3395355.

43 S. H. Ali, M. Heydarzadeh, S. Dusmez, X. Li, A. S. Kamath, and B. Akin, "Lifetime estimation of discrete IGBT devices based on Gaussian process," *IEEE Trans. Ind. Appl.*, vol. 54, no. 1, pp. 395–403, Jan. 2018. doi: 10.1109/TIA.2017.2753722.

44 A. Malik, A. Haque, I. A. Khan, K. V. S. Bharath, and S. Siddiqui, "Support vector data description (SVDD) based inverter fault diagnostic method," in *2022 Second International Conference on Power, Control and Computing Technologies (ICPC2T)*, Mar. 2022, pp. 1–6. doi: 10.1109/ICPC2T53885.2022.9776662.

45 T. Wang, H. Xu, J. Han, E. Elbouchikhi, and M. E. H. Benbouzid, "Cascaded H-bridge multilevel inverter system fault diagnosis using a PCA and multiclass relevance vector machine approach," *IEEE Trans. Power Electron.*, vol. 30, no. 12, pp. 7006 7018, Dec. 2015. doi: 10.1109/TPEL.2015.2393373.

46 M. Rigamonti, P. Baraldi, A. Alessi, E. Zio, D. Astigarraga, and A. Galarza, "An ensemble of component-based and population-based self-organizing maps for the identification of the degradation state of insulated-gate bipolar transistors," *IEEE Trans. Reliab.*, vol. 67, no. 3, pp. 1304–1313, Sep. 2018. doi: 10.1109/TR.2018.2834828.

47 W. Chen, L. Zhang, K. Pattipati, A. M. Bazzi, S. Joshi, and E. M. Dede, "Data-driven approach for fault prognosis of SiC MOSFETs," *IEEE Trans. Power Electron.*, vol. 35, no. 4, pp. 4048–4062, Apr. 2020. doi: 10.1109/TPEL.2019.2936850.

48 S. Afrasiabi *et al.*, "Photovoltaic array fault detection and classification based on T-distributed stochastic neighbor embedding and robust soft learning vector quantization," in *2021 IEEE International Conference on*

Environment and Electrical Engineering and 2021 IEEE Industrial and Commercial Power Systems Europe (EEEIC/I&CPS Europe), Sep. 2021, pp. 1–5. doi: 10.1109/EEEIC/ICPSEurope51590.2021.9584770.

49 R. S. Sutton and A. G. Barto, "Reinforcement learning: An introduction," *IEEE Trans. Neural Networks*, vol. 9, no. 5, pp. 1054–1054, Sep. 1998. doi: 10.1109/TNN.1998.712192.

50 Y. H. Tabrizi and M. N. Uddin, "Multi-agent reinforcement learning-based maximum power point tracking approach to fortify PMSG-based WECSs," *IEEE Trans. Ind. Appl.*, pp. 1–11, 2024. doi: 10.1109/TIA.2024.3429080.

51 A. Malik, A. Haque, V. S. B. Kurukuru, M. A. Khan, and F. Blaabjerg, "Overview of fault detection approaches for grid connected photovoltaic inverters," *e-Prime - Adv. Electr. Eng. Electron. Energy*, vol. 2, p. 100035, Apr. 2022. doi: 10.1016/j.prime.2022.100035.

52 M. Pecht and R. Jaai, "A prognostics and health management roadmap for information and electronics-rich systems," *Microelectron. Reliab.*, vol. 50, no. 3, pp. 317–323, Mar. 2010. doi: 10.1016/j.microrel.2010.01.006.

53 W. Bodewes, J. de Hoog, E. L. Ratnam, and S. Halgamuge, "Exploring the intersection of artificial intelligence and microgrids in developing economies: A review of practical applications," *Curr. Sustain. Energy Rep.*, vol. 11, no. 1, pp. 10–23, Mar. 2024. doi: 10.1007/s40518-024-00233-w.

54 S. Zhao and H. Wang, "Enabling data-driven condition monitoring of power electronic systems with artificial intelligence: Concepts, tools, and developments," *IEEE Power Electron. Mag.*, vol. 8, no. 1, pp. 18–27, Mar. 2021. doi: 10.1109/MPEL.2020.3047718.

55 Q. Chen and K. A. Folly, "Application of artificial intelligence for EV charging and discharging scheduling and dynamic pricing: A review," *Energies*, vol. 16, no. 1, p. 146, Dec. 2022. doi: 10.3390/en16010146.

56 A. Govind, K. Jayaswal, V. K. Tayal, and P. Kumar, "Simulation and real time implementation of shunt active power filter for power quality enhancement using adaptive neural network topology," *Electr. Power Syst. Res.*, vol. 228, p. 110042, Mar. 2024. doi: 10.1016/j.epsr.2023.110042.

3

Machine Learning Fundamentals

Ahteshamul Haque, Azra Malik, and Mansha Khursheed

Advance Power Electronics and Research Laboratory, Jamia Millia Islamia (A Central University), New Delhi, India

3.1 Introduction

Machine learning (ML) is a branch under artificial intelligence (AI) that facilitates the computers in learning and making decisions based on the data. The term "Machine Learning" was given in 1959 by an American inventor in artificial intelligence and computer gaming Arthur Samuel. Machine learning entails the creation of algorithms capable of identifying the trends and correlations in huge data sets, enabling enhanced systems' performance with time without overt programming. Machine learning is roughly categorized into the following: supervised learning, unsupervised learning, and semi-supervised learning, each having specific methodology and applications [1]. Other branches of machine learning such as deep learning and reinforcement learning also play a crucial role in power electronics applications. Deep learning allows for the exact modeling and prediction of complicated, non-linear behaviors in systems such as inverters and motor drives, which improves their control and efficiency [2]. Reinforcement learning, on the other hand, improves real-time decision-making processes, such as adaptive control techniques for energy management in smart grids and electric cars, by learning optimum policies through interaction with the environment [3]. Machine learning manifests tremendous benefits and has been utilized effectively in a variety of industries, including image classification, computer vision, voice recognition, and self-driving automobiles [4]. The growth of machine learning benefits power electronics, and the integration of the two has enormous potential. There can be various applications of machine learning including power module heatsink optimization, converter control, controlling wind energy conversion systems' maximum power point (MPP), detecting anomalies in inverters, and

Artificial Intelligence for Power Electronics, First Edition.
Edited by Ahteshamul Haque, Saad Mekhilef, and Azra Malik.

predicting remaining useful life (RUL) for super-capacitors. By incorporating machine learning, power-electronics-based systems gain self-awareness and flexibility, enhancing the system autonomy [5].

Power electronics is the study of how electrical power is controlled and converted using electronic equipment. It covers a wide range of applications, including renewable energy systems, power supply, electric vehicles, and motor drives [6]. The use of machine learning into power electronics has created new opportunities for improving the efficiency, reliability, and performance of these systems. Various ML techniques have been introduced in the existing power electronics systems such as grid connected system, fault detection and diagnosis, electric vehicle, forecasting applications, and energy storage as shown in Figure 3.1. Traditional control strategies for power electronic converters have weaknesses in terms of small-signal bandwidth, stability concerns, and efficiency compromise [7]. These strategies, such as linear pulse width modulation (PWM) control, have restricted small-signal bandwidth and require compromises to confirm stability under diverse loading conditions and variations in parameters [8]. Although free from small-signal bandwidth restrictions and stability issues,

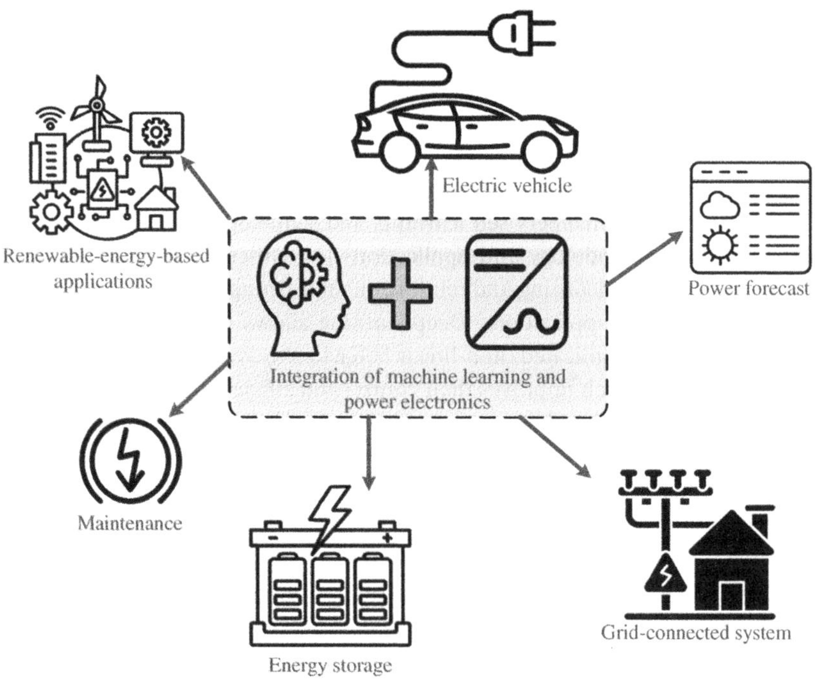

Figure 3.1 Various applications of machine learning in power electronics.

hysteresis control requires careful matching of sensing networks and can result in compromised efficiency due to unbalanced currents [9]. Furthermore, conventional control strategies need high-gain amplifiers, compromising output voltage accuracy. The conventional linear controllers in power electronic converters too have fundamental dynamic performance limitations [10]. These confinements have driven to the investigation of advanced control methods to improve dynamic performance and robustness. This is where machine-learning-based approaches play a key role in advanced control applications in power electronics. In a similar manner, these strategies are widely adopted in other application areas too for their inherent advantages. Following are the focal points of utilizing ML for improved performance and efficiency.

- Power electronic systems can improve productivity, reliability, fault detection, and energy management.
- ML permits the recognizable proof of troublesome designs and relationships through conventional strategies. Subsequently, an ideal execution is done.
- ML can reduce peak consumption, enabling the system to optimize energy management in power electronics and drive systems.
- ML is used in designing and controlling power converters such as buck, boost, and buck–boost converters.
- ML is adopted for fault diagnosis and identification. The ML Algorithms can analyze the converter's output waveform and detect deviations from the expected output [11].

An optimization-based evaluation of metaheuristic strategies for achieving desired waveform, control design, and optimum power quality is presented utilizing machine learning [12]. Other similar algorithms adopting neural network have been developed with precise network design, training technique, and other essential aspects [13]. This study addresses a wide range of power and non-power electronics applications. Using the approaches provided, many examples are shown [14]. An extensive and thorough description of machine-learning-based methods for MPP tracking in photovoltaic (PV) systems is presented [15]. Furthermore, machine learning methods are examined for PV systems with an emphasis on a wide range of PV applications [16]. Likewise, machine learning may be employed for maintenance in renewable-energy-based systems, for applications like preventive maintenance, fault diagnosis, etc. The most susceptible components in solar PV-based systems are power switching devices and capacitors. An overview of machine-learning-related approaches for condition monitoring of capacitors in PV is developed [17]. It focuses on proposing root-cause investigation for failures in different kinds of capacitors, particularly for PV systems. Another work, which outlines the method to monitor reliability in a power system using machine learning techniques, is described [18]. Additionally, it covers maintenance of PV systems and appropriate machine learning methods. Numerous failure identification approaches employing machine-learning-based classification algorithms that concentrate on inverter switch related anomalies in PV systems are explored [19, 20]. Further sections of this chapter delve deeply into the elements and characteristics of machine-learning-based methods. It will help the readers to grasp the fundamentals of machine learning before exploring the intricacies of various branches of machine learning as discussed in the upcoming chapters.

3.2 Key Components of Machine Learning

Algorithms and statistical models are used in the complex field of machine learning (ML) to help computers learn from data and make predictions or decisions. Each of the many essential elements that go into creating and implementing machine learning models is vital to the process as a whole. Machine Learning Algorithms are a set of rules or programs that can learn the hidden patterns from the data, predict the output, and improve the performance from experience on their own. The following are the main elements of machine learning:

3.2.1 Data Preparation

Data is the set of observations or measurements used to make a machine-learning model. In ML, Data is used to learn patterns and relationships between input variables and target outputs, which can be used for prediction or classification tasks [4]. Data can be labeled or unlabeled for ML algorithms. It is necessary to collect relevant and high-quality data from a variety of sources, such as databases, application programming interfaces (APIs), sensors, and simulation interfaces. The performance of the machine learning model is directly impacted by the quantity and quality of the data [21]. It includes removing noise and outliers from the data, assessing density to examine the distribution of the data, grouping the data according to similarity, translating high-dimensional data into low-dimensional data to help with feature reduction, and more. It could also be necessary to transform or normalize data in order to ensure consistency and standardize the range of various variables [22]. Data Normalization is used to transform the data into a common format for utilization in analytics and machine learning algorithms. Data normalization in Machine Learning can be applied to numerical and categorical data, and helps in reducing the complexity of the data. Further, normalization of data helps in improving model performance, addressing skewness in data, and improving model performance as well as data interpretability. Various techniques used for normalization are:

- **Min–Max Scaling**: Min–max scaling is the normalization technique that abstracts the information from the minimum value by subtracting the data value and then partitioning this value by the greatest and least value difference. Then, this information is converted into a range between 0 and 1.
- **Z-Score Normalization**: In Z-Score Normalization, information is converted by subtracting the mean from each data point into a standard normal distribution; it is then divided by the standard distribution. Z-Score Normalization is used where the information contains the normal distribution. It is the technique that makes a difference to make the information interpretable.
- **Decimal Scaling**: This normalization strategy subtracts the minimum value of each data point from the other data point and then divides it by the greatest and least value difference. By doing that, the information is converted into a range from 0 to 1. As it makes a difference in diminishing the information into a manageable range, therefore it is used for managing exceptionally expensive datasets.
- **Log Transformation**: This normalization method takes the Log of each data point to convert information into a logarithmic scale. The log transformation decreases the variety of the information; therefore, this method is used when managing information that includes a wide extent of values.

3.2.2 Feature Selection and Engineering

Once the data is prepared, it is important to analyze the data for any unnecessary information. In a dataset, not every feature (variable) has the same weight. Finding the most pertinent features that considerably increase the prediction potential of the model is the aim of feature selection. As a result, the data's dimensionality is decreased, the model performs better, and computing time is reduced.

Another critical stage in this phase is feature engineering, which is adding new features or changing current ones to enhance model performance. This procedure aids in identifying the relevant trend in the collected data. Higher-dimensional sets of features are sometimes obtained through feature extraction in a given application, which might demand more time and computational efforts. Consequently, various methods have been proposed to condense the feature-set into a more meaningful and computationally efficient feature-set aggregation. Numerous techniques for extracting features have been studied in the literature, such as time-domain format, frequency-domain, and time–frequency-domain techniques.

- **Time-Domain Format**: These methods involve using statistical techniques on the considered signal's time-series description. In accordance with overlap consideration, the signal is initially split into smaller time-series windows, and those windows are then subjected to a variety of statistics. These characteristics include determining the skewness, kurtosis, peak, entropy, variance, mean, standard deviation, and root-mean-square (RMS) values. [20].

- **Frequency Domain Format**: Frequency domain approaches begin with the translation of time domain signals to the corresponding signals in frequency domain [23]. This can be accomplished using algorithms such as the Fourier transform (FT), wavelet transform (WT), and fast Fourier transform (FFT). The transition in frequency domain simplifies computational analysis. Several frequency domain characteristics may be determined, such as energy, power spectral density (PSD), mean power, peak frequency, spectral kurtosis, spectral skewness, signal-to-noise ratio (SNR), spectral entropy, and Shannon entropy. [24]. Table 3.1 includes a variety of temporal and frequency domain characteristics.
- **Time–Frequency Domain Format**: These are frequently employed for the study of non-stationary data. They need to convert the time-domain data acquired in 1-D (dimension) to a 2-D time–frequency representation [25]. Many techniques can be adopted for this transformation like short-time Fourier transform (STFT) [26], Wigner–Ville distribution (WVD) [27], and Hilbert–Huang transform (HHT) [28].

3.2.3 Model Selection

One of the major criteria for working on an ML-based application is to analyze and select appropriate algorithm and model. It is essential to select the right ML algorithm with regards to power-electronics-based applications. The decision relies on the kind of task (regression, classification, clustering, etc.), the characteristics of the data, and the desired outcome. Popular algorithms include decision trees, random forests, support vector machines, and neural networks. For a complex model like neural network, defining the architecture parameters, including number of layers, number of neurons in each layer, and activation function , is an important step [29]. The architecture influences the model's ability to learn from the data and generalize to new, previously unknown data.

3.2.4 Model Training

In order for the selected algorithm to discover the underlying patterns, the prepared data must be fed into it during training. The model uses optimization strategies like gradient descent to modify its parameters in response to the input data. Until the model performs satisfactorily or converges to a stable state, training continues. A validation set is used to prevent overfitting, which occurs when a model performs well on training data but poorly on new data. This set is used to evaluate

Table 3.1 Various time domain, frequency domain, and time–frequency domain features.

	Feature Name	Description	Formula of Feature
Time-domain features	RMS	Root mean square value of the considered signal is determined through RMS	$RMS = \sqrt{\frac{1}{n}\sum_{i=1}^{n} S_i^2}$
	Mean	Mean represents the average value of a particular signal	$\mu = \frac{1}{n}\sum_{i=1}^{n} S_i$
	Variance	Variance describes the degree of dispersion in the considered data in proportion to the mean	$Var = \frac{1}{n}\sum_{i=1}^{n}(S_i - \mu)^2$
	Standard deviation	The standard deviation quantifies how near the provided data point is to the average value	$\sigma = \sqrt{\frac{1}{n}\sum_{i=1}^{n}(S_i - \mu)^2}$
	Entropy	Entropy provides the measure of "uncertainty" in the given data	$H = -\sum_{i=1}^{n} p(S_i)\log_{10} p(S_i)$
	Kurtosis	Signal shape or "tailedness" may be quantified through Kurtosis	$K = \frac{\frac{1}{n}\sum_{i=0}^{n}(S_i - \mu)^4}{\left(\sqrt{\frac{1}{n}\sum_{i=0}^{n}(S_i - \mu)^2}\right)^4}$
	Skewness	Skewness determines the asymmetric characteristics of the considered signal	$S_K = \frac{\frac{1}{n}\sum_{i=0}^{n}(S_i - \mu)^3}{\left(\frac{1}{n}\sum_{i=0}^{n}(S_i - \mu)^2\right)^3}$

(Continued)

Table 3.1 (Continued)

	Feature Name	Description	Formula of Feature
Frequency domain features	Power	It quantifies the signal strength across a specified time duration	$P = \lim_{n\to\infty} \frac{1}{2T} \int_{-T}^{T} \lvert S_i(t)\rvert^2 \, dt$
	Energy	Overall strength possessed by the signal is determined through energy	$E = \int_{-\infty}^{\infty} \lvert S_i(t)\rvert^2 dt$
	Mean frequency	It measures the mean frequency from the obtained power spectrum	$MF = \frac{\sum_{i=1}^{n} f_i P_i}{\sum_{i=1}^{n} P_i}$
	Signal-to-noise ratio (SNR)	SNR measures the distinctiveness of the desired signal from the disturbances and noise (unit is decibels)	$SNR_{dB} = 10 \log_{10} \left(\frac{P_{signal}}{P_{noise}} \right)$
	Power spectral density (PSD)	It quantifies the distribution of power in the frequency domain representation of signal	$PSD = \lim_{T\to\infty} \frac{1}{T} \int_{-\infty}^{\infty} \left[\int_{-\infty}^{\infty} S_{i_T}(t-\tau)\, S_{i_T}(t)\, dt \right] e^{-i2\pi f\tau} d\tau$
	Spectral skewness (SS)	SS measures the degree of symmetry nearby the spectral centroid (μ_f)	$SS = \frac{\sum_{i=b_1}^{b_2} (f_i - \mu_f)^3 s_i}{(\mu_s)^3 \sum_{i=b_1}^{b_2} s_i}$
			$\begin{bmatrix} f_i \rightarrow \textit{Frequency belonging to bin i} \\ s_i \rightarrow \textit{Spectral value at bin i} \\ \mu_s \rightarrow \textit{Spectral spread} \\ b_1, b_2 \rightarrow \textit{Band edges} \end{bmatrix}$
	Spectral Kurtosis (SK)	The degree of flatness of a given spectrum about its centroid is measured through SK	$SK = \frac{\sum_{i=b_1}^{b_2} (f_i - \mu_f)^4 s_i}{(\mu_s)^4 \sum_{i=b_1}^{b_2} s_i}$

the model's performance and adjust the hyperparameters like learning rate and number of iterations, that are not directly learned from the training set.

3.2.5 Model Evaluation

Model validation is important in the process of development that aids in identifying the data model for our data and uplifts its performance. Model validation is a set of processes that ensure the proper functioning of the ML model. It includes assessing its design objectives and its usefulness for end users. There are two methods of evaluation in ML:

- **Hold Out Method**: This is a technique that divides data into two sets, one test is used for testing and another test is used for training. It offers simplicity, adaptability, and speed in evaluating distinctive algorithm techniques.
- **Cross-Validation Method**: Cross-validation is a technique that divides the dataset into multiple samples to measure a model's accuracy. The training is done on one subset of data, and it evaluates on another complementary subset. The three main methods of cross-validation are *k*-fold cross validation, stratified cross-validation, and leave-one-out cross validation (LOOCV).

Different metrics are used to evaluate the machine learning model's performance, and these metrics change depending on the application. The standard metrics for classification problems include F1 score, accuracy, precision, and recall. For regression problems, common metrics include mean squared error and

Table 3.2 Model performance metrics.

Regression Performance Parameters		Classification Performance Parameters	
Metric	**Formula**	**Metric**	**Formula**
Mean absolute error (MAE)	$\frac{1}{n}\sum \mid (y-\hat{y}) \mid$	Accuracy	$\frac{TP+TN}{TP+TN+FP+FN}$
Mean square error (MSE)	$\frac{1}{n}\sum (y-\hat{y})^2$	Precision	$\frac{TP}{TP+FP}$
Root mean square error (RMSE)	$\sqrt{\frac{1}{n}\sum (y-\hat{y})^2}$	Error	$\frac{FP+FN}{TP+TN+FP+FN}$
R-square (R^2)	$1-\frac{\sum (y-\hat{y})^2}{\sum (y-\bar{y})^2}$	F1	$\frac{TP}{TP+FN}$
Mean absolute percentage error (MAPE)	$\frac{1}{n}\sum \left\lvert \frac{y-\hat{y}}{y} \right\rvert$	Recall	$\frac{TP}{TP+FN}$

mean absolute error. These metrics help in determining the model's effectiveness and how well it achieves the desired objectives. These metrics are further detailed in Table 3.2. The samples that have been correctly predicted to be positive or negative are labeled as True Positive (TP) and True Negative (TN). False-Negative (FN) and False-Positive (FP) samples, on the other hand, are those that are falsely predicted as negative and positive, respectively. For regression-problem-related metrics, n is the prediction samples number, y is the actual value, and $\hat{y}$ is the predicted value.

Further, confusion matrix is an $N \times N$ matrix, where N is the number of classes. It measures the performance of machine learning classification problems where the output can be two or more classes. It is useful in measuring the precision-recall, Specificity, Accuracy, area under the curve (AUC), etc.

3.3 Fundamental Concepts and Definitions

In machine learning, parameter values are learned through model training, whereas in conventional techniques (i.e. using pre-set algorithms), a computer is given both the model and its parameters to accomplish an objective. Labeled data are instances having a meaningful "label," "class," or "tag" that are instructive. Unlabeled data lacks explanation and consists of merely row data with missing "label" or "tag" assignment. The ML classes or categories have already been specified in the introduction section and are further explained in Figure 3.2. Supervised learning algorithms use labeled data to map new inputs to the actual target output. Unsupervised learning algorithm, on the other hand, evaluates an unlabeled dataset with unknown desired values in order to get insights through identifying hidden complex patterns. Semi-supervised algorithms analyze datasets with labeled and unlabeled samples. These algorithms are intended to take advantages of both supervised and unsupervised approaches. Another branch of machine learning called "deep learning" models uses intricate patterns in data by employing many-layered neural networks, hence the name "deep" neural networks.

In general, there are two types of supervised learning methods: regression and classification. Regression-based methods include predicting numerical output variables, whereas classification methods put the output in one of the defined categories or classes. Similarly, unsupervised learning strategies are classified as dimensionality reduction (also known as embedding) or clustering [18]. For example, anomaly detection to detect fake data by identifying outliers or unusual patterns in a dataset comes under clustering.

Semi-supervised learning-based models function as active learners. It can be further classified in two categories: Generative Adversarial Networks (GANs) and

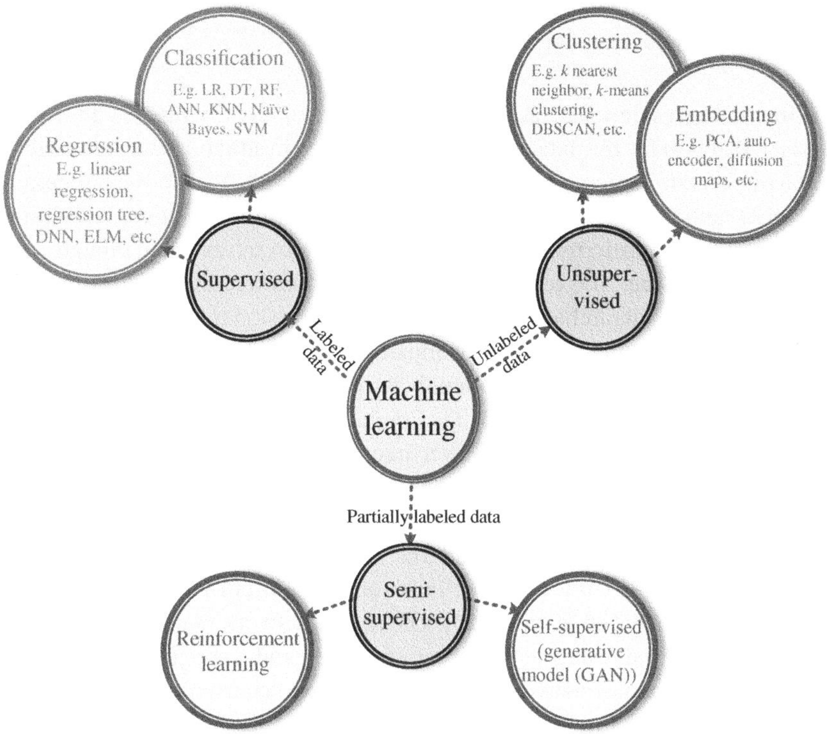

Figure 3.2 Various types of machine learning algorithms.

reinforcement learning. GANs employ deep neural learning methods to detect and interpret trends in input data. The generative model may then be used to generate new data samples that look similar to those in the training dataset. In reinforcement techniques, if a model completes a task properly, it is rewarded. Reinforcement learning aims to construct a model that maximizes rewards through an iterative approach [30]. Reinforcement learning is appropriate for an interactive or adaptive environment in which a model can evolve itself based on expert-defined policies, such as playing a game or driving autonomous vehicles. In this section, relevant ML techniques will be discussed for categories like regression and classification.

3.3.1 Regression

Regression analysis is a set of statistical methods used for estimating the relationships among variables. With an emphasis on the relationship between a dependent variable and one or more independent variables (predictors), it encompasses a wide range of modeling and analysis approaches.

- **Linear Regression**: The goal of this method is to identify the straight hyperplane that best fits the data. It is frequently applied in situations where variables have linear correlations, and regularization methods like Elastic-Net, Ridge, and LASSO help prevent overfitting. It is rigid in determining the optimal solution for complicated patterns and non-linear connections in variables, though.
- **Regression Tree**: The regression tree and decision tree used in classification share the same hierarchical structure, but the regression tree accepts numerical values as input. In addition to maximizing learning gain, the branching technique teaches non-linear correlations between variables. This approach is prone to overfitting issues even though it is simple to use and resistant to outlier.
- **Deep Neural Network**: Multi-layer neural networks, often known as deep neural networks, are extensively employed in many fields. In fact, DNNs might be utilized as classifiers as well as regression techniques because of their capacity to recognize intricate patterns [31]. Non-linear activation functions and hidden layers connecting the input and the output learn the non-linear relationships between features [25]. The neural network method performance can be enhanced using a variety of strategies and approaches, as well as sophisticated neural-network-based models like Recurrent Neural Network (RNN) and Convolutional Neural Network (CNN). Unlike other techniques, developing an effective neural network model with DNNs requires a thorough understanding of the neural network's parameters tuning.
- **Extreme Learning Machine (ELM)**: There are several uses for Extreme Learning Machines (ELM) in the data-driven methodology. ELM is utilized in classification, regression, clustering, feature extraction, sparse representation, and compression. Instead of using a gradient-based backpropagation process to update the network weighted values, this feedforward network randomly gives values to the network's bias and weight terms [32]. This type of algorithm's primary benefits include a quicker training period and superior interpolation outcomes. Conversely, even when compared to simple MLP models, the accuracy results of ELM are not encouraging.
- **Gradient Boosting Regression**: Gradient boosting regression utilized for regression tasks is an ensemble learning technique. It builds models sequentially, attempting with every iteration to fix the mistakes of the previous versions. A powerful prediction model is created by combining the models. While boosting concentrates on training new models to correct the residual errors of the combined ensemble of already-learned models, the ensemble method combines many weak learners to generate a strong learner. Gradient

Descent updates the model parameters iteratively with the objective of minimizing the loss function.

3.3.2 Classification

Many classification techniques have been presented in literature as listed below

- **Logistic Regression (LR)**: For binary classification tasks, in which an output is assigned to either class (0 or 1), LR is frequently utilized. Using a logistic function and hypothesis, a threshold is established in this approach to designate which instances will be labeled to which class. The hypothesis calculates the probability of occurrences to produce data, which will then be fitted into the logarithm function to create the sigmoid, an S-shaped curve. Next, new inputs class is determined using the logarithm function. Logistic regression can be applied to multiclass classification issues by using the one versus all technique, even though it performs better in binary classification tasks.
- **Artificial Neural Network (ANN)**: ANN, which is a class under supervised learning, takes inspiration from the human brain architecture. It consists of a network of neurons mimicking the biological neuron network, and further, imitates the human brain for learning [33]. Key components under ANN are layers, weights, biases, activation function, loss function, etc. Various structures of ANN include feedforward neural network (FNN), convolutional neural network (CNN), radial basis function network (RBFN), recurrent neural network (RNN), and transformer networks. A general RBFN structure is shown in Figure 3.3.

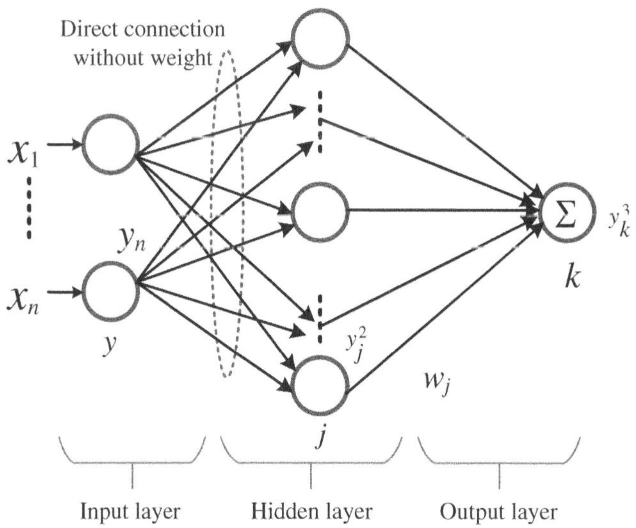

Figure 3.3 General architecture of RBFN.

ANNs are now the preferred models for many cutting-edge classification applications across a wide range of disciplines due to advancements in deep learning. The performance and adaptability of ANNs make them indispensable for modern machine learning, regardless of their challenges.

- ***K*-Nearest Neighbor (KNN)**: KNN is a simple, instance-based, non-parametric learning technique and is widely employed in machine learning for classification applications. A data point is classified according to the *k* nearest neighbors' majority class within the feature space [34]. The hyperparameter "*k*" in KNN denotes the number of nearest neighbors to take into account when classifying a sample. The method finds the *k* nearest points after computing the distance (usually Euclidean) between data point and every other point in the dataset. Although KNN is simple to use and intuitive, the curse of dimensionality can make it difficult to process huge datasets and make it computationally expensive.
- **Decision Trees (DTs)**: DTs have been considered for many classification applications. DT comprises an inverted decision tree with a root at the top that divides into multiple branches below [35]. The tree makes its final decision about the result at the point where the branch stops dividing into other branches. Moreover, it is prone to overfitting, particularly in a complex tree with several branches and conditions, which hinders its ability to generalize effectively to unknown inputs. Techniques including regularization, boosting, and bagging are typically employed in the DT to prevent overfitting issues.
- **Random Forest (RF)**: RF and the DT classifier are extremely similar, with RF employing many decision trees, rather than DT's single tree. This method can be employed to classify large amounts of data or quantify the significance of each feature in the selected plan of action. Because it is capable of being more precise and solve the overfitting problem of DTs, RFs are frequently preferred over decision trees in various applications. Nevertheless, because of its intricate structure and inherent slower speed compared to other models, this technique is not advised for use in real-time prediction.
- **Naïve Bayes (NB)**: NB is a highly effective classification algorithm that extends Bayes' theorem, operating under the assumption that each feature is independent to model the relationships between inputs and outputs. It is based on an extension of the Bayes theorem. The Bayes theorem evaluates an event's likelihood based on past occurrences [36]. NB is straightforward to implement and can efficiently predict labels for new data. Moreover, when domain knowledge supports the independence of features, it often outperforms other classification algorithms like logistic regression, especially with smaller datasets. However, in practice, it is difficult to obtain data with completely independent features; also, the method assigns zero probability and fails to classify inputs that were not followed up on during the training phase.

- **Support Vector Machine (SVM)**: The fundamental concept of Support Vector Machines (SVM) involves mapping data into a higher-dimensional space to identify an optimal hyperplane that separates different classes. In essence, support vectors represent coordinates in this new *n*-dimensional space. Although SVM is frequently utilized for binary classification tasks, it can be computationally intensive and slow when dealing with large datasets.

3.3.3 Clustering

Clustering techniques aim to group similar instances into the same cluster based on shared properties. These methods are widely used beyond machine learning, including in fields such as pattern recognition, image analysis, statistical analysis, and data compression.

- ***K*-Means Clustering**: Among the most recognized clustering algorithms is *K*-means, which is one of the simplest and most intuitive. It partitions instances into *k* clusters or centroids with the same variance. By deciding the number of clusters (K), the algorithm iteratively refines the clusters to minimize a measure known as inertia, adjusting the positions of the centroids accordingly. Due to its simplicity and scalability to large datasets, *K*-means is extensively applied in various domains.
- **Density-Based Spatial Clustering of Applications with Noise (DBSCAN)**: DBSCAN is commonly adopted in machine learning and data mining, and identifies core points in areas of high density and expands clusters within a definite radius, typically using Euclidean distance. Areas of low density are classified as outliers. The main challenges in DBSCAN include choosing appropriate clustering attributes, handling noise with unpredictable densities, and managing significant variances in the number of boundary points in different directions of the clusters. An expert defines the minimum number of points required to form a dense region and the maximum distance between points in the same region. Although DBSCAN is a popular clustering method, it performs poorly with widely spaced or high-dimensional datasets.
- **Gaussian Mixture Models (GMMs)**: Assumes that the data comes from a combination of multiple Gaussian distributions. It employs the Expectation–Maximization (EM) algorithm to estimate the factors and allot data points to clusters using probabilistic models.

3.3.4 Embedding Algorithms

In many scenarios, particularly within the realm of big data, having a vast number of features or variables in a dataset can complicate the task of understanding their interrelationships. Training a model with the entire dataset may lead to overfitting, where the model fails to generalize well to new, unknown data. To address this, Embedded Algorithms (EAs) can be utilized to derive new features from the data, preserving essential information before applying advanced machine learning models. EAs can also be employed directly for prediction. Two common types of embedding algorithms include:

- **Principal Component Analysis (PCA)**: PCA targets to lower the dimensionality of large datasets while retaining as much variance as possible [21]. It projects data onto principal components or axes and uses techniques like Singular Value Decomposition (SVD) to achieve dimensionality reduction or make predictions by leveraging the eigenvectors of the data's covariance matrix.
- **Auto-Encoder**: This cutting-edge technique utilizes neural networks to compress data. Auto-encoders are effective in applications like data compression. They learn to encode the input data into a lower-dimensional representation and reconstruct it as closely as possible to its original form, filtering out noise through its architecture. An auto-encoder typically comprises an encoder, a decoder, and a bottleneck. The encoder compresses the data into a compressed form, the decoder reconstructs the input by minimizing noise, and the bottleneck stores the encoded data. The encoder of a properly trained auto-encoder can be utilized for dimensionality reduction, or the entire model can be applied in tasks such as anomaly detection.

3.4 Machine Learning (ML) Applications in Power Electronics

ML applications in power electronics are discussed in detail in this section.

3.4.1 ML Application in Forecasting

With the growing need for renewable energy, machine learning plays an increasingly important role in solar and wind energy forecasting, paving the way to a more sustainable and dependable energy future. In the field of renewable energy, the use of machine learning to forecast solar and wind energy is a revolutionary development. Machine learning approaches have the ability to accurately anticipate energy production by utilizing large volumes of historical data and advanced algorithms. As the number of grid-integrated PV systems increases, precise power generation forecasts are becoming more critical. This growth in PV systems is driven by factors like decreasing investment costs, incentives, building regulations, and directives [4]. With this trend expected to continue, the integration of more PV

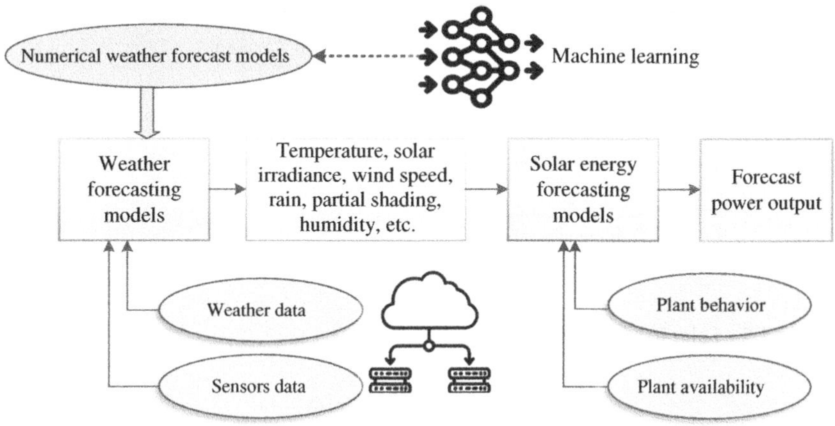

Figure 3.4 A general framework for solar PV generation forecast.

systems into the power grid may lead to stability issues due to abrupt and intermittent climate changes. Research has leveraged load forecasts to improve system operation and examined various forecasting techniques to enhance precision in power grids. Weather, being a chaotic system, significantly impacts forecasting accuracy. A general step-by-step framework for solar power generation forecast can be illustrated in Figure 3.4. Long-term forecasts, such as seasonal predictions, are challenging to achieve. This challenge has led to the development of advanced techniques using statistical and stochastic models to support long-term planning and offer a thorough understanding of distributed energy resources and loads.

Early research on grid management considered the distributed generation primarily concentrated on load prediction rather than predicting the outputs of these energy resources. An approach utilizing ANN combined with discrete wavelet transform (DWT) for data preprocessing is illustrated in Figure 3.5 [37]. This model determines the solar irradiation based on sunshine duration and mean temperature. Similar methodologies have been developed for forecasting the output of solar PV panels, emphasizing simplicity and avoiding complex calculations [33]. These models provide 24-hour ahead output predictions using an enhanced back-propagation method to improve forecast accuracy. Additionally, a study comparing various solar PV energy forecasting strategies utilizing ML is discussed in [38]. Specifically, an evaluation and differentiation between support vector regression (SVR) and ANN for energy production estimation from a Florida-based solar PV system is presented [39]. This investigation emphasizes on forecasting energy generation at intervals of 15 minutes, 1 hour, and 24 hours ahead. A hierarchical approach inculcating these ML algorithms is established using 15-minute average power measurements collected over the year.

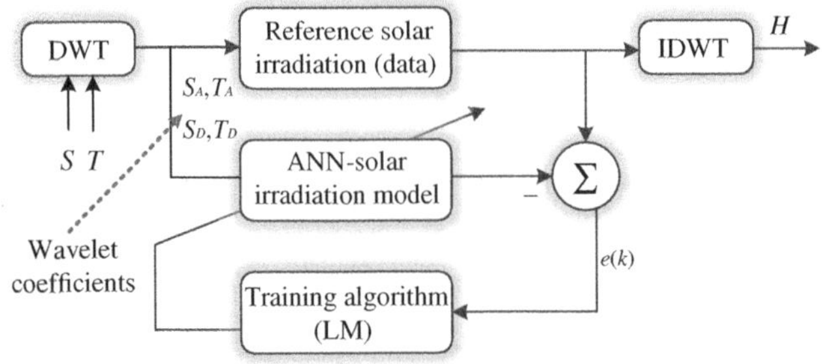

Figure 3.5 Solar irradiance prediction using ANN.

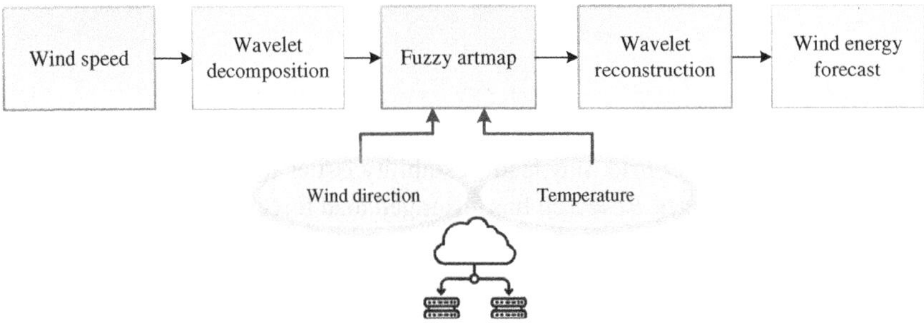

Figure 3.6 Wind energy forecast utilizing Fuzzy artmap method.

Machine learning, a sophisticated branch of AI, is adept at identifying patterns in data, making it ideal for forecasting wind energy outputs. Traditionally, statistical methods were used for short-term wind energy forecasts, but recent trends show a shift toward data-driven techniques for both short-term and long-term predictions [40]. A notable study introduced a hybrid machine learning technique combining wavelet transform-based data filtering with the fuzzy ARTMAP network as shown in Figure 3.6 [41]. This approach, tested against other forecasting methods, demonstrated significant improvements in prediction accuracy. Applied to data from the North Cape wind farm in PEI, Canada, the model also accurately predicted wind power output, effectively mitigating biases from turbulence and shear at new locations.

3.4.2 ML Application in Control

In power electronics control applications, machine learning (ML) is mainly used for regression tasks to map system inputs and outputs nonlinearly. These tasks are crucial for regulating systems to achieve desired performance. Traditional methods have limitations such as requiring extensive control knowledge, being

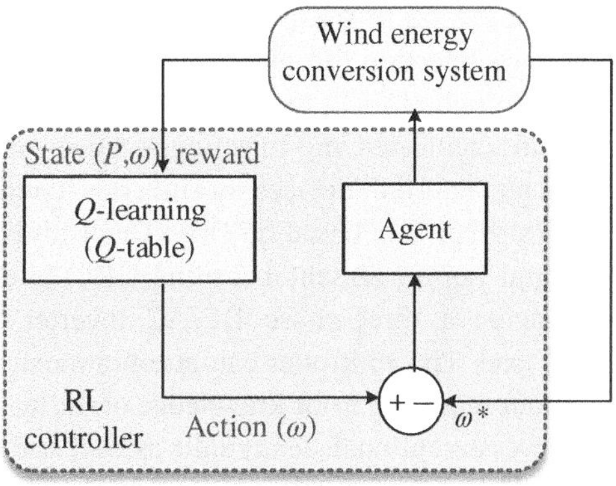

Figure 3.7 Reinforcement learning-based MPPT for WECS.

time-consuming for complex systems, and lacking adaptability to changing conditions. ML techniques like Artificial Neural Networks (ANN) and Reinforcement Learning (RL) address these issues by enhancing performance and adaptability, providing more robust and efficient control solutions. While there have been significant progressions in PV technology, such as reduced costs and improved cell efficiency, the limited energy conversion efficiency remains a barrier to widespread adoption. Effective maximum power point tracking (MPPT) is essential to addressing this issue, given the dependence of solar power generation on factors like solar irradiance and temperature. To maximize energy extraction from PV arrays, a control unit with a proficient MPPT strategy is crucial. Improving MPPT capability is the most cost-effective method to enhance overall PV system efficiency. A comprehensive review of various MPPT techniques for solar PV panels is discussed in [15]. RL-based control is effectively applied in MPPT for renewable energy systems, especially wind energy conversion system (WECS) [42]. An intelligent real-time MPPT algorithm using RL leverages online learning to create an optimal control strategy stored in the Q-table as illustrated in Figure 3.7. This table contains state transition probabilities that maximize power output when specific actions are taken based on the current system state. ANN-based controllers for PV inverters are found to produce output with very low total harmonic distortion (THD) [11]. In [43], a space vector pulse-width modulation (SVPWM) technique for a three-level voltage-fed inverter is executed using an FNN. The neural network takes sampled command phase voltages as input and generates pulse width modulation patterns for SVPWM as output.

Typically, RBFNs provide better generalization than FNNs, along with quicker training and execution times. An example of RBFN application includes its role in managing the DC-link voltage and AC line voltage in a three-phase induction generator, as illustrated in [44]. Anti-islanding protection is vital for identifying irregularities and disconnecting distributed generators (DGs) from the grid, using active, passive, and hybrid methods [45]. ML-based approaches enhance detection through enabling control in real-time conditions. Low voltage ride through (LVRT) maintains grid stability during abnormalities, keeping PV systems connected and injecting reactive current for voltage recovery, achievable through external devices or inverter controller modifications [46]. Another control algorithm based on Deep Deterministic Policy Gradient (DDPG)-assisted Integral Reinforcement Learning (IRL) is developed [47]. It is responsible for managing a three-phase DC/AC inverter that supplies a resistive–inductive (RL) load. The controller can autonomously adjust its control gains in real-time without requiring prior knowledge of the system model. The proposed algorithm achieves exceptional steady state as well as dynamic responses while maintaining relatively low computational complexity.

3.4.3 ML Application in Electric Vehicle (EV)

ML has found several applications in the field of electric vehicles (EVs), enhancing aspects such as energy optimization, battery management, charging infrastructure, autonomous driving, and predictive maintenance. These advancements help in improving the efficiency, safety, and overall performance of EVs.

- **Energy Optimization**: ML approaches are also employed to optimize energy consumption in electric vehicles (EV). By anticipating driving behavior and road conditions, machine learning algorithms can dynamically change power distribution between the battery and the electric motor. This assures effective energy consumption and increases the range of EVs. For example, reinforcement learning may be used to develop optimal driving methods that reduce energy usage over time [48].
- **Charging Infrastructure**: ML is helpful in increasing the efficiency of EV charging facilities. Smart charging stations utilize machine learning algorithms to control energy distribution, shorten charging time, and balance the strain on the power grid [6]. By forecasting peak consumption periods and dynamically modifying billing rates, ML contributes to grid stability and efficient use of available resources. An ML-based EV charging management application is employed to direct EVs to charging stations in order to reduce load variation, charging costs, voltage fluctuations, and power losses, while taking into account traditional, and fast charging, along with vehicle-to-grid (V2G) technologies. The findings demonstrate the high efficiency with long, and short term memory (LSTM) for EV management to confirm enhanced accuracy [49].
- **Battery Management**: Battery management systems (BMS) are one of the essential applications of machine learning (ML). Accurate state-of-charge (SOC) and state-of-health (SOH) estimates are critical to the lifespan and

reliability of EV batteries. SOC and SOH are predicted using ML methods like as neural networks and support vector machines, which evaluate trends in battery usage data, temperature, and charging cycles. This allows for more accurate forecasts and improved battery life management [50].

3.4.4 ML Application in Maintenance

Maintenance activities like condition monitoring, fault diagnosis, and predicting the remaining useful life (RUL) are crucial for enhancing reliability. Machine-learning-based intelligent fault classification techniques have shown high accuracy in diagnosing converter faults in grid-connected PV systems [5]. The methodology for machine-learning-based fault diagnosis in power electronic converters related to grid-connected PV systems is illustrated in Figure 3.8. Artificial Neural Networks (ANNs) are employed to identify and classify power switch faults in multilevel H-bridge inverters. Discrete Wavelet Transform (DWT) is used to extract features such as signal power and energy [22]. Radial Basis Function Networks (RBFNs) and supervised-learning-based Probabilistic Neural Networks (PNNs) are suggested for diagnosing faults in diode-clamped multilevel inverter [51]. A modified Convolutional Neural Network (CNN)

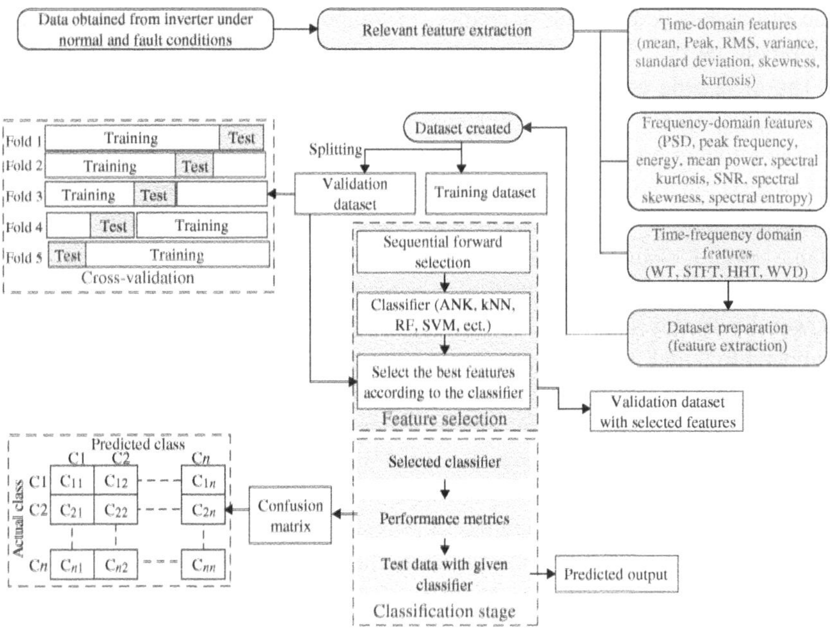

Figure 3.8 Framework for classification in case of inverter faults.

with Global Average Pooling (GAP) is utilized for diagnosing inverter switch faults [25].

Machine-learning-based monitoring techniques, such as ANN regression, fuzzy systems, and neo-fuzzy neuron (NFN) models, are utilized for assessing the health of capacitors [52]. Capacitor health estimation methods, including Recursive Least Squares (RLS) and Support Vector Regression (SVR), are compared using offline training [53]. Supervised-learning-based ANN approaches are also explored for monitoring the health of electrolytic capacitors [54]. Batteries play a critical role in grid-connected PV systems, and their performance needs to be maintained [55]. Diagnostic techniques for battery health monitoring, such as state of charge (SOC), state of health (SOH), and state of function (SOF), have been studied extensively [56]. Machine learning techniques, including Bayesian regression, relevance vector machines (RVM), and adaptive Gaussian mixture models (AGMM), are applied for battery fault analysis and predicting the remaining useful life (RUL) [57]. Reliability analysis focuses on estimating the lifespan of power electronic devices and predicting their remaining useful life (RUL) in the event of a failure [58]. Traditional analytical methods like Monte Carlo or Markov analyses are often employed, but they may face challenges in achieving rapid convergence [59]. Machine-learning-based lifetime estimation addresses these limitations by accounting for sudden faults or changes in component stress due to ride-through operations [60].

The increasing adoption of wind power in electricity grids is driven by the growing capacity of individual turbines and the shift from onshore to offshore locations to harness greater wind energy potential [61]. This shift has heightened the impact on the power grid and increased maintenance and repair costs following failures. Enhanced reliability and resilience are essential to withstand extreme disturbances from the grid and environment. Industrialists and manufacturers prioritize reliability in wind power generation to ensure high power availability [62]. Factors such as moisture, dust, vibration, chemicals, temperature, and high voltage contribute to failures. Traditional statistical methods have proven inadequate, prompting a shift to a physics-of-failure-based approach, focusing on specific failure phenomena. A machine-learning-based condition monitoring approach for wind turbines uses supervisory control and data acquisition (SCADA) data and neural networks to predict gearbox temperature and analyze anomalies. SVM-based fault diagnostic techniques and machine learning methods detect structural health issues in wind turbine blades as illustrated in Figure 3.9 [63]. High-speed shaft bearings, critical components, are monitored using vibration-based fault detection and unsupervised learning techniques.

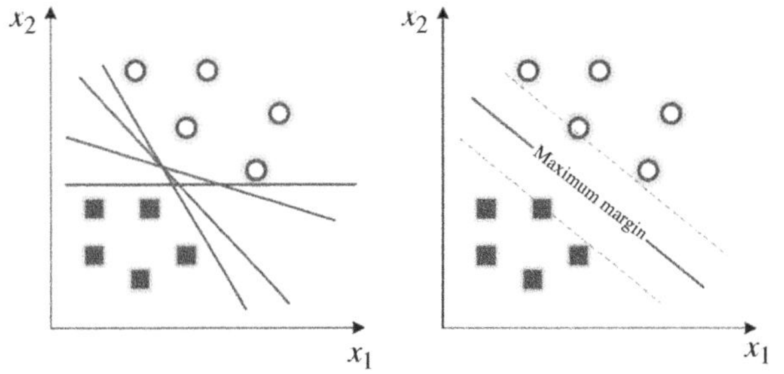

Figure 3.9 SVM technique for wind turbine condition monitoring.

3.5 Case Study

In this chapter, two interesting case examples are presented that highlight the many uses of ML in power electronics field. In the first case study, the PV generation is predicted and optimized through the use of a regression-based methodology. The second case study explores a classification-based fault detection technique for grid-connected photovoltaic systems, demonstrating how ML can detect and classify faults to ensure operational stability. These illustrations show the power electronics industry's numerous applications for machine learning as well as its innovative approaches.

3.5.1 Regression (PV Generation Forecasting)

The required data for solar PV generation forecasting is gathered from *Kaggle* website, with a file named "house_generation_10kw.csv." This generation dataset for a single house of solar PV capacity 10 Kw contains 97,333 entries having 16 columns for two years. These columns include the data pertaining to parameters like current, voltage, frequency, inverter temperature, power, and daily generation, and all the column values are non-null. The forecasting is implemented in MATLAB editor, with various steps detailed here through code:

- **Loading the Dataset**: Write the following code in the editor to load the dataset.

```
% Load the dataset
data = readtable('house_generation_10kw.csv', 'Delim-
iter', '\t');
```

- **Data Preprocessing**:

```
% Convert Tiempo to datetime format
data.Tiempo = datetime(data.Tiempo, 'InputFormat',
'dd.MM.yyyy HH:mm:ss');
```

```
% Extract relevant time features
data.HOUR = hour(data.Tiempo);
data.DAY = day(data.Tiempo);
data.MONTH = month(data.Tiempo);

% Handle missing values if any
data = rmmissing(data);
```

- **Plotting the Required Data**:

```
% Plot power over time
figure;
plot(data.Tiempo, data.power);
xlabel('Date Time');
ylabel('Total Power');
title('Total Power Over Time');
```

- **Relevant Feature Selection**:

```
% Select features and target for model training
features=data(:, {'V1MPPT_V_','V2MPPT_V_','V3MPPT_V_',
'I1MPPT_A_', 'I2MPPT_A_', 'I3MPPT_A_', 'HOUR', 'DAY',
'MONTH'});
target = data.power;
```

- **Training the Model**: Here, basic linear regression model is utilized, and the required code is detailed here:

```
% Convert table to array for model training
X = table2array(features);
y = target;

% Split data into training and testing sets
cv = cvpartition(height(data), 'HoldOut', 0.2);
X_train = X(training(cv), :);
y_train = y(training(cv), :);
X_test = X(test(cv), :);
y_test = y(test(cv), :);

% Train a Linear Regression model
mdl = fitlm(X_train, y_train);

% Predict on test data
y_pred = predict(mdl, X_test);
```

```
% Evaluate the model
mae = mean(abs(y_test - y_pred));
mse = mean((y_test - y_pred).^2);
r2 = mdl.Rsquared.Ordinary;

fprintf('MAE: %.2f\n', mae);
fprintf('MSE: %.2f\n', mse);
fprintf('R-squared: %.2f\n', r2);

figure;
plot(y_test, 'o');
hold on;
plot(y_pred, '*');
xlabel('Sample Index');
ylabel('Power');
title('Actual vs Predicted Values');
legend('Actual', 'Predicted');
hold off;
```

- **Predict with New Data**:

```
% Predict power generation for a new input value
Xnew = [112, 127, 124, 0.1, 0.1, 0.1, 9, 11, 1];
% New input feature value
predictedY = predict(mdl, Xnew)
```

Once the code is run on MATLAB, following results are obtained as provided in Figures 3.10 and 3.11:

MAE: 122.90
MSE: 50099.82
R-squared: 0.99
predictedY = 119.0743

The results show that the basic linear model is working well for the prediction task, since the *R*-squared value is highly close to 1. The predicted value is 119.0743, while the actual value was 121, so it is observed that the predicted value is quite close to the actual value.

3.5.2 Classification (Inverter Fault Classification)

In this case study, grid-connected PV system is considered, and mainly inverter faults are studied for classification purpose. For data acquisition step, the PV

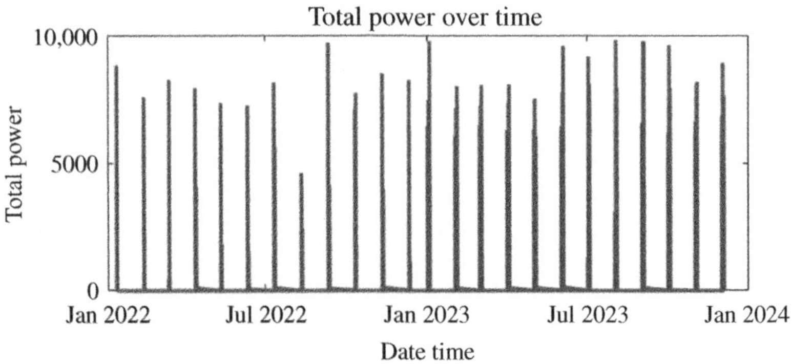

Figure 3.10 The graph of total power over time.

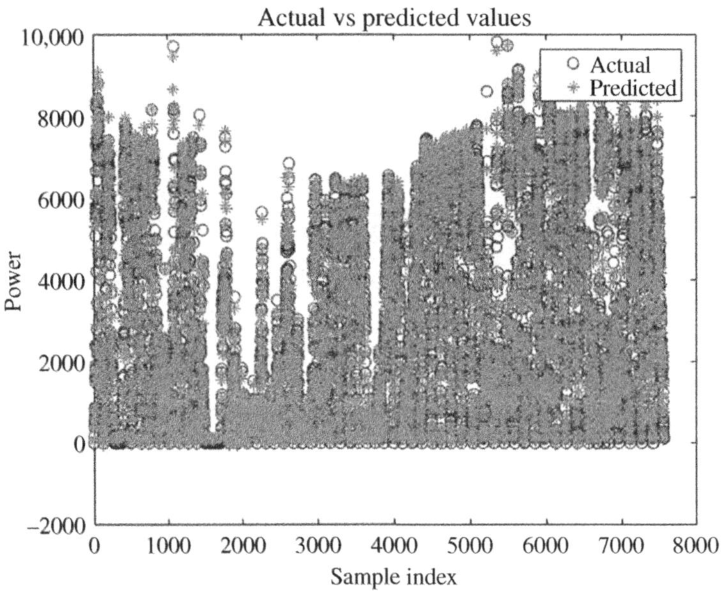

Figure 3.11 The graph of actual versus predicted values.

system is simulated through injecting various faults and collecting the respective data. Overall, there are five classes for classification namely – (i) No-fault, (ii) Switch open, (iii) Switch short, (iv) Filter fault, and (v) Capacitor fault. Once the fault data is acquired, it needs to be pre-processed through utilizing wavelet transform. For this purpose, discrete wavelet transform (DWT) is utilized, and mother wavelet Daubechies-5 (db5) is considered for decomposing the signals,

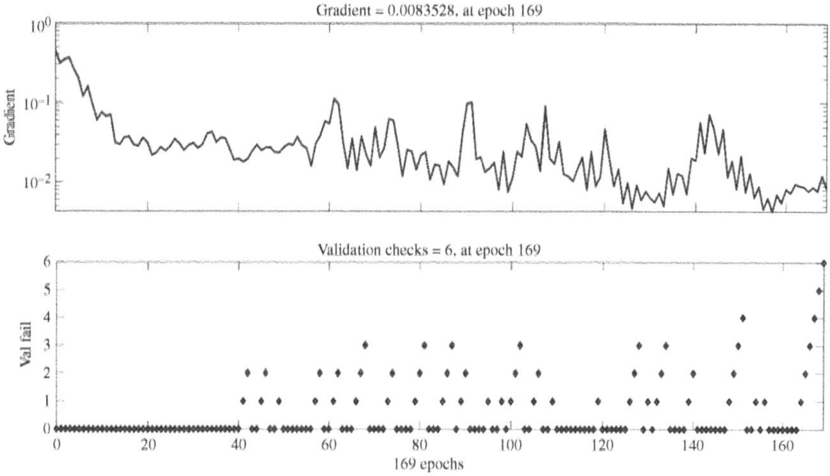

Figure 3.12 Training curves for the ANN-based inverter fault classification.

and their reconstruction [20]. After that various features are calculated like SNR, Peak, Entropy, total harmonic distortion (THD), power generated, Kurtosis, and Skewness. These features are then fed to the ANN for classification purpose using "nprtool" from MATLAB.

Once the "nprtool" command is provided to the command window, the "nprtool" window is opened in MATLAB. The input and output data files are then imported, and training is started after providing certain parameters. These parameters include number of layers, and training, testing, and validation dataset ratio. The training is stopped at 169 epochs, and the training curve is obtained as shown in Figure 3.12. Further the best validation performance curve is shown in Figure 3.13, at epoch 163.

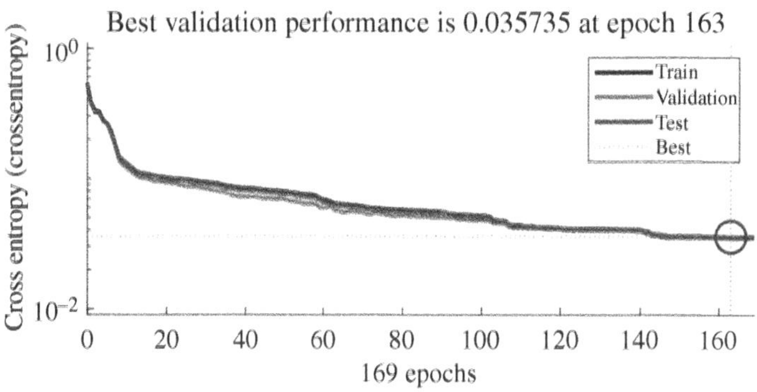

Figure 3.13 Validation performance curve.

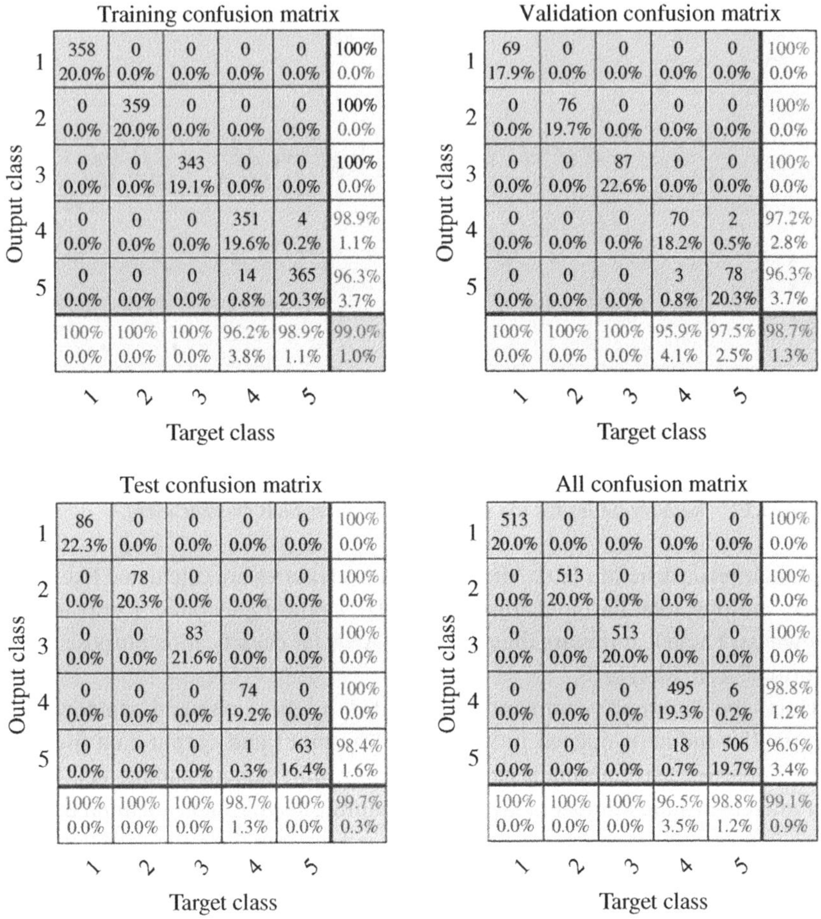

Figure 3.14 Confusion matrix for various datasets.

The overall confusion matrix for training, testing, and validation dataset is illustrated in Figure 3.14, where the training accuracy is observed to be around 99%, testing accuracy is 99.7%, validation accuracy is 98.7%, and overall accuracy is 99.1%, which is very high as compared to the traditional models utilized for fault classification purpose. In addition, receiver operating characteristic (ROC) curve is shown for all the datasets in Figure 3.15, where it can be seen that it is quite close to 1, which further highlights the preciseness of the utilized ML algorithm.

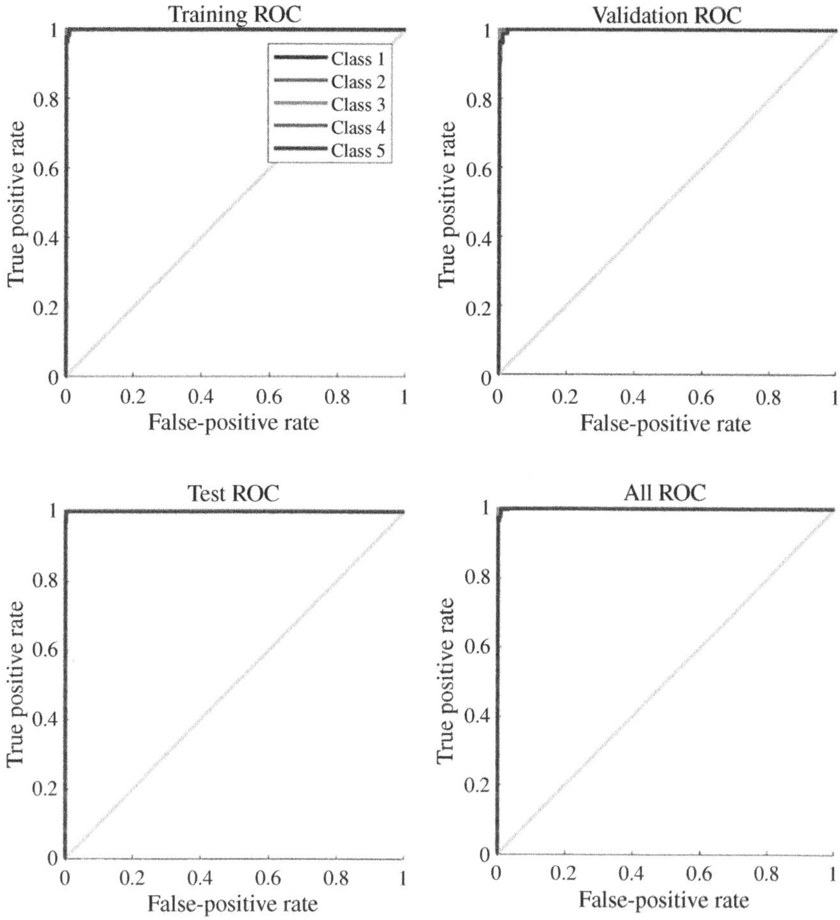

Figure 3.15 ROC curves for training, testing, validation, and overall dataset.

3.6 Challenges

To fully harness the potential of ML power electronics systems must overcome a number of obstacles. Assuring excellent data quality, which entails precisely gathering thorough data on voltage, current, temperature, and other relevant variables, is a significant challenge. It is important to ensure that the data must be carefully pre-processed, cleaned, transformed, and organized before it is fed into ML models. The quality of this input data has a major impact on the reliability and precision of ML predictions. Furthermore, model interpretability is critical since power electronics frequently work in high-stakes scenarios, where comprehending ML decision-making processes is important. To gain the trust, it is vital to construct interpretable and explainable models, while black-box models might be troublesome because they provide too little insight into the prediction process. There's an additional layer of complexity in meeting real-time operational requirements. Strong real-time data processing infrastructures and extremely efficient algorithms are required for power electronics systems in order to respond to dynamic conditions rapidly and precisely. Moreover, the operating environment's inherent complexity and variability, which includes varying load conditions, environment influences, and data noise, must be dealt with appropriate ML integration. Reliable operation demands algorithms to be established that are robust to these variations. In order to maintain effectiveness, ML models must be continuously updated and monitored as systems change and new data becomes available. Retraining and recalibration may also be necessary. These challenges must be addressed to fully harness ML's potential in enhancing the performance and reliability of power electronics systems.

3.7 Future Research Directions

ML will play a major role in various applications in the future by improving system resilience, efficiency, and flexibility to create smarter, more intelligent power electronics systems. Further, future research direction may involve following key aspects:

- To create robust machine learning models and derive valuable insights for improved decision-making, leverage *modern data analytics*.
- Develop effective algorithms that can handle *real-time processing* and low latency responses while adhering to the strict time limits of power electronics systems.
- To take use of both machine learning and classical control techniques' advantages for real-time operations and predictive maintenance, *utilizing hybrid models* can be solution.
- *Enhance ML model robustness* and resilience so they can adjust to changing circumstances and continue to maintain high performance in spite of environmental instabilities.
- Develop *autonomous power electronics systems* with machine learning (ML) that can self-diagnose, self-heal, and self-optimize to minimize the need for human involvement.

3.8 Conclusion

An introduction to the Machine Learning (ML) techniques in power electronics is given in this chapter. As we can see across a wide variety of applications, power electronics is instrumental. The significance of using ML technologies lies in its accuracy feature. A glimpse of various Machine Learning methods implementation in power electronics applications has been given in this chapter. A few case studies are also detailed to provide a better picture of how ML can be useful, and fast in solving various power-electronics-related problems. The future of the journey of machine learning in power electronics is driving faster. The integration of Machine Learning in Power Electronics is poised to drive further innovation.

Exercises

3.1 How do supervised, unsupervised, and reinforcement learning differ in their application to power electronics?

3.2 How does machine learning improve the efficiency and reliability of power electronics systems?

3.3 What are the advantages and limitations of using supervised learning for predictive maintenance in power electronics systems?

3.4 In what ways can machine learning help in controlling power converters and optimizing energy conversion?

3.5 What challenges exist in integrating machine learning with power electronics, and how can they be addressed?

3.6 How does machine learning contribute to fault detection and diagnostics in power electronic systems?

3.7 What are the key evaluation metrics used to assess machine learning models in power electronics?

3.8 What is the importance of cross-validation in evaluating machine learning models in power electronics applications?

3.9 How do accuracy, precision, recall, and F1 score apply to classification tasks in power electronics?

3.10 Why is it essential to consider both training and validation datasets when evaluating the performance of machine learning models in power electronics?

3.11 How can ROC-AUC help in evaluating the accuracy of ML models?

3.12 How can overfitting be detected in machine learning models used for power electronics, and what methods help mitigate it?

References

1 O. H. Abu-Rub, A. Y. Fard, M. F. Umar, M. Hosseinzadehtaher, and M. B. Shadmands, "Towards intelligent power electronics-dominated grid via machine learning techniques," *IEEE Power Electron. Mag.*, vol. 8, no. 1, pp. 28–38, 2021. doi: 10.1109/MPEL.2020.3047506.

2 A. Malik, A. Haque, K. V. Satya Bharath, and S. Padmanaban, "Data driven fault classification technique for grid connected PV inverter," in *IECON Proceedings (Industrial Electronics Conference)*, vol. 2021, Oct. 2021. doi: 10.1109/IECON48115.2021.9589347.

3 Y. Zeng *et al.*, "Multiagent deep reinforcement learning-aided output current sharing control for input-series output-parallel dual active bridge converter," *IEEE Trans. Power Electron.*, vol. 37, no. 11, pp. 12955–12961, 2022. doi: 10.1109/TPEL.2022.3181243.

4 V. S. B. Kurukuru, A. Haque, M. A. Khan, S. Sahoo, A. Malik, and F. Blaabjerg, "A review on artificial intelligence applications for grid-connected solar photovoltaic systems," *Energies*, vol. 14, no. 15, p. 4690, Aug. 2021. doi: 10.3390/en14154690.

5 A. Malik, A. Haque, V. S. B. Kurukuru, M. A. Khan, and F. Blaabjerg, "Overview of fault detection approaches for grid connected photovoltaic inverters," *e-Prime Adv. Electr. Eng. Electron. Energy*, vol. 2, no. 2021, p. 100035, 2022. doi: 10.1016/j.prime.2022.100035.

6 S. Mateen, M. Amir, A. Haque, and F. I. Bakhsh, "Ultra-fast charging of electric vehicles: A review of power electronics converter, grid stability and optimal battery consideration in multi-energy systems," *Sustain. Energy, Grids Netw.*, vol. 35, p. 101112, Sep. 2023. doi: 10.1016/j.segan.2023.101112.

7 Y. Yu, G. Konstantinou, B. Hredzak, and V. G. Agelidis, "Operation of cascaded H-bridge multilevel converters for large-scale photovoltaic power plants under bridge failures," *IEEE Trans. Ind. Electron.*, vol. 62, no. 11, pp. 7228–7236, 2015. doi: 10.1109/TIE.2015.2434995.

8 S. Mateen, A. Haque, F. S. Zehgeer, and F. I. Bakhsh, "Current stress minimization in dual active bridge converter for ultra-fast charging of electric vehicles," in *2023 11th National Power Electronics Conference (NPEC),* Dec. 2023, pp. 1–6. doi: 10.1109/NPEC57805.2023.10385015.

9 R. Davoodnezhad, D. G. Holmes, and B. P. McGrath, "A novel three-level hysteresis current regulation strategy for three-phase three-level inverters," *IEEE Trans. Power Electron.*, vol. 29, no. 11, pp. 6100–6109, 2014. doi: 10.1109/TPEL.2013.2295597.

10 M. A. Khan, A. Haque, and V. S. B. Kurukuru, "Droop based Low voltage ride through implementation for grid integrated photovoltaic system," in *2019 International Conference on Power Electronics, Control and Automation (ICPECA),* Nov. 2019, pp. 1–5. doi: 10.1109/ICPECA47973.2019.8975467.

11 S. Zhao, F. Blaabjerg, and H. Wang, "An overview of artificial intelligence applications for power electronics," *IEEE Trans. Power Electron.*, vol. 36, no. 4, pp. 4633–4658, 2021. doi: 10.1109/TPEL.2020.3024914.

12 M. R. G. Meireles, P. E. M. Almeida, and M. G. Simões, "A comprehensive review for industrial applicability of artificial neural networks," *IEEE Trans. Ind. Electron.*, vol. 50, no. 3, pp. 585–601, 2003. doi: 10.1109/TIE.2003.812470.

13 B. K. Bose, "Neural network applications in power electronics and motor drives—An introduction and perspective," in *IECON Proceedings (Industrial Electronics Conference)*, vol. 54, no. 1, pp. 25–27, 2008. doi: 10.1109/IECON.2008.4757921.

14 B. K. Bose, "Artificial intelligence techniques in smart grid and renewable energy systems -Some example applications," *Proc. IEEE*, vol. 105, no. 11, pp. 2262–2273, 2017. doi: 10.1109/JPROC.2017.2756596.

15 M. Seyedmahmoudian *et al.*, "State of the art artificial intelligence-based MPPT techniques for mitigating partial shading effects on PV systems – A review," *Renew. Sustain. Energy Rev.*, vol. 64, pp. 435–455, 2016. doi: 10.1016/j.rser.2016.06.053.

16 A. Mellit and S. A. Kalogirou, "Artificial intelligence techniques for photovoltaic applications: A review," *Prog. Energy Combust. Sci.*, vol. 34, no. 5, pp. 574–632, 2008. doi: 10.1016/j.pecs.2008.01.001.

17 H. Soliman, H. Wang, and F. Blaabjerg, "A review of the condition monitoring of capacitors in power electronic converters," *IEEE Trans. Ind. Appl.*, vol. 52, no. 6, pp. 4976–4989, 2016. doi: 10.1109/TIA.2016.2591906.

18 L. Duchesne, E. Karangelos, and L. Wehenkel, “Recent developments in machine learning for energy systems reliability management,” *Proc. IEEE*, vol. 108, no. 9, pp. 1656–1676, 2020. doi: 10.1109/JPROC.2020.2988715.

19 K. V. S. Bharath, A. Haque, and M. A. Khan, “Condition monitoring of photovoltaic systems using machine leaming techniques,” in *2018 2nd IEEE International Conference on Power Electronics, Intelligent Control and Energy Systems (ICPEICES)*, Oct. 2018, pp. 870–875. doi: 10.1109/ICPEICES.2018.8897413.

20 V. S. Bharath Kurukuru, A. Haque, R. Kumar, M. A. Khan, and A. K. Tripathy, “Machine learning based fault classification approach for power electronic converters,” in *9th IEEE International Conference on Power Electronics, Drives & Energy Systems* 2020, 2020. doi: 10.1109/PEDES49360.2020.9379365.

21 K. Dhibi *et al.*, “Reduced kernel random forest technique for fault detection and classification in grid-tied PV systems,” *IEEE J. Photovoltaics*, vol. 10, no. 6, pp. 1864–1871, 2020. doi: 10.1109/JPHOTOV.2020.3011068.

22 V. S. B. Kurukuru, A. Haque, A. K. Tripathi, and M. A. Khan, “Condition monitoring of IGBT modules using online TSEPs and data-driven approach,” *Int. Trans. Electr. Energy Syst.*, vol. 31, no. 8, pp. 1–24, 2021. doi: 10.1002/2050-7038.12969.

23 W. Caesarendra and T. Tjahjowidodo, “A review of feature extraction methods in vibration-based condition monitoring and its application for degradation trend estimation of low-speed slew bearing,” *Machines*, vol. 5, no. 4, p. 21, 2017. doi: 10.3390/machines5040021.

24 V. S. Bharath Kurukuru, F. Blaabjerg, M. A. Khan, and A. Haque, “A novel fault classification approach for photovoltaic systems,” *Energies*, vol. 13, no. 2, p. 308, 2020. doi: 10.3390/en13020308.

25 W. Gong, H. Chen, Z. Zhang, M. Zhang, and H. Gao, “A data-driven-based fault diagnosis approach for electrical power DC-DC inverter by using modified convolutional neural network with global average pooling and 2-D feature image,” *IEEE Access*, vol. 8, pp. 73677–73697, 2020. doi: 10.1109/ACCESS.2020.2988323.

26 W. K. Lu and Q. Zhang, “Deconvolutive short-time fourier transform spectrogram,” *IEEE Signal Process Lett.*, vol. 16, no. 7, pp. 576–579, 2009. doi: 10.1109/LSP.2009.2020887.

27 W. Martin and P. Flandrin, “Wigner-Ville spectral analysis of nonstationary processes,” *IEEE Trans. Acoust.*, vol. 33, no. 6, pp. 1461–1470, 1985. doi: 10.1109/TASSP.1985.1164760.

28 Z. K. Peng, P. W. Tse, and F. L. Chu, “A comparison study of improved Hilbert-Huang transform and wavelet transform: Application to fault diagnosis for rolling bearing,” *Mech. Syst. Signal Process.*, vol. 19, no. 5, pp. 974–988, 2005. doi: 10.1016/j.ymssp.2004.01.006.

29 Z. Liu, C. Xiao, T. Zhang, and X. Zhang, "Research on fault detection for three types of wind turbine subsystems using machine learning," *Energies*, vol. 13, no. 2, pp. 1–21, 2020. doi: 10.3390/en13020460.

30 C. Li, Y. A. Chen, C. Jin, R. Sharma, and J. Kleissl, "Online PV smart inverter coordination using deep deterministic policy gradient," *Electr. Power Syst. Res.*, vol. 209, p. 107988, Aug. 2022. doi: 10.1016/j.epsr.2022.107988.

31 G. Seo, S. Yoon, J. Song, E. Srivastava, and E. Hwang, "Label-free fault detection scheme for inverters of PV systems: Deep reinforcement learning-based dynamic threshold," *Appl. Sci.*, vol. 13, no. 4, p. 2470, 2023. doi: 10.3390/app13042470.

32 S. M. Miraftabzadeh, M. Longo, F. Foiadelli, M. Pasetti, and R. Igual, "Advances in the application of machine learning techniques for power system analytics: A survey," *Energies*, vol. 14, no. 16, p. 4776, Aug. 2021. doi: 10.3390/en14164776.

33 M. Ding, L. Wang, and R. Bi, "An ANN-based approach for forecasting the power output of photovoltaic system," *Procedia Environ. Sci.*, vol. 11, no. PART C, pp. 1308–1315, 2011. doi: 10.1016/j.proenv.2011.12.196.

34 M. Manohar, E. Koley, Y. Kumar, and S. Ghosh, "Discrete wavelet transform and kNN-based fault detector and classifier for PV integrated microgrid," *Lect. Notes Netw. Syst.*, vol. 38, pp. 19–28, 2018. doi: 10.1007/978-981-10-8360-0_2.

35 D. P. Mishra, S. R. Samantaray, and G. Joos, "A combined wavelet and data-mining based intelligent protection scheme for microgrid," *IEEE Trans. Smart Grid*, vol. 7, no. 5, pp. 2295–2304, 2016. doi: 10.1109/TSG.2015.2487501.

36 Q. Liu, S. Liu, Q. Dai, X. Yu, D. Teng, and M. Wei, "Data-driven approaches for diagnosis of incipient faults in cutting arms of the roadheader," *Shock Vib.*, vol. 2021, p. 8865068, 2021. doi: 10.1155/2021/8865068.

37 A. Mellit, "Artificial intelligence technique for modelling and forecasting of meteorological data: A survey," *Atmos. Turbul. Meteorol. Model. Aerodyn.*, vol. 1, no. 1, pp. 293–328, 2011.

38 D. Su, E. Batzelis, and B. Pal, "Machine learning algorithms in forecasting of photovoltaic power generation," in *SEST 2019 – 2nd Int. Conf. Smart Energy Syst. Technol.*, 2019. doi: 10.1109/SEST.2019.8849106.

39 Z. Li, S. M. Mahbobur Rahman, R. Vega, and B. Dong, "A hierarchical approach using machine learning methods in solar photovoltaic energy production forecasting," *Energies*, vol. 9, no. 1, p. 55, 2016. doi: 10.3390/en 9010055.

40 S. Rajagopalan and S. Santoso, "Wind power forecasting and error analysis using the autoregressive moving average modeling," in *2009 IEEE Power & Energy Society General Meeting. PES'09*, pp. 1–6, 2009. doi: 10.1109/PES.2009. 5276019.

41 A. U. Haque, P. Mandal, J. Meng, and M. Negnevitsky, "Wind speed forecast model for wind farm based on a hybrid machine learning algorithm," *Int. J. Sustain. Energy*, vol. 34, no. 1, pp. 38–51, 2015. doi: 10.1080/14786451.2013.826224.

42 L. Avila, M. De Paula, I. Carlucho, and C. Sanchez Reinoso, "MPPT for PV systems using deep reinforcement learning algorithms," *IEEE Lat. Am. Trans.*, vol. 17, no. 12, pp. 2020–2027, 2019. doi: 10.1109/TLA.2019.9011547.

43 Y. Zhang, "Neural network algorithm with reinforcement learning for parameters extraction of photovoltaic models," *IEEE Trans. Neural Netw. Learn. Syst.*, vol. 34, pp. 1–11, 2021. doi: 10.1109/TNNLS.2021.3109565.

44 M. Zhang, P. I. Gómez, Q. Xu, and T. Dragicevic, "Review of online learning for control and diagnostics of power converters and drives: Algorithms, implementations and applications," *Renew. Sustain. Energy Rev.*, vol. 186, p. 113627, Oct. 2023. doi: 10.1016/j.rser.2023.113627.

45 K. Hu, Z. Liu, Y. Yang, F. Iannuzzo, and F. Blaabjerg, "Ensuring a reliable operation of two-level IGBT-based power converters: A review of monitoring and fault-tolerant approaches," *IEEE Access*, vol. 8, pp. 89988–90022, 2020. doi: 10.1109/ACCESS.2020.2994368.

46 H. Bai, X. Wang, and F. Blaabjerg, "Passivity enhancement in renewable energy source based power plant with paralleled grid-connected VSIs," vol. 53, no. 4, pp. 1–9, 2017. doi: 10.1109/TIA.2017.2685363.

47 C. Xiang *et al.*, "A novel deep deterministic policy gradient assisted learning based control algorithm for three-phase DC/AC inverter with an RL load," *IEEE J. Emerg. Sel. Top. Power Electron.*, vol. 11, no. 6, p. 1, 2022. doi: 10.1109/JESTPE.2022.3174638.

48 T. Mazhar *et al.*, "Electric vehicle charging system in the smart grid using different machine learning methods," *Sustainability*, vol. 15, no. 3, p. 2603, 2023. doi: 10.3390/su15032603.

49 M. Shibl, L. Ismail, and A. Massoud, "Electric vehicles charging management using machine learning considering fast charging and vehicle-to-grid operation," *Energies*, vol. 14, no. 19, p. 6199, 2021. doi: 10.3390/en14196199.

50 W. He *et al.*, "Data-driven estimation of battery state-of-health with formation features," *J. Micromechanics Microengineering*, vol. 34, no. 7, p. 075004, Jul. 2024. doi: 10.1088/1361-6439/ad520c.

51 M. Bhattacharya, S. Saha, D. Khan, and T. Nag, "Wavelet based component fault detection in diode clamped multilevel inverter using probabilistic neural network," in *2017 2nd International Conference for Convergence in Technology I2CT 2017*, vol. 2017, pp. 1163–1168, Jan. 2017. doi: 10.1109/I2CT.2017.8226310.

52 K. W. Lee, M. Kim, J. Yoon, S. Bin Lee, and J. Y. Yoo, "Condition monitoring of DC-link electrolytic capacitors in adjustable-speed drives," *IEEE Trans. Ind. Appl.*, vol. 44, no. 5, pp. 1606–1613, 2008. doi: 10.1109/TIA.2008.2002220.

53 C. Bhargava, V. K. Banga, and Y. Singh, "An intelligent prognostic model for electrolytic capacitors health monitoring: A design of experiments approach," *Adv. Mech. Eng.*, vol. 10, no. 10, pp. 1–11, 2018. doi: 10.1177/168781401878 1170.

54 V. A. Sankaran, F. L. Rees, and C. S. Avant, "Electrolytic capacitor life testing and prediction," in *Conference Record – IAS Annual Meeting (IEEE Industry Applications Society)*, vol. 2, pp. 1058–1065, 1997. doi: 10.1109/ias.1997.628992.

55 M. F. Samadi and M. Saif, "State-space modeling and observer design of Li-ion batteries using Takagi-Sugeno fuzzy system," *IEEE Trans. Control Syst. Technol.*, vol. 25, no. 1, pp. 301–308, 2017. doi: 10.1109/TCST.2016.2549270.

56 J. Yu, "Health degradation detection and monitoring of lithium-ion battery based on adaptive learning method," *IEEE Trans. Instrum. Meas.*, vol. 63, no. 7, pp. 1709–1721, 2014. doi: 10.1109/TIM.2013.2293234.

57 D. Zhou, Z. Li, J. Zhu, H. Zhang, and L. Hou, "State of health monitoring and remaining useful life prediction of lithium-ion batteries based on temporal convolutional network," *IEEE Access*, vol. 8, pp. 53307–53320, 2020. doi: 10.1109/ACCESS.2020.2981261.

58 A. Sangwongwanich and F. Blaabjerg, "Monte Carlo simulation with incremental damage for reliability assessment of power electronics," *IEEE Trans. Power Electron.*, vol. 36, no. 7, pp. 7366–7371, 2021. doi: 10.1109/TPEL.2020.3044438.

59 A. Sangwongwanich, Y. Yang, and S. Member, "Mission profile-oriented control for reliability and lifetime of photovoltaic inverters," *IEEE Trans. Ind. Appl.*, vol. 56, no. 1, pp. 601–610, 2019. doi: 10.1109/TIA.2019.2947227.

60 Y. Yang, A. Sangwongwanich, and F. Blaabjerg, "Design for reliability of power electronics for grid-connected photovoltaic systems," *CPSS Trans. Power Electron. Appl.*, vol. 1, no. 1, pp. 92–103, 2016.

61 C. Magazzino, M. Mele, and N. Schneider, "A machine learning approach on the relationship among solar and wind energy production, coal consumption, GDP, and CO_2 emissions," *Renew. Energy*, vol. 167, pp. 99–115, 2021. doi: 10.1016/j.renene.2020.11.050.

62 A. Arcos Jiménez, L. Zhang, C. Q. Gómez Muñoz, and F. P. García Márquez, "Maintenance management based on machine learning and nonlinear features in wind turbines," *Renew. Energy*, vol. 146, pp. 316–328, 2020. doi: 10.1016/j.renene.2019.06.135.

63 C. Q. Gómez Muñoz, F. P. García Márquez, and J. M. Sánchez Tomás, "Ice detection using thermal infrared radiometry on wind turbine blades," *Meas. J. Int. Meas. Confed.*, vol. 93, pp. 157–163, 2016. doi: 10.1016/j.measurement.2016.06.064.

4

Data Collection and Pre-processing

Manauwar Hussain, Suwaiba Mateen, and Ahteshamul Haque

Advance Power Electronics and Research Laboratory, Jamia Millia Islamia (A Central University), New Delhi, India

4.1 Introduction

Data is the backbone of Artificial Intelligence (AI), much like how experiences and observations are to human learning. AI models learn from data by identifying patterns, relationships, and trends. The better the data, the more accurately the model can generalize from it to make predictions or decisions. Applications of AI are highly dependent on data, making the quality and privacy of data essential concerns. Data is crucial for AI because it helps these systems learn and make accurate predictions. AI models use large amounts of data, often labeled by humans, to identify patterns and relationships. The effectiveness of AI systems depends heavily on the quality, amount, and variety of the data they are trained on. High-quality, diverse data leads to better performance and more capable AI [1].

Good data is vital for artificial intelligence because it affects how well AI models perform. High-quality data helps AI make accurate predictions and reliable outcomes, building trust and confidence among users [2]. It helps in ensuring accuracy, completeness, and consistency. Inaccuracies in data can lead to AI models learning incorrect patterns, leading to poor predictions or decisions. Similarly, incomplete data can skew the training process. For example, in predicting the remaining useful life (RUL) of components in power electronics, missing operational data can lead to an inaccurate understanding of wear and tear. Another issue is the inconsistent data. Data should be consistent across sources and time periods. Inconsistent data, such as different units of measurement or time stamps that don't align, can confuse the model and reduce its performance. Ensuring data quality also involves identifying and correcting biases in the data. An example of bias is if the training data of an AI system to predict faults predominantly comes from

Artificial Intelligence for Power Electronics, First Edition.
Edited by Ahteshamul Haque, Saad Mekhilef, and Azra Malik.

a specific type of equipment or operating condition, the model may not generalize well to other types or conditions, leading to biased predictions. It is important to mitigate bias in a system so as to reduce unfair treatment of certain groups or individuals [2]. Another important factor in data preparation is the role of human oversight. Much of AI's success hinges on labeled data, where human experts annotate data points to provide the correct output labels. This labeling process must be rigorous to avoid introducing errors.

The data collection and pre-processing come with many challenges. In some domains, especially emerging ones, data might be scarce or expensive to collect. This can limit the effectiveness of AI models. Techniques like transfer learning (where models trained on related tasks are adapted to new ones) [3] and synthetic data generation are best suited to address this challenge. Also, as AI applications move toward real-time decision-making, the need for rapid data processing becomes critical. This requires efficient data pipelines that can handle large volumes of data with minimal latency [4]. In case of data collection and pre-processing it is important to focus on data privacy and ethical considerations. AI systems increasingly rely on personal and sensitive data; hence, privacy becomes a significant concern. Beyond privacy, there is the ethical use of data [5]. For instance, using data collected without consent or repurposing data for unintended uses can lead to ethical breaches. In power electronics, this might involve ensuring that operational data from equipment is used only for its intended purpose and not for unrelated analyses without proper consent.

Data is not just the fuel for AI, but the critical foundation that determines its success. In this chapter the importance of data in AI for power electronics, data collection and pre-processing techniques, data quality and challenges associated with it are analyzed. Data Annotation and labeling methods are discussed in detail to understand the importance of labeling and organizing data effectively. To understand the real-time case study for pre-processing of data, data smoothing and detecting outliers in MATLAB is shown in the chapter. As AI continues to evolve, the importance of robust data practices keeps on growing, making it an area of ongoing focus and innovation.

4.2 Data Collection in Power Electronics

Data collection for AI applications in power electronic systems is often driven by the need to solve the problem of ensuring that relevant data are collected. This process is challenging due to several factors. Firstly, the sheer volume of data that needs to be managed is immense, making it difficult to handle, store, and process effectively. Secondly, the diversity of data sources adds another layer of complexity, as data may come from various sensors, devices, and systems, each with its own format and standards. Integrating these diverse data streams into a cohesive dataset requires significant effort and sophisticated tools. Finally, ensuring that the collected data includes representative samples is crucial. Representative samples are necessary to ensure that the AI models trained on this data can generalize well and perform accurately in real-world scenarios. This involves careful planning and strategy to capture data that reflects the various conditions and variations that the AI system might encounter in actual operation. Together, these challenges highlight the complexity and importance of effective data collection strategies in developing reliable AI applications for power electronic systems [6].

In power electronic systems employing AI, ensuring high-quality data collection hinges on several critical factors. First and foremost, understanding the specific data requirements is essential. This involves precisely identifying the types of data necessary to train and optimize AI algorithms for their intended applications within power electronics. Secondly, determining the appropriate sources of data is crucial. These sources may include various sensors, monitoring devices, and operational logs that provide the relevant information needed for AI systems to make informed decisions and predictions. Additionally, assessing the quantity of data required is vital. While gathering sufficient data volume is important for robust model training and validation, it is equally essential to ensure that the data collected is representative and covers the full spectrum of conditions and scenarios relevant to power electronic operations. By meticulously addressing these aspects – data requirements, sources, and quantity – developers can enhance the reliability and effectiveness of AI applications in power electronic systems, thereby optimizing performance and facilitating more informed decision-making processes [7].

4.2.1 Types of Data in Power Electronics

In power electronics, various types of data are collected and utilized for different applications, ranging from system monitoring and control to predictive maintenance and optimization. The data can be broadly categorized based on its source, nature, and the specific application within the field of power electronics. The five primary types of data include electrical data, thermal data, fault and diagnostic data, control data, and performance metrics.

Electrical data include parameters like voltage, current, power, frequency, and power factor. These parameters are essential for understanding the operational state of the system. This data can be for real-time monitoring, fault detection, and performance analysis. For example, deviations in voltage or current levels can indicate potential issues like overloading or component failure.

Thermal data includes parameters such as temperature and heat flux. This data is critical for assessing the thermal performance of power electronic components [8]. Monitoring these parameters helps in implementing thermal management strategies, ensuring that components operate within safe limits to prolong their lifespan.

The third category of data in power electronics is the fault and diagnostic data. This data are the fault codes that identify and record specific faults in system. This can include over-voltage, over-current, short circuits, and thermal overloads. Similarly, error logs and diagnostic messages help to identify and troubleshoot issues [9]. This data allows operators to detect problems early and take corrective actions before a failure occurs, thereby minimizing downtime and maintenance costs.

The control data includes parameters like setpoints and control signals. Setpoints refers to the target values for key parameters like voltage, or current, which the control system uses to regulate the operation. The data on signals sent by the control system to adjust the operation of various components, such as changing the duty cycle is referred to as the control signals [10].

Performance metrics such as efficiency and Total Harmonic Distortion (THD) are key indicators of how well a power electronic system is operating. Efficiency measures how effectively the system converts input power to output power, while THD reflects the quality of the electrical output. These metrics are used for system optimization, energy management, and ensuring compliance with power quality standards. High efficiency and low THD are desirable for reducing energy losses and minimizing disturbances in the power grid [11].

4.2.2 Data Sources

For AI to effectively improve systems, it must integrate data from diverse sources. The fusion of sensor data with data acquisition systems, historical records, and external factors enables AI to build a comprehensive understanding of the system. This holistic approach allows for more accurate predictions, optimized control, and proactive maintenance, leading to a reliable and efficient system. The most common data sources in power electronics are:

- **Sensors and Measurement Instruments**: Sensors are the primary source of real-time data in power electronics. These include voltage sensors, current sensors, temperature sensors, and other measurement instruments. Electrical and thermal data are collected from these sensors. They provide raw data [12] that AI models need to monitor system health, detect anomalies, and make predictions.
- **Diagnostic and Monitoring Systems**: These systems continuously monitor the health of components and systems, logging faults, errors, and other diagnostic information. The fault and diagnostic data is collected by these systems [13]. It is essential for AI models focused on predictive maintenance, fault detection, and root cause analysis.

- **Historical Maintenance Records**: Maintenance logs, service records, and historical performance data provide a long-term view of system behavior and component reliability. It helps AI models to understand long-term trends and patterns, which is critical for predictive analytics and lifecycle management. AI can use this data to predict when maintenance is needed and to extend the lifespan of components [14].
- **Simulation and Modeling Tools**: Simulation software and modeling tools generate synthetic data that mimics real-world operations of systems under various conditions. Performance metrics can be collected from these tools. Simulation data is used to train AI models as shown by authors in [15], especially when real-world data is scarce or difficult to obtain. It helps AI understand how systems might behave under different scenarios, which is vital for robust model training.

4.2.3 Data Collection Methods

The data collection methods include real-time data collection, historical data collection, online and offline data collection [16]. Direct measurement using sensors and measurement instruments is the most common approach, providing real-time data on critical parameters such as voltage, current, and temperature. The precision and responsiveness of sensor-based data collection make it indispensable for real-time monitoring and control, ensuring that AI systems can react promptly to any changes in operating conditions. This data is often complemented by data logging systems, which store large volumes of historical data for trend analysis and predictive maintenance. For geographically dispersed or remote systems, such as solar farms [17], remote monitoring and telemetry are used to centralize data collection and reduce the need for on-site inspections. These methods involve collecting data from remote systems and transmitting it to a central location for processing and analysis. Simulation and modeling tools generate synthetic data, especially when real-world data is scarce, enabling the testing of various scenarios and the training of AI models. Simulation tools, such as those provided by platforms like MATLAB Simulink, Typhoon HIL, etc., allow engineers to model complex power electronic circuits and observe how they would perform under different loads, environmental conditions, or fault scenarios.

In industrial settings, SCADA (Supervisory Control and Data Acquisition) systems play a crucial role, integrating real-time data collection with control functions for complex processes. These systems gather data from various sensors and devices within a power electronic system, providing real-time monitoring and control capabilities [18]. SCADA systems not only collect data but also allow operators to control system parameters remotely, ensuring that AI-driven decision-making is integrated with the actual operation of the system. Embedded systems within power electronic devices, such as microcontrollers and Digital Signal Processors (DSPs), also contribute by collecting and processing data locally, essential for real-time control applications. These systems are embedded directly into devices like inverters, converters, and motor drives, where they collect data locally and process it in real-time [19]. Cloud-based data collection has become increasingly important, allowing for scalable data storage, advanced analytics, and remote accessibility, often facilitated by Internet of Things (IoT) devices. For example, an IoT-enabled energy meter might send usage data to a cloud platform, where AI algorithms analyze it to optimize energy consumption. Cloud-based data collection is particularly beneficial for systems that require large-scale data analysis, real-time monitoring, and remote accessibility [20]. Finally, event-based data collection focuses on capturing detailed data during specific conditions, such as faults or transient events, which is crucial for diagnosing and resolving issues. Together, these methods ensure that AI systems in power electronics have access to the high-quality, diverse data needed to optimize performance, predict failures, and improve overall system reliability.

4.3 Data Quality and Challenges

Data quality is a critical aspect of power electronic systems utilizing AI. The accuracy, completeness, and consistency of the data used for training and testing AI models have a direct impact on the performance and effectiveness of these systems. When data quality is low, it can result in biased, inaccurate, or irrelevant outcomes, which in turn negatively affect decision-making processes based on AI outputs within power electronics [21]. Ensuring high-quality data is therefore essential for AI systems in power electronics to deliver reliable and valuable results. This process involves several key activities, including data cleansing, validation, enrichment, and management. Data cleansing ensures that any errors or inconsistencies are removed from the dataset, while validation confirms that the data meets the required standards and is accurate. Data enrichment involves enhancing the dataset with additional relevant information, and effective data management ensures that the data remains organized and accessible [22]. AI applications in power electronics demand data that is not only high-quality but also relevant, representative, and reliable. This high-quality data is crucial for these AI systems to produce optimal outcomes. Moreover, the consistency of data quality over time is vital, necessitating ongoing monitoring and maintenance of the data. This ongoing process helps to ensure that the data used for training and validation remains of high quality, thereby maintaining the performance of the AI systems in power electronics [3]. Data quality in power electronic

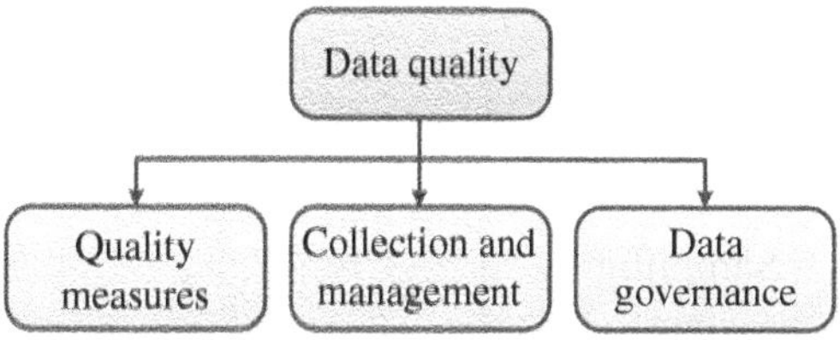

Figure 4.1 Challenging elements of data quality.

systems utilizing AI is a multidimensional concept that encompasses several crucial factors, including accuracy, completeness, consistency, and timeliness. Accuracy refers to the correctness of the data, ensuring that the information used by AI models is precise and error-free. Completeness involves having all necessary data available without any missing elements, which is essential for comprehensive analysis and decision-making. Consistency ensures that the data remains uniform and reliable across different datasets and over time, preventing discrepancies that could lead to faulty conclusions. Lastly, timeliness pertains to the relevance of the data being up-to-date and available when needed, which is critical for real-time applications and decision-making processes in power electronics. Together, these dimensions of data quality are vital for the effective performance and reliability of AI systems in power electronic applications [23]. The challenging elements of data quality in power electronic systems utilizing AI are shown in Figure 4.1. These generally are divided into three categories by the authors in [24] as: quality measures, collection and management, and data governance.

4.4 Data Pre-processing Techniques

Data pre-processing is the process of getting the dataset ready for usage after using Exploratory Data Analysis (EDA) to understand the subtleties of dataset and the key concerns in it. In an ideal world, the dataset would be perfect and without any problems. Unfortunately, real-world data will always present some issues that need to be addressed. Inconsistencies such as sensor noise, missing measurements, and varying signal scales often happen in the real world and need to be adjusted to make the data more useful and understandable. This process of cleaning and solving most of the issues in the data is called as the data preprocessing step. By addressing these issues, such as filtering out noise, handling missing data, and normalizing signal scales, it is ensured that the data used is clean, relevant, and suitable for analysis, ultimately leading to more accurate and reliable results [25]. A data pre-processing pipeline is shown in Figure 4.2.

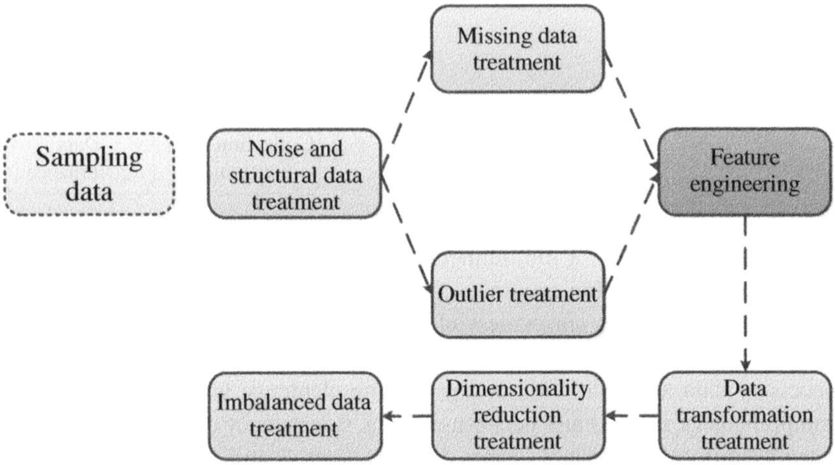

Figure 4.2 Data pre-processing pipeline.

4.4.1 Data Cleaning

Data cleaning is an essential part of data pre-processing, involving the detection and correction of bad and inaccurate observations to improve data quality. This process includes identifying and handling incomplete, inaccurate, duplicated, irrelevant, or null values in the dataset. Depending on the problem domain and project goals, the problematic data points can be either modified or deleted [26]. Some common issues in power electronics data include:

- **Noisy Data**: Noisy data refers to contains errors, inconsistencies, or irrelevant information, which can negatively impact the performance of AI models [27]. For instance, if a dataset has a column for "voltage" with negative values, these observations don't make sense and should be deleted or set to null. Another example is removing irrelevant data. For example, in case of predicting the efficiency of a power inverter, data about the color of the equipment is irrelevant and should be removed. Outliers, although sometimes valid, can also be considered noise. While some outliers may represent important variations, others might be noise that should be excluded from the analysis.
  A common technique to handle noisy data is binning. Sort the values, divide them into "bins" (equal-sized buckets), and then apply the mean or median to each bin, smoothing the data [28]. This method reduces the impact of noise and outliers, creating a more stable and reliable dataset for AI modeling. By applying techniques like binning and carefully filtering out irrelevant or erroneous data, the quality of the dataset is improved.

- **Missing Data**: Missing data is another common issue [29]. Most machine learning models can't handle missing values, so it is needed to fill in these gaps through a process called imputation. One solution is to remove the observation if the dataset is large and a few missing records won't affect the overall distribution, or if most attributes of the observation are null, making it meaningless.
  Another approach is to use a global constant like "NA" or 0 to fill the gap, especially if predicting the missing value is difficult, or to use the mean or median of the attribute. Alternatively, backward/forward fill methods can be applied by using the previous or next value to fill the missing data.
  A more robust approach involves using machine learning algorithms to predict missing values. For example, the kNN Imputer finds the k-nearest instances to the missing value instance and uses their mean for the missing attribute, or regression can be used to predict missing values based on other attributes. Authors in [30] utilized a variant of kNN (kNN-ST: kNN Spatio-Temporal) for predicting missing data in environmental crowd sensing. Choosing the right technique depends on the problem and the type of missing data, such as Missing Completely at Random (MCAR), Missing at Random (MAR), or Missing Not at Random (MNAR). Python's sklearn library provides tools like the KNN Imputer for this purpose [25].
- **Structural Errors**: Structural errors include typos and inconsistencies in data values. For instance, for a dataset about capacitor brands, it may appear as "ABB," "abb," or "A.B.B." These issues need to be fixed before using the data in AI model. An easy fix is to transform all words to lowercase. Addressing these errors usually requires manual intervention rather than automated techniques.
  By following these data cleaning steps, High-quality dataset is ensured that leads to more accurate and reliable analysis and modeling results [25].

4.4.2 Data Transformation

Data transformation is a critical step in the preprocessing phase that converts data from one format to another. Some algorithms require transformed data for optimal performance, and skipping this process can lead to poor model performance or even bias. For example, the kNN model uses distance measures to find the nearest neighbors to a given record. If one feature in the model has a much higher scale than others, it will dominate the distance calculation, creating bias [31]. Key techniques to handle this issue include transformation for categorical variables, such as ordinal and nominal variables. Ordinal variables have a defined order, like efficiency ratings (low, medium, high), and can be mapped to numbers using a function or the Label Encoder class in sklearn. Nominal variables, like types of power converters (buck,

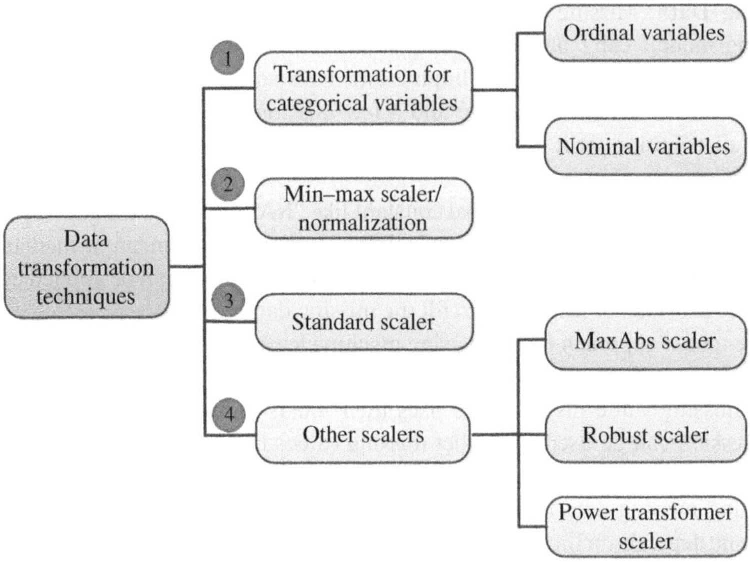

Figure 4.3 Different data transformation techniques.

boost, buck–boost), have no inherent order and are typically handled using One Hot Encoding, which transforms one column into multiple columns representing each unique value [32]. To avoid multicollinearity, one of these columns can be dropped. Data-transformation techniques can be illustrated in Figure 4.3.

For continuous data, methods like Min–Max Scaler (normalization) and Standard Scaler (z-score normalization or standardization) are commonly used. The Min–Max Scaler scales data between a predefined range, usually 0 to 1, and is useful when data doesn't follow a normal distribution, though it is sensitive to outliers. The Standard Scaler transforms data so that the mean is zero and the standard deviation is one, working best with data that follows a normal distribution and is not sensitive to outliers. Other scaling techniques include the MaxAbs Scaler, which scales data within the range of −1 to 1 by dividing each value by the absolute maximum value, and the Robust Scaler, which removes the median and scales data using the interquartile range (IQR), making it robust to outliers. The Power Transformer Scaler adjusts data to more closely follow a normal distribution, useful for data with varying variance (heteroscedasticity) [33]. Depending on the specific problem, different scalers can improve results, and while Min–Max and Standard Scalers are the most common, exploring other techniques can also be beneficial [34].

4.4.3 Feature Engineering

Feature engineering process involves applying domain knowledge to transform raw data into meaningful inputs that better capture the underlying patterns and relationships within the system. By doing so, the model's ability to make accurate predictions and optimizations is significantly improved. In this context, feature engineering often involves the manual creation of new features by transforming existing ones [35]. For example, in a dataset that includes raw measurements such as voltage, current, and temperature, one might create new features like power, efficiency ratios, or temperature gradients. These derived features can provide the model with more context and insight into the operational state of the system. Transformations could include mathematical operations like logarithms, exponentiations, or ratios, which can help to normalize data, highlight trends, or reduce the impact of noise. Additionally, engineers might apply domain-specific knowledge to create custom features that reflect the unique characteristics of power electronic systems, such as harmonic distortion measures, switching frequencies, energy losses, intrusion detection in system [36].

This tailored approach ensures that AI models are not only accurate but also interpretable and relevant to the practical needs of power electronics applications. Some basic examples that can be applied to the dataset to potentially increase model's performance are [37]:

- **Decompose a DateTime**: If the data is obtained from different power devices, the original datetime might be too complex for a model to understand. This datetime can be replaced with simpler features like "Hour of Day," "Day of the Week," or "Month of the Year" to help the model find patterns related to time.
- **Reframe Numerical Quantities**: This approach can be used if, for example, the data obtained on the power output of devices is throughout the year. Instead of just using the raw power output, new features that capture seasonal variations like "Summer Output," "Winter Output," "Spring Output," and "Fall Output" can be created. This way, the model can better understand how power output changes with seasons.
- **Using Domain Knowledge**: These examples show how one can use their understanding of power electronics to create new features that help the model make better predictions. The goal is to create features that are more informative and easier for the model to understand.

4.4.4 Data Reduction Techniques

In power electronics, data reduction can be likened to the process of simplifying a complex electrical system while maintaining its core functionalities and performance characteristics [37]. This involves using various techniques such as component selection, circuit simplification, sampling techniques, signal compression methods, and data binning or discretization to remove noise and eliminate redundant components, resulting in a more efficient and manageable system.

Data reduction works by identifying and eliminating unnecessary components and processes through the aforementioned techniques. Each technique optimizes the system in its own way. The benefits of data reduction in this field include improved efficiency in system analysis, processing, and storage, reduced processing time and computational requirements, enhanced interpretability of system data, mitigation of issues related to high dimensionality, and improvements in model accuracy and performance [38]. Some commonly used data reduction techniques are:

- **Dimensionality Reduction**: These include two major techniques for data reduction: Principal Component Analysis (PCA) and Linear Discriminant Analysis (LDA). PCA is a statistical technique that transforms the original dataset into a smaller set of uncorrelated variables called principal components [39], whereas LDA reduces dimensionality by focusing on maximizing the separation between different classes in the data [40].
- **Feature Selection**: Feature selection includes techniques such as filter methods, wrapper methods, and embedded methods [41]. In filter methods the relevance of each feature in the dataset based on statistical measures like correlation or mutual information is evaluated. Features that have little or no correlation with the target variable is removed, simplifying the dataset and reducing computational load. The wrapper methods involve using a machine learning model to evaluate the importance of each feature by iteratively adding or removing features and assessing model performance. This method is computationally intensive but leads to better feature subsets. The embedded methods, on the other hand, perform feature selection during the model training process, where algorithms like decision trees or LASSO regression automatically select the most important features [42]. A detailed comparison of various feature selection methods explanation of various techniques utilized by these methods is shown in Figure 4.4 [43].
- **Sampling**: Sampling can be random or stratified [44]. Random sampling involves selecting a subset of the data randomly, which can reduce the dataset size while maintaining a representative sample. And stratified sampling divides the data into strata or groups based on certain characteristics (e.g. operating conditions, load levels) and then samples from each group proportionally. This ensures that the reduced dataset remains balanced and representative of all conditions.

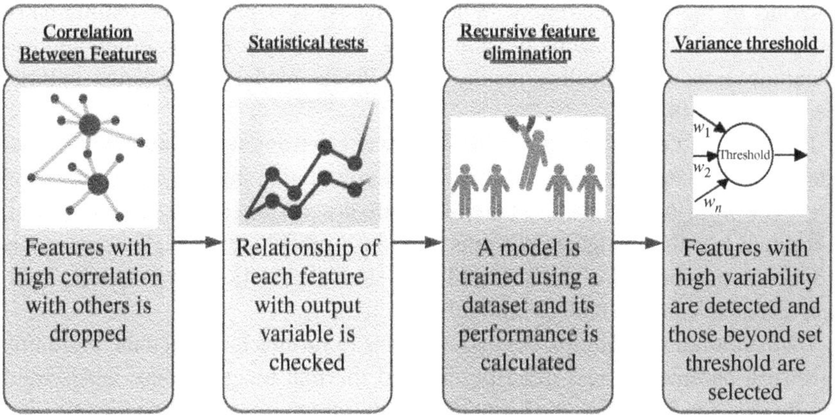

Figure 4.4 Feature selection procedure in data pre-processing.

- **Data Aggregation**: Data aggregation techniques, such as averaging or summarizing data over specific intervals, help in reducing the number of data points while preserving the overall trends [45]. For example, instead of storing every millisecond reading from a voltage sensor, data might be averaged over seconds or minutes, significantly reducing the dataset size while still capturing the essential information. In time-series data common in power electronics, data points can be aggregated over specific time windows to reduce the volume of data [46].

4.5 Data Annotation and Labeling

Data annotation means labeling and organizing data so AI can use it effectively. In the field of power electronics, this involves using a large amount of high-quality data to build AI models. This data needs to be labeled accurately to help companies improve their AI systems, making them work better [47]. Accurate labeling is essential because, in power electronics, AI needs to distinguish between different types of data such as electrical signals, voltage levels, current measurements, thermal images, or a mix of these. Once the data is correctly identified, the AI can use this information to perform various tasks, such as detecting faults, predicting energy needs, optimizing performance, and ensuring system reliability.

For example, when AI receives data from a power electronics system, it must understand what type of data it is dealing with to process it correctly. If the data pertains to voltage levels, the AI will analyze it differently than if it were thermal images. This precise data labeling ensures that the AI model can interpret and respond to the data accurately. Whether the AI is deployed for maintenance purposes, energy management, automated control systems, or other applications, accurate data labeling results in a reliable model that delivers optimal performance [48].

In the context of machine learning for power electronics, data labeling involves marking raw data such as signal waveforms, component images, and operational logs with informative labels. These labels are crucial for training the AI model, helping it learn to recognize patterns and make predictions [49]. This process is applied in numerous use cases, including system diagnostics, power quality analysis, predictive maintenance, and fault detection.

Data annotation encompasses adding labels to various forms of data, including electrical parameters, text descriptions, and thermal images. These labels train AI models for tasks like fault prediction, autonomous grid management, and more. The role of humans in this process, often referred to as the "Human in the Loop," is vital. Human intelligence is essential for verifying, validating, and correcting the AI model's outcomes, ensuring efficiency and enabling continuous improvement.

4.5.1 Methods for Data Labeling

Data labeling can be done by three ways: Manually, automated, and semi-automated [50]. These methods are discussed in detail, and their comparison based on different aspects is shown in Table 4.1.

Manual labeling refers to the usual way of creating a training dataset by manually labeling data. This means experts look at each piece of data and label it by hand. It works very well when it's important to avoid mistakes. For example, having skilled engineers label power grid data by hand to build a model for predicting faults ensures the data is reliable. It can capture rare or unusual cases and, thereby, ensures high-quality data. Manual labeling can also be done by external workers (like temporary workers or contractors), known as crowdsourcing. This requires lot of time and effort and is expensive to hire professional data labelers. Automated data labeling means that humans are not involved in the data labeling process. Instead, machine learning models train themselves by learning labeling rules from data samples and then apply these rules to unlabeled data. The procedure is fast, cost effective, consistent, and is highly scalable. One mistake in automated data labeling can increase the probability of future errors. Semi-automated data labeling combines the strengths of automated labeling with human expertise. In this process, machine learning rapidly labels the data, and then human experts review and correct any mistakes made by the algorithm. This approach speeds up the data labeling process while maintaining high data quality. Examples of tools that enable this include Labelbox and SuperAnnotate.

Table 4.1 Comparison Among Different Data-Labeling Methods.

Aspect	Manual Labeling	Automated Labeling	Semi-automated Labeling
Accuracy	High, depends on expert knowledge	Varies, depends on rule quality and model accuracy	Moderate to high
Cost	High	Low	Moderate
Scalability	Low	High	Moderate
Speed	Slow	Fast	Moderate
Human involvement	Full, requires expert annotation	Minimal	Partial
Use case suitability	Best for small datasets or when accuracy is critical	Best for very large datasets where speed and scalability are key	Suitable for medium-to-large datasets, when some human validation is needed
Error handling	Errors can be minimized with careful annotation	Errors depend on the accuracy of the models	Errors reduced through active learning and iteration
Adaptability	High	Limited	Adaptable
Training requirement	Requires domain experts	Requires initial setup and model training	Requires both experts and trained models
Examples	Labeling fault conditions, efficiency states manually	Rule-based labeling of operational states	Active learning for fault detection, label propagation for performance prediction

4.6 Case Study: Data Smoothing and Detecting Outliers

Outlier identification finds data points that differ noticeably from the rest of the data, while data smoothing refers to methods for removing undesired noise from data. Data smoothing and outlier detection are essential steps in data preprocessing, especially when dealing with noisy datasets. They help in making the data more suitable for analysis or modeling by reducing noise and identifying or removing anomalous data points. In this chapter MATLAB software is used for data cleaning and detecting local extrema. Smoothing and detrending processes are used to eliminate noise and remove polynomial trends, allowing for a clearer analysis of underlying patterns.

4.6.1 Moving Window Methods

In order to statistically represent a neighborhood of points in the data, moving window approaches are strategies to analyze the data in smaller batches at a time. Frequently used for data smoothing, the moving average method moves a window along the data and calculates the mean of the points inside each frame. By doing this, small differences between data points may be eliminated. Take measurements of the wind speed, for instance, which were made every minute for roughly three hours. To reduce gusts of wind that are traveling at a high pace, the movmean function within a 10-minute interval is used. The resulting plot is shown in Figure 4.5.

```
load windData.mat
mins = 1:length(speed);
window = 10;
meanspeed = movmean(speed,window);
plot(mins,speed,mins,meanspeed)
axis tight
legend("Measured Wind Speed","Average Wind Speed over 10 min Window")
xlabel("Time")
ylabel("Speed")
```

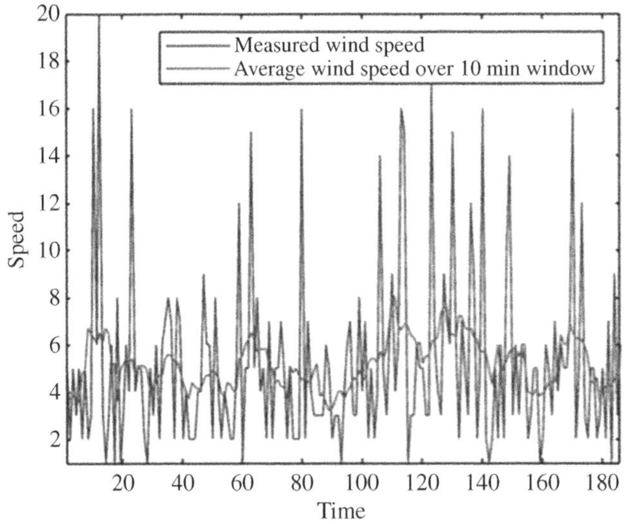

Figure 4.5 Plot for measured wind speed and average wind speed over 10 minutes window.

```
medianspeed = movmedian(speed,window);
plot(mins,speed,mins,medianspeed)
axis tight
legend("Measured Wind Speed","Median Wind Speed over 10 min Window")
xlabel("Time")
ylabel("Speed")
```

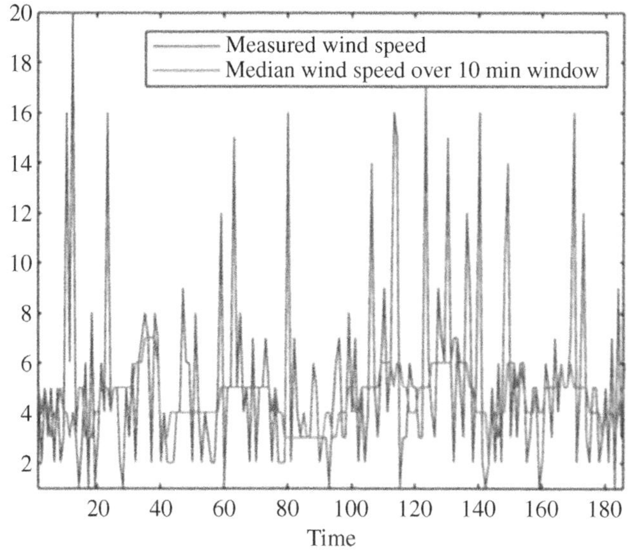

Figure 4.6 Plot for measured wind speed and median wind speed over 10 minutes window.

Similarly, movmedian function may be used to find the median wind speed across a sliding window. To do this enter the following commands are entered on editor window in MATLAB. The resulting waveforms are shown in Figure 4.6.

Now, it can be seen that moving window smoothing is not appropriate for all data. Make a sinusoidal signal, for instance, and add random noise. The corresponding plot is depicted in Figure 4.7.

Use a moving mean with a window size of 4 to smooth the noisy data as shown in Figure 4.8.

The moving mean accurately captures the overall form of the data, but it is not very good at capturing the valleys or local minima. The mean is not a very good approximation to the valley spots because in each window, there are two larger neighbors surrounding those sites. The shorter peaks are completely eliminated by the mean when the window size is increased. The resulting plot is shown in

```
t = 1:0.2:15;
A = sin(2*pi*t) + cos(2*pi*0.5*t);
Anoise = A + 0.5*rand(1,length(t));
plot(t,A,t,Anoise)
axis tight
legend("Original Data","Noisy Data")
```

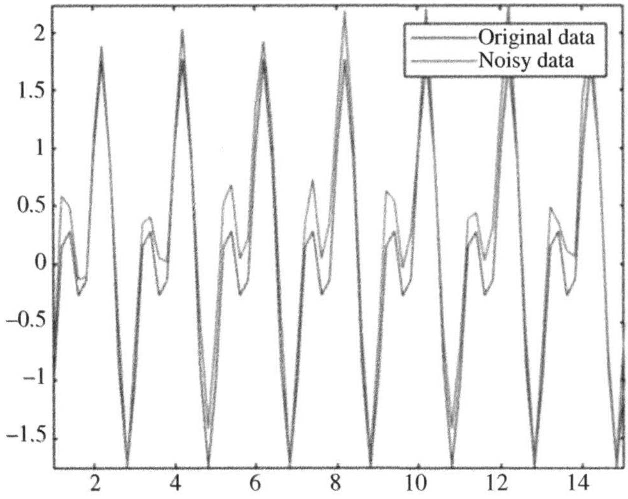

Figure 4.7 Plot for original and noisy data.

Figure 4.9. Alternative methods of smoothing could be taken into consideration for this kind of data.

4.6.2 Common Smoothing Methods

Many smoothing alternatives, including the well-liked Savitzky–Golay approach for signal processing, are offered via the smoothdata function. By default, smoothdata selects the method's best-guess window size based on the available data. The smooth data using this approach is shown in Figure 4.10.

Another smoothing technique that works especially well when the data contains both noise and outliers is the Robust Lowess approach. Introduce an anomaly into the turbulent data, then employ Robust Lowess to even out the data, hence removing the anomaly. The plot for smooth data is shown in Figure 4.11.

```
window = 4;
Amean = movmean(Anoise,window);
plot(t,A,t,Amean)
axis tight
legend("Original Data","Moving Mean - Window Size 4")
```

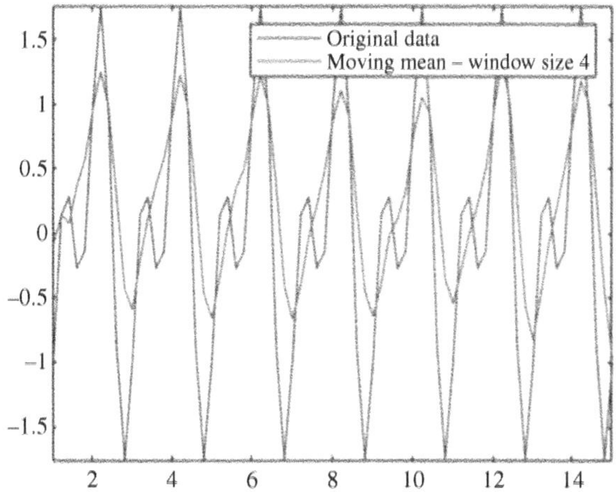

Figure 4.8 Plot for original data and moving mean for window size 4.

```
Amean = movmean(Anoise,5);
plot(t,A,t,Amean)
axis tight
legend("Original Data","Moving Mean - Window Size 10")
```

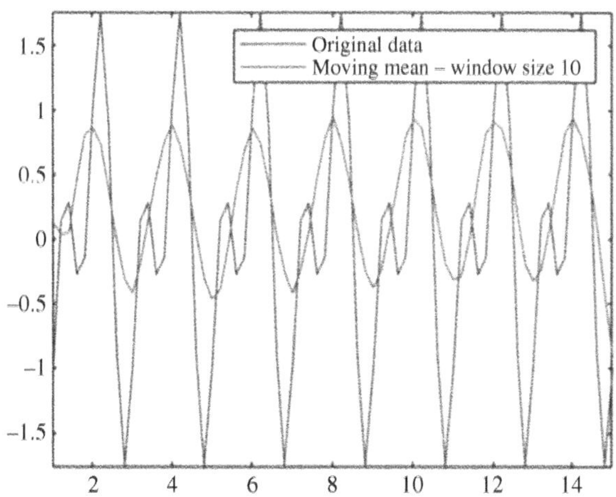

Figure 4.9 Plot for original data and moving mean for window size 10.

```
[Asgolay,window] = smoothdata(Anoise,"sgolay");
plot(t,A,t,Asgolay)
axis tight
legend("Original Data","Savitzky-Golay","location","best")
```

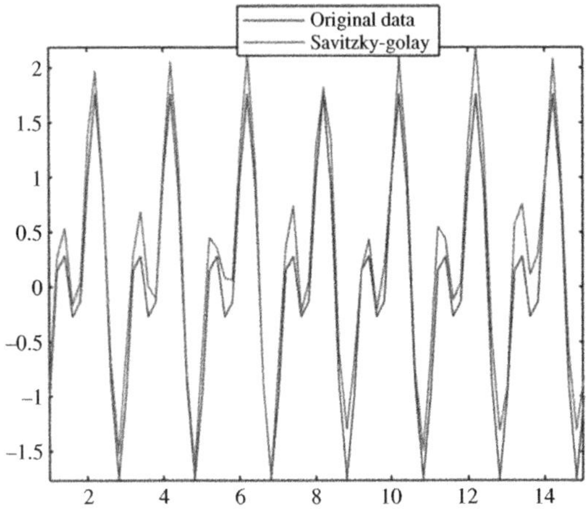

Figure 4.10 Plot for original data and smooth data utilizing Savitzky–Golay approach.

```
Anoise(36) = 20;
Arlowess = smoothdata(Anoise,"rlowess",10);
plot(t,Anoise,t,Arlowess)
axis tight
legend("Noisy Data","Robust Lowess")
```

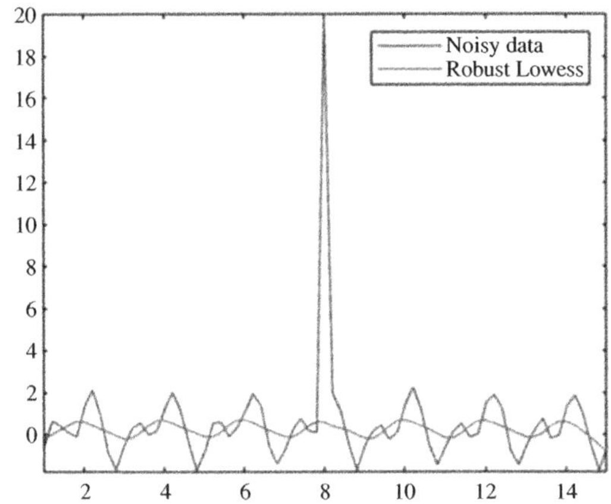

Figure 4.11 Plot for noisy data and smooth data utilizing Robust Lowess approach.

4.6.3 Detecting Outliers

Data processing outcomes and other computed values might be severely distorted by outliers in the data. For instance, deceptive peaks or valleys can be obtained if an attempt to smooth data with outliers using a moving median is performed. Figure 4.12 shows the plot for moving median.

By utilizing the filloutliers function along with offering a fill technique, one can replace outliers in the data. For instance, enter the value of the neighbor's value exactly next to the outlier in Anoise. The plot for noisy data with outlier and filled outlier is shown in Figure 4.13.

Alternatively, the rmoutliers function can be used to exclude outliers from the data. Take away Anoise's outlier, for instance. Figure 4.14 shows the noisy data with and without outlier.

4.6.4 Nonuniform Data

Different data sets have different point spacing, which might have an impact on data processing techniques. For the data in Airreg, create a datetime vector with erratic sampling times. Samples were taken hourly for the first 2 days, and then every minute for the next 30. This is represented by the time vector. Now Figure 4.15 shows the insightful results.

```
Amedian = smoothdata(Anoise,"movmedian");
plot(t,Anoise,t,Amedian)
axis tight
legend("Noisy Data","Moving Median")
```

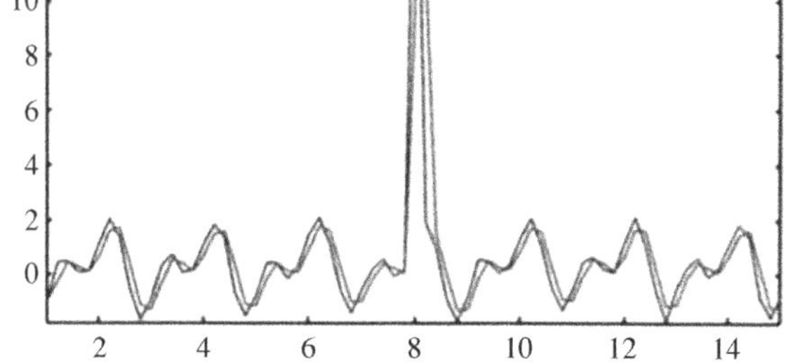

Figure 4.12 Plot for noisy data and moving median.

```
Afill = filloutliers(Anoise,"next");
plot(t,Anoise,t,Afill,"o-")
axis tight
legend("Noisy Data with Outlier","Noisy Data with Filled Outlier")
```

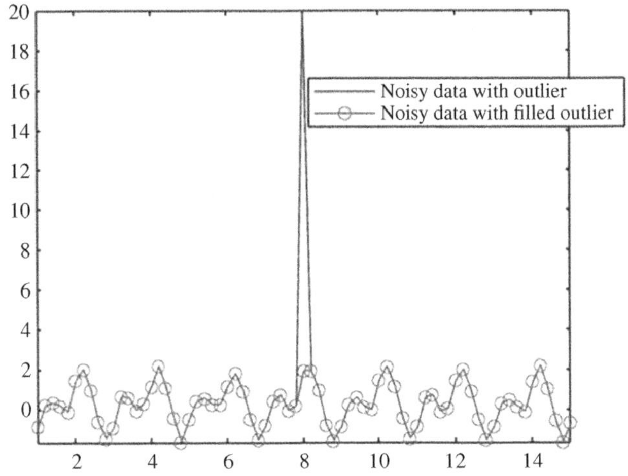

Figure 4.13 Plot for noisy data with outlier and with filled outlier.

```
Aremove = rmoutliers(Anoise);
plot(t,Anoise,t(~TF),Aremove,"o-")
axis tight
legend("Noisy Data with Outlier","Noisy Data with Outlier Removed")
```

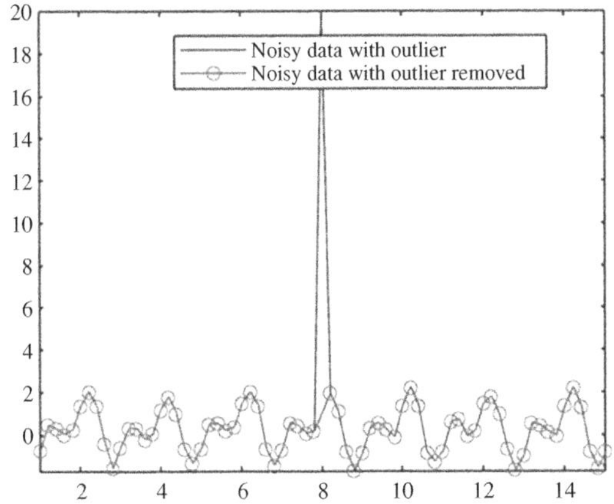

Figure 4.14 Noisy data with and without outlier.

```
t0 = datetime(2024,1,1,1,1,1);
timeminutes = sort(t0 + minutes(1:30));
timehours = t0 + hours(1:48);
time = [timeminutes timehours];
Airreg = rand(1,length(time));
plot(time,Airreg)
axis tight
```

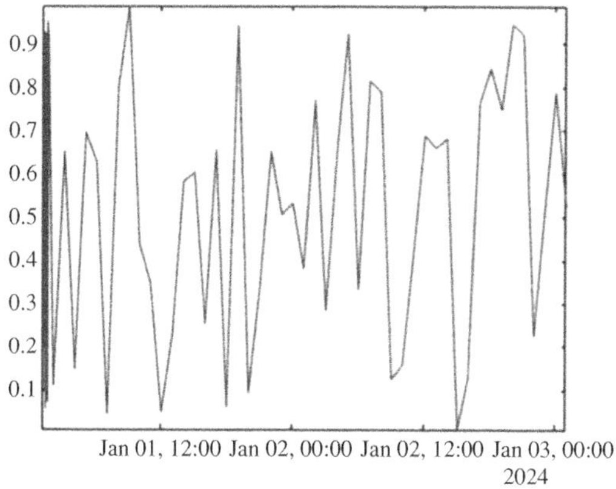

Figure 4.15 DateDatetime vector with erratic sampling time.

Smoothdata smooths by default with respect to numbers that are equally spaced, in this example, 1, 2, … , 78. Smoothing does not remove noise from the first half hour of data because integer time stamps do not match the sampling of the points in Airreg as provided in Figure 4.16.

The sample points to a number of MATLAB® data processing functions, such as filloutliers, movmean, and smoothdata, can be provided, to make sure that the data is treated in accordance with its sampling units and frequencies. Use the SamplePoints name-value parameter with the time stamps in time in Airreg to eliminate the high-frequency volatility in the first half-hour of data. The plot for smoothed data with sampled points is shown in Figure 4.17.

Data preprocessing is essential for accurate and meaningful analysis, involving several techniques to prepare data for further study. Data cleaning is one of these key techniques, which focuses on identifying, correcting, or removing bad or missing data to enhance data quality. Detecting local extrema and abrupt changes is crucial for uncovering significant trends or shifts within the data. Smoothing and detrending processes are used to eliminate noise and remove polynomial trends, allowing for a clearer analysis of underlying patterns. Scaling adjusts the data's range, making it easier to compare across different datasets.

```
Adefault = smoothdata(Airreg,"movmean",4);
plot(time,Airreg,time,Adefault)
axis tight
legend("Original Data","Smoothed Data with Default Sample Points")
```

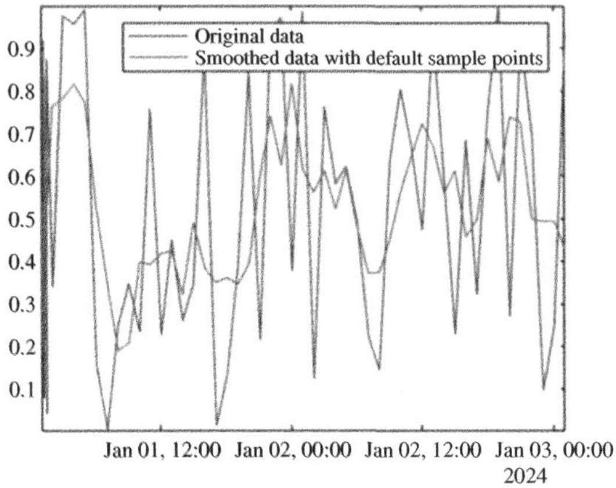

Figure 4.16 Plot of original data and smoothed data with default sample points.

```
Asamplepoints = smoothdata(Airreg,"movmean", ...
    hours(3),"SamplePoints",time);
plot(time,Airreg,time,Asamplepoints)
axis tight
legend("Original Data","Smoothed Data with Sample Points")
```

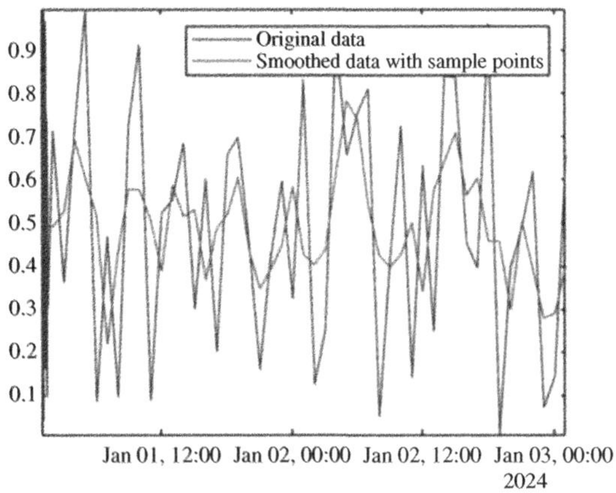

Figure 4.17 Plot for original data and smoothed data with sample points.

4.7 Challenges and Limitations

The challenges and limitations in data collection and preprocessing for AI in power electronics are multifaceted, requiring careful consideration and advanced techniques to overcome. Power electronics systems generate vast amounts of data [51], but the quality of this data is often inconsistent because of issues like sensor inaccuracies, communication noise, and data corruption. For a grid-tied inverter, the data from voltage and current sensors is often noisy due to electromagnetic interference. This noise can cause incorrect readings, leading to inaccurate models if not properly pre-processed. Techniques like filtering and smoothing are compulsory to ensure better data quality and integrity. Other than the issue of poor quality of data, less quantity of data also hinders the pre-processing of data. For example, in case of a predictive maintenance of a power converter, the number of failure events is usually low, leading to an imbalanced dataset. AI models trained on such datasets may become biased toward normal operating conditions, reducing their ability to accurately predict faults. Similarly, in this case if the faults are incorrectly labeled because of absence of a domain expert, the AI model misclassifies these states and leads to unreliable predictions.

The real-time data processing is only possible if the pre-processing technique is fast and efficient and can handle high-speed data streams without introducing significant delays. In a real-time voltage regulation system, data must be collected, pre-processed, and fed into an AI model almost instantaneously. Any delay in processing can lead to instability in the system, making real-time data processing a critical challenge. Ensuring privacy and security of collected data is very important. Cybersecurity threats can compromise data integrity, leading to incorrect AI model outcomes. In smart grid applications, data from distributed energy resources (DERs) is collected and processed by AI systems. If this data is intercepted or tampered with, the AI model's predictions for energy distribution can be skewed, potentially leading to grid instability. Besides these challenges, it is necessary to have synchronization of data sources for accurate analysis. Especially in a multi-inverter system, data from various inverters might be collected at different times or with different sampling rates [52]. Misalignment of this data can lead to incorrect AI model predictions, necessitating careful synchronization during preprocessing.

4.8 Conclusion

Data collection for AI integration with power electronics entails the methodical gathering of information from many sources. To gather data, a variety of sensors are used in the power electronic systems. Regular data collection via sensor networks results in continuous and comprehensive system information. The chapter aims for the readers to understand the importance of data in AI. Different types, sources, and collection techniques of data in power electronics are analyzed and discussed in detail. The emphasis on data quality and challenges with respect to it are discussed. Ensuring data accuracy, handling missing data, managing noisy data, dealing with outliers, data redundancy and relevance is important for integrating AI And power electronics. Different data pre-processing techniques for data cleaning, data transformation, feature engineering, and data reduction are discussed in detail. The importance of data annotation and different types of labeling methods are analyzed. These methods are compared with respect to different aspects for better understanding of the readers.

The chapter includes an example for data smoothing and detecting outliers in MATLAB software. It is an essential step in data preprocessing, especially when dealing with noisy datasets. They help in making the data more suitable for analysis or modeling by reducing noise and identifying or removing anomalous data points. The chapter shows step-by-step procedure for smoothing of data using Savitzky–Golay Filter approach and Robust Lowess approach. Similarly, a step wise procedure for detecting outliers is shown in the chapter. The challenges and limitations of data collection and its pre-processing are discussed.

In summary, the integration of AI with power electronics promises transformative advancements, enabling smarter, more efficient, and predictive systems. However, the foundation of successful AI applications lies in the quality and preparation of the data that feeds these models. Data collection and preprocessing are critical stages that can significantly influence the performance and reliability of AI systems.

Exercises

4.1 Why is data considered the backbone of Artificial Intelligence (AI)?

4.2 Explain how data quality affects the performance of AI models.

4.3 List and define three essential characteristics of high-quality data that influence AI model outcomes.

4.4 Provide an example of how incomplete data could negatively impact AI predictions in power electronics.

4.5 Discuss why data privacy is a concern in AI applications.

4.6 Identify two main challenges associated with data collection for AI.

4.7 Explain how simulation tools like MATLAB Simulink and Typhoon HIL assist in generating data for AI models.

4.8 What advantages does cloud-based data collection offer for AI applications in power electronics?

4.9 How can historical maintenance records contribute to predictive analytics in AI applications?

4.10 What types of data sources are commonly used in AI applications for power electronics, and why are they important?

4.11 Define data pre-processing and explain its importance in preparing real-world data for analysis.

4.12 Explain how decomposing a Date Time feature might improve model understanding in a time-based dataset. Provide an example.

4.13 Define sampling and describe the difference between random sampling and stratified sampling.

4.14 Define the concept of "Human in the Loop" in data annotation. Why is it important?

4.15 Explain Min–Max scaling and Standard scaling. In what situations would you use each?

4.16 Explain how noise affects datasets in power electronics and provide two examples of noisy data.

4.17 Imagine you are tasked with designing a data collection system for a power electronics setup with diverse sensors. What initial steps would you take to ensure effective data integration and collection?

References

1 S. Zhao, F. Blaabjerg, and H. Wang, "An overview of artificial intelligence applications for power electronics," *IEEE Trans. Power Electron.*, vol. 36, no. 4, pp. 4633–4658, Apr. 2021. 10.1109/TPEL.2020.3024914.

2 E. Ferrara, "Fairness and bias in artificial intelligence: A brief survey of sources, impacts, and mitigation strategies," *Sci*, vol. 6, no. 1, p. 3, Dec. 2023. 10.3390/sci6010003.

3 S. Fayaz, S. Z. Ahmad Shah, N. M. ud din, N. Gul, and A. Assad, "Advancements in data augmentation and transfer learning: A comprehensive survey to address data scarcity challenges," *Recent Adv. Comput. Sci. Commun.*, vol. 17, no. 8, pp. 14–35, Nov. 2024. 10.2174/0126662558286875231215054324.

4 L. Alzubaidi *et al.*, "A survey on deep learning tools dealing with data scarcity: Definitions, challenges, solutions, tips, and applications," *J. Big Data*, vol. 10, no. 1, p. 46, Apr. 2023. 10.1186/s40537-023-00727-2.

5 E. Zaidan and I. A. Ibrahim, "AI governance in a complex and rapidly changing regulatory landscape: A global perspective," *Humanit. Soc. Sci. Commun.*, vol. 11, no. 1, p. 1121, Sep. 2024. 10.1057/s41599-024-03560-x.

6 V. Mnih *et al.*, "Human-level control through deep reinforcement learning," *Nature*, vol. 518, no. 7540, pp. 529–533, Feb. 2015. 10.1038/nature14236.

7 J. P. Daries *et al.*, "Privacy, anonymity, and big data in the social sciences," *Commun. ACM*, vol. 57, no. 9, pp. 56–63, Sep. 2014. 10.1145/2643132.

8 M. Rosu *et al.*, "Power electronics and drive systems," in *Multiphysics Simulation by Design for Electrical Machines, Power Electronics, and Drives*. Wiley, 2018, pp. 251–281. 10.1002/9781119103462.ch6.

9 A. Moradzadeh, B. Mohammadi-Ivatloo, K. Pourhossein, and A. Anvari-Moghaddam, "Data mining applications to fault diagnosis in power electronic systems: A systematic review," *IEEE Trans. Power Electron.*, vol. 37, no. 5, pp. 6026–6050, May 2022. 10.1109/TPEL.2021.3131293.

10 K. Sozański, *Digital Signal Processing in Power Electronics Control Circuits*. London: Springer, 2017. 10.1007/978-1-4471-7332-8.

11 A. Kalair, N. Abas, A. R. Kalair, Z. Saleem, and N. Khan, "Review of harmonic analysis, modeling and mitigation techniques," *Renew. Sustain. Energy Rev.*, vol. 78, pp. 1152–1187, Oct. 2017. 10.1016/j.rser.2017.04.121.

12 M. Ohtsu and D. G. Aggelis, "Sensors and instruments," in *Acoustic Emission Testing*. Cham: Springer, 2022, pp. 21–44. 10.1007/978-3-030-67936-1_3.

13 S. Rahimpour, O. Husev, D. Vinnikov, N. V. Kurdkandi, and H. Tarzamni, "Fault management techniques to enhance the reliability of power electronic converters: An overview," *IEEE Access*, vol. 11, pp. 13432–13446, 2023. 10.1109/ACCESS.2023.3242918.

14 S. Zhao and H. Wang, "Enabling data-driven condition monitoring of power electronic systems with artificial intelligence: Concepts, tools, and developments," *IEEE Power Electron. Mag.*, vol. 8, no. 1, pp. 18–27, Mar. 2021. 10.1109/MPEL.2020.3047718.

15 Y. Peng, S. Zhao, and H. Wang, "A digital twin based estimation method for health indicators of DC–DC converters," *IEEE Trans. Power Electron.*, vol. 36, no. 2, pp. 2105–2118, Feb. 2021. 10.1109/TPEL.2020.3009600.

16 A. P. Srivastava, "AI-powered data collection and analysis," in *Utilizing AI Tools in Academic Research Writing*. IGI Global Scientific Publishing, 2024, pp. 114–140. 10.4018/979-8-3693-1798-3.ch008.

17 S. Shapsough, M. Takrouri, R. Dhaouadi, and I. A. Zualkernan, "Using IoT and smart monitoring devices to optimize the efficiency of large-scale distributed solar farms," *Wirel. Netw.*, vol. 27, no. 6, pp. 4313–4329, Aug. 2021. 10.1007/s11276-018-01918-z.

18 W. Zhang, H. Zhang, L. Fang, Z. Liu, and C. Ge, "A secure revocable fine-grained access control and data sharing scheme for SCADA in IIoT systems," *IEEE Internet Things J.*, vol. 9, no. 3, pp. 1976–1984, Feb. 2022. 10.1109/JIOT.2021.3091760.

19 F. A. Silva, "Control circuits in power electronics: Practical issues in design and implementation [Book News]," *IEEE Ind. Electron. Mag.*, vol. 11, no. 1, pp. 84–85, Mar. 2017. 10.1109/MIE.2017.2652783.

20 A. Haque, K. V. S. Bharath, and S. Mateen, "Smart grid concept and technologies for smart cities," in *Smart Cities: Power Electronics, Renewable Energy, and Internet of Things*. Boca Raton: CRC Press, 2023, pp. 96–129. 10.1201/9781032669809-5.

21 W. Liang *et al.*, "Advances, challenges and opportunities in creating data for trustworthy AI," *Nat. Mach. Intell.*, vol. 4, no. 8, pp. 669–677, Aug. 2022. 10.1038/s42256-022-00516-1.

22 A. Gharaibeh *et al.*, "Smart cities: A survey on data management, security, and enabling technologies," *IEEE Commun. Surv. Tutorials*, vol. 19, no. 4, pp. 2456–2501, 2017. 10.1109/COMST.2017.2736886.

23 L. L. Pipino, Y. W. Lee, and R. Y. Wang, "Data quality assessment," *Commun. ACM*, vol. 45, no. 4, pp. 211–218, Apr. 2002. 10.1145/505248.506010.

24 A. Halevy *et al.*, "Goods," in *Proceedings of the 2016 International Conference on Management of Data*, Jun. 2016, pp. 795–806. 10.1145/2882903.2903730.

25 N. Azevedo, "Data preprocessing techniques: 6 steps to clean data in machine learning," *Scalable Path*, 2021. https://www.scalablepath.com/data-science/data-preprocessing-phase (accessed Aug. 29, 2024).

26 X. Ding, H. Wang, G. Li, H. Li, Y. Li, and Y. Liu, "IoT data cleaning techniques: A survey," *Intell. Converg. Networks*, vol. 3, no. 4, pp. 325–339, Dec. 2022. 10.23919/ICN.2022.0026.

27 S. Zhao, Y. Peng, F. Yang, E. Ugur, B. Akin, and H. Wang, "Health state estimation and remaining useful life prediction of power devices subject to noisy and aperiodic condition monitoring," *IEEE Trans. Instrum. Meas.*, vol. 70, pp. 1–16, 2021. 10.1109/TIM.2021.3054429.

28 R. A. Davis, A. J. Charlton, J. Godward, S. A. Jones, M. Harrison, and J. C. Wilson, "Adaptive binning: An improved binning method for metabolomics data using the undecimated wavelet transform," *Chemom. Intell. Lab. Syst.*, vol. 85, no. 1, pp. 144–154, Jan. 2007. 10.1016/j.chemolab.2006.08.014.

29 Y.-S. Sim, J.-S. Hwang, S.-D. Mun, T.-J. Kim, and S. J. Chang, "Missing data imputation algorithm for transmission systems based on multivariate imputation with principal component analysis," *IEEE Access*, vol. 10, pp. 83195–83203, 2022. 10.1109/ACCESS.2022.3194545.

30 N. Marchang and R. Tripathi, "KNN-ST: Exploiting spatio-temporal correlation for missing data inference in environmental crowd sensing," *IEEE Sens. J.*, vol. 21, no. 3, pp. 3429–3436, Feb. 2021. 10.1109/JSEN.2020.3024976.

31 P. Zhao and L. Lai, "Analysis of KNN density estimation," *IEEE Trans. Inf. Theory*, vol. 68, no. 12, pp. 7971–7995, Dec. 2022. 10.1109/TIT.2022.3195870.

32 M. K. Dahouda and I. Joe, "A deep-learned embedding technique for categorical features encoding," *IEEE Access*, vol. 9, pp. 114381–114391, 2021. 10.1109/ACCESS.2021.3104357.

33 V. N. G. Raju, K. P. Lakshmi, V. M. Jain, A. Kalidindi, and V. Padma, "Study the influence of normalization/transformation process on the accuracy of supervised classification," in *2020 Third International Conference on Smart Systems and Inventive Technology (ICSSIT)*, Aug. 2020, pp. 729–735. 10.1109/ICSSIT48917.2020.9214160.

34 L. B. V. de Amorim, G. D. C. Cavalcanti, and R. M. O. Cruz, "The choice of scaling technique matters for classification performance," *Appl. Soft Comput.*, vol. 133, p. 109924, Jan. 2023. 10.1016/j.asoc.2022.109924.

35 H. Liu, *Feature Engineering for Machine Learning and Data Analytics*. CRC Press, 2018. 10.1201/9781315181080.

36 A. Ghubaish, Z. Yang, A. Erbad, and R. Jain, "LEMDA: A novel feature engineering method for intrusion detection in IoT systems," *IEEE Internet Things J.*, vol. 11, no. 8, pp. 13247–13256, Apr. 2024. 10.1109/JIOT.2023.3328795.

37 Y. Gong, G. Liu, Y. Xue, R. Li, and L. Meng, "A survey on dataset quality in machine learning," *Inf. Softw. Technol.*, vol. 162, p. 107268, Oct. 2023. 10.1016/j.infsof.2023.107268.

38 M. Ashraf *et al.*, "A survey on dimensionality reduction techniques for time-series data," *IEEE Access*, vol. 11, pp. 42909–42923, 2023. 10.1109/ACCESS.2023.3269693.

39 J. Zheng, Z. Yang, and Z. Ge, "Deep residual principal component analysis as feature engineering for industrial data analytics," *IEEE Trans. Instrum. Meas.*, vol. 73, pp. 1–10, 2024. 10.1109/TIM.2024.3420267.

40 F. Nie, H. Chen, S. Xiang, C. Zhang, S. Yan, and X. Li, "On the equivalence of linear discriminant analysis and least squares regression," *IEEE*

Trans. Neural Netw. Learn. Syst., vol. 35, no. 4, pp. 5710–5720, Apr. 2024. 10.1109/TNNLS.2022.3208944.

41 G. Chandrashekar and F. Sahin, "A survey on feature selection methods," *Comput. Electr. Eng.*, vol. 40, no. 1, pp. 16–28, Jan. 2014. 10.1016/j.compeleceng.2013.11.024.

42 Y. Guo, W. Wang, and X. Wang, "A robust linear regression feature selection method for data sets with unknown noise," *IEEE Trans. Knowl. Data Eng.*, pp. 1–1, 2021. 10.1109/TKDE.2021.3076891.

43 A. Kaur, K. Guleria, and N. Kumar Trivedi, "Feature selection in machine learning: Methods and comparison," in *2021 International Conference on Advance Computing and Innovative Technologies in Engineering (ICACITE)*, Mar. 2021, pp. 789–795. 10.1109/ICACITE51222.2021.9404623.

44 P. A. De los Santos, R. J. Burke, and J. M. Tien, "Progressive random sampling with stratification," *IEEE Trans. Syst. Man Cybern. Part C Appl. Rev.*, vol. 37, no. 6, pp. 1223–1230, Nov. 2007. 10.1109/TSMCC.2007.905818.

45 A. Samanta and T. G. Nguyen, "Quality-driven energy-efficient big data aggregation in WBANs," *IEEE Sensors Lett.*, vol. 6, no. 8, pp. 1–4, Aug. 2022. 10.1109/LSENS.2022.3192620.

46 T. Zhu, J. Li, H. Gao, and Y. Li, "Data aggregation scheduling in battery-free wireless sensor networks," *IEEE Trans. Mob. Comput.*, vol. 21, no. 6, pp. 1972–1984, Jun. 2022. 10.1109/TMC.2020.3035671.

47 R. Munro, *Human-in-the-Loop Machine Learning: Active Learning and Annotation for Human-Centered AI*. Manning, 2021.

48 S. Li, T. Liu, J. Tan, D. Zeng, and S. Ge, "Trustable co-label learning from multiple noisy annotators," *IEEE Trans. Multimed.*, vol. 25, pp. 1045–1057, 2023. 10.1109/TMM.2021.3137752.

49 J. Lin, T. Yu, and Z. J. Wang, "Rethinking crowdsourcing annotation: Partial annotation with salient labels for multilabel aerial image classification," *IEEE Trans. Geosci. Remote Sens.*, vol. 60, pp. 1–12, 2022. 10.1109/TGRS.2022.3191735.

50 K. Pykes, *What Is Data Labeling And Why Is It Necessary for AI?* Datacamp, 2024.

51 K. L. Tsui, Y. Zhao, and D. Wang, "Big data opportunities: System health monitoring and management," *IEEE Access*, vol. 7, pp. 68853–68867, 2019. 10.1109/ACCESS.2019.2917891.

52 Q. Meng and Z. Hou, "Data-driven multi-inverter cooperative control for voltage tracking and current sharing in islanded AC microgrids," *Trans. Inst. Meas. Control*, vol. 41, no. 11, pp. 3145–3157, Jul. 2019. 10.1177/0142331218818653.

5

Fuzzy Logic and Metaheuristic Methods in Power Electronics

Fatima Shabir Zehgeer and Ahteshamul Haque

Advance Power Electronics and Research Laboratory, Jamia Millia Islamia (A Central University), New Delhi, India

5.1 Introduction

The popularity of the renewables sources of energy is immensely increasing due to the environmental benefits, energy security, economic growth, sustainability, and the reduction in the energy cost. Power electronics has a key role in renewable energy systems as they are important for converting, controlling, and conditioning the electric energy generated from the renewable sources. Power electronics converters find application in energy conversion, energy storage, grid integration and stability, electrification of electric vehicles, and power quality improvement [1]. Ever since the invention of the thyristors in 1950s, power electronics has revolutionized the control and conversion of energy. Thus, advancements are being made in the development of the various topologies of the converters and the control schemes [2].

AI methods (Expert methods, Fuzzy methods, Metaheuristics methods, machine learning) have found increased popularity in the area of power electronics and have resulted in increased efficiency, reduced cost, optimal control and design [3]. Fuzzy and metaheuristics methods are important in power electronics as they can handle complex, non-linear, and dynamic systems efficiently. These methods have evolved immensely and resulted in the efficient control, design optimization, and performance enhancement of the power electronic system [4].

Fuzzy logic methods are not dependent on the complex computations and mathematical models and thus offer great advantage. Fuzzy logic methods, based on

Artificial Intelligence for Power Electronics, First Edition.
Edited by Ahteshamul Haque, Saad Mekhilef, and Azra Malik.

If–then rules, mimic the human decision making by handling the fuzzy or uncertain information. Basic steps in the fuzzy logic involve [5]:

- Fuzzification of the input crip values using the membership function
- Rule Evaluation (Inference) in which If–then rules are applied
- Aggregation of all rules to give a fuzzy output set
- Defuzzification which involves the conversion of fuzzy output into the crip value

Fuzzy logic methods have the better adaptability, robustness, and simplified control. Fuzzy logic methods have found the application in the control of DC–DC converters, inverter control, MPPT in solar inverters, motor drives, diagnostics, fault detection, improvement in power quality, converter design, and energy management [6].

Metaheuristic methods tend to find the most optimal solutions and can be categorized into two groups:

- Trajectory-based methods
- Population-based methods.

Population-based methods take inspiration from nature and include evolutionary algorithms and algorithms based on swarm intelligence. Metaheuristic algorithms are used to optimize the design and control parameters of the power electronics system [7]. Metaheuristic methods find applications in design optimization of converters, harmonic minimization, parameter tuning, and energy management in microgrids [8].

Combined fuzzy and metaheuristic methods create a more powerful and efficient system. Fuzzy logic tuned by metaheuristic algorithms such as Genetic algorithm (GA) or particle swarm optimization (PSO) can be used to optimize the parameters of the fuzzy logic controller [9]. The major applications of the fuzzy logic and metaheuristic methods in power electronics are shown in Figure 5.1.

This chapter aims to explain the major application of the fuzzy logic and the metaheuristic methods in power electronics. This chapter will help to comprehend.

- How fuzzy logic methods are applied to power electronics.
- Fundaments of the metaheuristic methods and various optimization techniques.
- Application of the various metaheuristic methods the power electronics.
- Hybrid Approaches: Combined application of fuzzy-logic-based metaheuristic methods.
- Understanding the real-world examples which focus on the practical application of fuzzy and metaheuristic methods in power electronics.

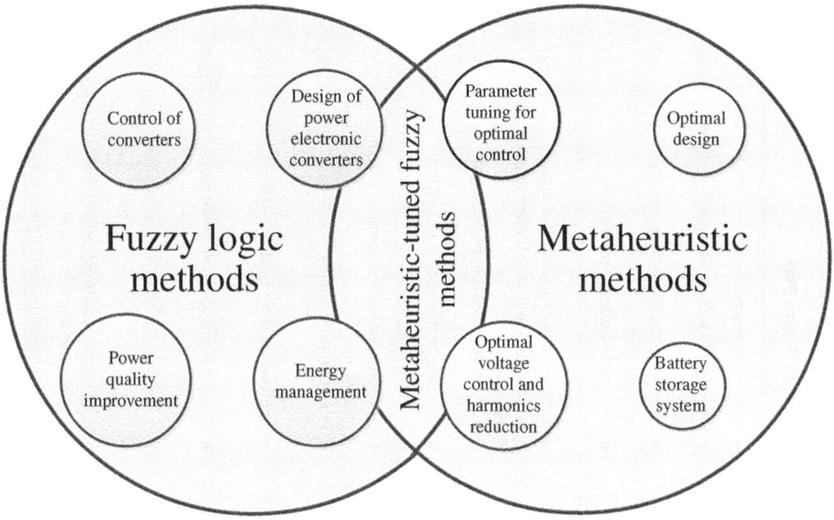

Figure 5.1 Fuzzy logic and metaheuristic methods in power electronics.

5.2 Applications of Fuzzy Logic Methods in Power Electronics

The fuzzy logic methods have been extensively used in power electronic methods for the control and design of the converters. Fuzzy logic has also been proposed for power quality improvement. Additionally, faults in converters have been detected with the fuzzy logic methods. Various applications of fuzzy logic methods in power electronics are summarized in Table 5.1.

5.2.1 Converter Control

A fuzzy-logic-based controller is presented in [10] for maximizing the power output. The maximum power point tracking (MPPT)-based fuzzy logic control (FLC) results in a robust PV system with reduced system harmonics. The FLC is implemented for controlling the switching of bidirectional converter. The input to the fuzzy logic controller is the voltage difference between the input and the load voltage. The triangular membership function (MF) is used for FLC. The fuzzy logic controller is connected to the pulse width modulation (PWM) generator. The gate pulses are generated as per the fuzzy rules. Thus, the fuzzy-logic-based controls the switching devices of the converter in order to maintain the constant load voltage and track maximum power point (MPP).

Table 5.1 Application of Fuzzy Logic Methods in Power Electronics.

Fuzzy Logic Methods	Application	Advantages	Limitations/Future Work	Reference
Fuzzy logic for converter control	MPPT control of PV module	• Better Adaptability	• Performance needs be further improved	[10]
	FLC of resonant converter	• Reduced current stress • Voltage gain is high	• Unregulated voltage for open-loop control	[11]
	FLC of LLC resonant converter	• Zero voltage switching is achieved	• High power LLC resonant converters to be developed	[12]
Fuzzy logic for improving power quality	Controlling FACTS devices like STATCOM	• High accuracy • Increased dynamic stability	• Increased number of oscillations	[13]
	FLC of shunt active power filter	• Reduced Total Harmonic Distortion (THD)	• Integration with advanced control methods is required to further reduce harmonics	[14]
	FLC of multilevel inverter	• Enhanced power factor • Cost effective method	• Active power needs to be managed	[15]
Fuzzy inference system for design of power electronic devices	DC–DC converter design	• Smooth response and small overshoot	• Extensive mathematical modelling	[16]
	High-frequency inductor design	• Optimization results in trade-off between parameters	• Time-consuming and complex design process	[17]
Fuzzy inference system for fault detection	Fault detection in five level inverter	• Accurate diagnosis	• Detection time is more	[18]
	Open circuit switch fault detection in boost converter	• Rapid fault detection	• Analysis for other fault types to be done	[19]
Fuzzy logic for power flow control	Power flow controller for Rotor Side Converter (RSC)	• Constant stator power output is maintained for different wind speeds • time response is better than conventional control method	• Developed voltage to be further analyzed	[20]
	Control of active and reactive power flow in HVDC System	• Enhancement in stability of transmission line	• With the advancement in the HVDC systems, the control method faces challenges and further work is to be done	[5]

In [21] another highly efficient fuzzy-logic-based intelligent MPPT control for the grid connected converter is presented. The fuzzy controller decreases the fluctuations in voltage, current, and power. Additionally, it is more precise and efficient than the conventional MPPT techniques like Incremental Conductance (IC) Method.

Resonant boost converters are an integral part of the electrical vehicle charging system. The resonant converter topology proposed in [11] is controlled with fuzzy logic. The resultant control method results in higher gain with less transformer turns ratio. It also reduces the current stress on transformer. The resonant LLC converter in [12] is also controlled with fuzzy logic. The fuzzy logic controller has three inputs. With five triangular MF, there are 25 fuzzy rules. The fuzzy controller controls the switching frequency of MOSFETS in the LLC converter so as to achieve the zero-voltage switching (ZVS).

FLC has been implemented for the voltage control of the resonant LLC converters in [22]. The fuzzy logic controller has three inputs: Normalized reference voltage, error between the input and reference voltage, and integral of the error. The reference switching frequency and reference switching frequency variation are the two outputs. Thus, even if there are variations in the input voltage, FLC ensures to maintain the constant output voltage.

5.2.2 Power Quality Improvement

Flexible AC transmission system (FACTS) is extensively developed for improving the voltage profile and enhancing power flow capacity. Several power-electronics-based FACTS controllers such as static VAR compensator (SVC) and unified power flow controllers (UPFC) are used. In [13], FLC is proposed to control the UPFC to enhance the transmission line stability. Forty-nine if–then fuzzy rules are used to develop the fuzzy logic controller. Two error signals are fed as inputs to the two fuzzy logic controllers connected converters of UPFC. The FLCs effectively control the voltage and regulates the power flow in a transmission line. Mamdani fuzzy controller for mitigating the power quality has also been proposed in [23]. The proposed fuzzy logic controller consists four main components namely.

- Fuzzifier
- Rule base
- Interface
- Defuzzifier.

Compared with the conventional proportional integral (PI) control, the fuzzy control for the FACT device results in more efficient reactive power compensation. In order to mitigate the power quality issues, the fuzzy logic controller has also

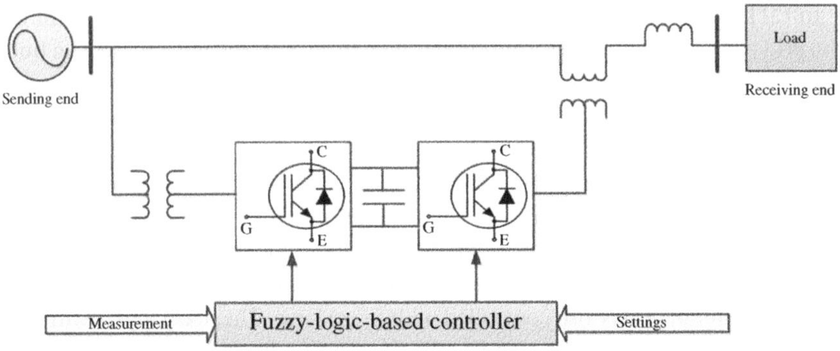

Figure 5.2 Fuzzy-logic-based controller for enhancing transmission line stability.

been implemented in [24] in which fuzzy logic controller controls the combination of dynamic voltage explorer (DVR) and STATCOM. The unified power flow fuzzy controller is shown in Figure 5.2.

In [14] FLC-based active filter has been employed for reduction in total harmonic distortion and improve the PQ characteristics. The fuzzy control employs triangular membership functions with the Mamdani-Type in operator. The fuzzy logic controller ensures the minimum ripple in dc link voltage. In smart grid, multi-level inverter (MLI) topology is employed for enhancing power quality. Fuzzy logic controller based on Mamdani-type method controls the MLI by controlling the DC link voltage [15]. The Gaussian membership function (MF) is able to perform better as compared to other membership functions such as triangular or trapezoidal membership function.

5.2.3 Fault Detection in Power Electronic Converters

The fuzzy logic has found application in the fault detection for the boost converters [18]. By controlling the input voltage, the output voltage is regulated to a desired value. By analyzing the input current, the fuzzy-based control can effectively detect the open circuit switch fault. Fuzzy-logic-based fault detection is also used in [19] to diagnose the open circuit switch fault in MLI. The fuzzy logic determines the various rules that link the input and the output variables. The fuzzy controller classifies the fault, and hence it can detect the fault under various open circuit fault conditions.

For the Multiterminal High-Voltage Direct Current Transmission (MTHVDC), the fuzzy logic is also implemented for fault detection [25]. The fault is detected in AC or DC side. Using the triangular membership function, the DC voltage is fuzzified into high, low, and medium membership functions. After fuzzification, the fault section is identified. The output 1 indicates fault section, while as 0 indicates no fault condition.

5.2.4 Power Flow Control

To control the power flow the grid, converters such as rectifiers and inverters are used. Converters on rotor side are controlled with the fuzzy logic. In [20] fuzzy logic is utilized to control the slip power at the rotor side converter (RSC). The FLC feeds the reference sine wave to the PWM signal generators. The IGBTS of the RSC are then triggered with the PWM gate pulses. For the High-Voltage Transmission applications, the fuzzy logic has been implemented to control the Voltage Source Converters (VSC) [5]. The Defuzzification Boolean Relations (DBR) Fuzzy logic controller controls the power between the converter stations of the High-Voltage Direct Current (HVDC) system.

5.2.5 Design of Power Electronic Devices

Various AI-based techniques have been proposed in literature for designing the high-frequency power electronic components such as high-frequency inductors and high-frequency transformers [26]. Fuzzy-logic-based design has been presented in [17] of the optimal design of high-frequency DC bias inductor. The objective was to optimize the core window fill factor and reduce the magnetic saturation. The optimization considers the trade-off between various parameters, but the overall design is complex. The DC–DC converter has been designed using fuzzy logic in [16] for charging of lead acid battery used in solar cells. The fuzzy logic controller utilizes the trapezoidal membership function for this design.

5.3 Applications of Metaheuristic Methods in Power Electronics

Metaheuristic optimization techniques like differential algorithm (DE), particle swarm optimization, and genetic algorithm have been applied to the power electronic converters. The methods successfully resulted in the efficient and cost-effective design of the converters and the high-frequency magnetic components [27]. The metaheuristic techniques have also been implemented for the control applications and the power quality improvement [28]. The application of the metaheuristic methods to the power electronic converters is explained in this section.

5.3.1 Power Quality Improvement

Genetic algorithm (GA) is implemented in [29] to seven-level inverter in order to minimize the harmonic. The genetic algorithm has also been implemented in [30] to optimize the induction and the switching turn on currents of the Dual Active Bridge converter used for the electric vehicle charging. With the GA timed modulation, the losses have decreased and system efficiency has increased. PSO has been implemented in [31] to improve voltage in a solar photo voltaic system. A modified metaheuristic algorithm-based on the adaptive wavelength and population size has been proposed in [32] for active and reactive power improvement in microgrid.

5.3.2 Control

For the optimization, PSO technique draws the inspiration from the movement of birds. The PSO algorithm begins with the definition of the problem. Then population which consists of the specific number of particles is initialized randomly. Each particle in the population is properly designated. In the main loop, velocity and position of each particle is updated, and current position is compared with the past position till the best and then the best position $g(t)$ is obtained.

Based on the PSO, the global maximum power point (GMPP) is tracked under different irradiance in [33]. Also, in [34], maximum power point is tracked with PSO algorithm. The voltage and current of the PV system is given to the PSO-based controller. The optimization parameter for the PSO algorithm is the duty cycle.

An improved PSO algorithm for obtaining MPP has been proposed in [35]. The Leader Particle Swarm optimization algorithm with five parameter tuning results in the faster convergence even under different shading conditions. In [36] GA results in optimal control of DC–DC converter.

The bacterial foraging algorithm has been implemented in [37] for reducing distortions in current and voltage. The PI controller parameters are optimized in accordance with the Bacterial Foraging Optimization (BFO) algorithm. Active power P and reactive power Q are the two inputs to the fuzzy-based PI controller. The control scheme is employed for generating the gate pulses for the IGBT-based grid-side inverter. The BFO algorithm-based control for the inverters in distributed generation is shown in Figure 5.3.

5.3.3 Optimal Design of the Power Electronic Converters

Metaheuristic methods are extensively used for designing the power electronic devices, especially the High-Frequency (HF) components. With the metaheuristic algorithms, HF inductors and HF transformers have been optimally designed.

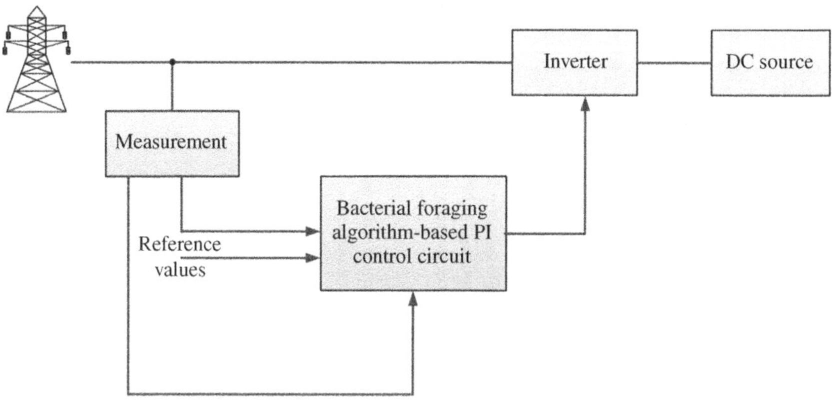

Figure 5.3 Bacterial foraging optimization (BFO)-based PI control of grid-side inverter.

In [38], the arm inductor of a multi-modular converter has been designed while optimizing the core area product. This optimization is based on Non-dominated Sorted Genetic algorithm, NSGA-II.

A comparative study for designing a spiral inductor using Artificial Bee Colony (ABC) optimization and the Genetic algorithm has been done in [39]. The study shows that ABC is more robust and converges faster. The genetic algorithm results in designing the optimally sized inductor. For designing the ferrite core inductor differential evolution has been implemented in [40]. Table 5.2 details the metaheuristic methods' applications in power electronics area.

5.4 Hybrid Approaches: Fuzzy Logic and Metaheuristic Methods in Power Electronics

In literature, the fuzzy logic methods have been used in conjunction with the metaheuristic methods. The various parameters of the fuzzy controller are optimized with the metaheuristic algorithms. This section explains the combined application of fuzzy and metaheuristic methods in power electronics.

5.4.1 Control of the Power Electronic Converters

In [42] Genetic algorithm is employed in order to optimize the various parameters of fuzzy-logic-based controller. DC–DC boost converter is then accordingly controlled. The error voltage and the difference in error are the two inputs to the fuzzy logic controller. Various FLC parameters are optimized with the genetic algorithm, and it results in a more efficient control.

Table 5.2 Metaheuristic Methods in Power Electronics.

Application Area in Power Electronics		Metaheuristic Algorithm	Optimization Parameter	Advantage	Limitations/Future Work	Reference
Control	Grid-side inverters for distributed generation	Bacterial Foraging Optimization (BFO)	• PI controller gains • (Kp and Ki)	• Reduced THD • Control response is better	• To be implemented for other PI controllers	[37]
	MPPT control for PV systems	Particle Swaram Optimization (PSO)	• Duty Cycle, D	• High efficiency • Steady-state performance	• Implementation is more complicated than traditional methods	[34]
		Leader Particle Swarm Optimization (LPSO)	• Constants • Scaling factors	• Less oscillations, • MPPT tracking under critical conditions	• Increased computational burden due to five parameter tuning	[35]
	Voltage regulator	Simulated annealing	• PID controller parameters • Derivative (DD) controller parameters	• Versatile	• Optimal solution not guaranteed	[41]
Power quality improvement	Modulation for dual-active bridge converter	Genetic algorithm II (NSGA-II)	• Inductor current, • turn on current	• Less losses	• Future work on real-time modulation	[30]
	H bridge converter	Genetic algorithm	• Switching angle • Voltage	• Minimized THD	—	[29]
Design	Ferrite core inductors	Differential evolution	• Inductor current	• Realistic saturation curves	• Deviations in values from test conditions	[40]
	Spiral inductors	Artificial bee colony optimization	• Geometric parameters	• Implementation is easy	• Slow convergence	[39]
	Arm inductor for multi-modular converter	Non-dominated Sorted Genetic Algorithm (NSGA-II)	• Core window fill factor	• Good adaptability	• Investigate more magnetic materials	[38]

Uncertain weather conditions result in partial shading which affects the accurate tracking of the MPP of the PV module. To address the issue, a hybrid optimization problem based on ant colony fuzzy method (ACO) is proposed in [43]. Firstly, ant colony optimization algorithm determines the period in which power is maximum. In the second step, the voltage and current corresponding to that period are given as input to the fuzzy-logic-based controller. Then the fuzzy rules determine the maximum power point. The ant fuzzy optimization (AFO) algorithm has the high efficiency of about 98.7%.

5.4.2 Power Quality Improvement

For mitigating the power quality issues, a power electronic device called Unified Power Quality Conditioner (UPQC) is used. Fuzzy logic has been used for generating the reference signals for the UPQC. Ant colony optimization is used to find an optimal membership function for the optimal FLC [44]. The triangular membership function with seven fuzzy rules is used. The proposed fuzzy–ACO method results in lower voltage sag with reduced total harmonic distortion to 0.13%. A fuzzy PSO for reducing THD has been implemented in [45].

5.4.3 Power Flow Control

The ant colony optimization technique has been implemented in [46] in order to optimize the parameters of the fuzzy logic controller. The fuzzy-logic-based controller controls the power flow for the electric vehicle. The flow of power is controlled between the PV battery, electric motor, and the supercapacitor. The ant colony optimization-based fuzzy control results in the optimal power flow to the energy storage devices like PV battery and the supercapacitor. The inputs to the fuzzy controller are as:

- Output from the PI control.
- State of Charge (SOC) of PV battery and the Supercapacitor.

The optimal output current signal from the fuzzy controller is used as reference for the generation of the gate pulses.

In [47] Genetic algorithm has been used in order to optimize the membership function of the fuzzy logic controller for battery charging. The optimal duty ratio is the output of the GA-tuned fuzzy controller, and it is given as input to the PWM generator. The PWM-generated pulses are used to control the switching on and switching off the MOSFETS of the DC–DC Converter.

Another hybrid method based on differential evolution-based fuzzy logic charging of the electric vehicles has been proposed in [48]. The converters in the system are designed for the optimal power flow between Vehicle and grid (V2G). There are

Table 5.3 Hybrid Approach: Fuzzy Logic and Metaheuristic Methods in Power Electronics.

Hybrid Approach Application		Metaheuristic Method	Advantages	Limitation/Future Work	References
Control	FLC-GA control of DC–DC boost converter	Genetic algorithm	• Robust and Stable • less overshoot	• Integration with other Optimization techniques	[42]
	AFO-based MPPT control for PV array	Ant colony optimization	• High Accuracy • Better Adaptability	• High cost • Complex hardware implementation	[43]
Improvement in power quality	Ant colony optimization for Unified Power Quality Conditioner (UPQC)	Ant colony optimization	• Reduced Voltage Sag • Reduction in THD	• Compensation for imbalance in voltage and reactive power is required.	[44]
	FLC-PSO for PV-based active filter	Particle swarm optimization	• Power factor is enhanced • Reduced THD	–	[45]
Energy management and power flow control	AFO for the electric vehicle charging system	Ant colony optimization	• Less power • High efficiency	• Advancement in MPPT control required.	[46] [50]
	FLC-GA control for battery charging	Genetic algorithm	• Reduced Computational time • Stable	• Steady-state error needs to be minimized.	[47]
	FLC-DE for vehicle to gird charging systems	Differential evolution	• Suitable for real V2G system	• Imprecise measurement	[48]
Design	FLC-PSO for increasing lifetime of power electronic devices	Particle swarm optimization	• Degradation of the components is reduced. • Reduced current overshoot.	• Time taken to reach the desired value is more	[49]

two inputs to the Fuzzy logic system and three membership functions per input. With the proposed fuzzy differential algorithm method, each component of the system draws the optimal current and the SOC for the batteries is also balanced.

5.4.4 Design

Various fuzzy–metaheuristic hybrid methods have been proposed for the optimal design of the power electronic components. For the optimal design, various factors like reliability and lifetime are considered. Hybrid Fuzzy–Particle Swarm optimization method has been implemented in [49] for designing the power electronic components used in industrial drives. The life of the power electronic devices is enhanced when optimally designed. The fuzzy–PSO-tuned controller consists of two objective functions:

- The motor in the drive should attain the desired speed.
- Increase the life of power electronics stage.

The PSO algorithm optimizes the membership function of the FLC by finding optimal positions. The optimization technique ensures that low temperature of the power electronic components is maintained so as to increase the life time of the power electronic elements. The comparison among hybrid approaches utilizing fuzzy logic and metaheuristic methods is detailed in Table 5.3.

5.5 Case Studies and Real-World Examples

Some examples have been presented in this subsection which illustrate the application of the fuzzy and metaheuristic methods in power electronics. A DC–DC converter has been controlled with the fuzzy logic method. Particle swarm optimization (PSO) has been implemented in the second case to find the optimal duty ratio for DC–DC converter. The switching signals generated form the optimal duty ratio result in maximizing the power output. The MPPT control of the PV system has been achieved with the fuzzy logic in the third case. All simulations have been carried in MATLAB Simulink.

5.5.1 Fuzzy Logic Control of the Buck Converter

In this example, a fuzzy logic DC–DC buck converter has been simulated in MATLAB/Simulink. The buck converter steps down a 20 V DC input voltage into DC output voltage of 12 V. The switching device of this step-down converter, i.e. IGBT is controlled with the fuzzy logic controller.

The values of various parameters are listed in Table 5.4.

Table 5.4 Values of Parameters for Buck Converter.

Parameter	Value
V_{in}	20 V
V_{out}	12 V
Inductance, L	0.15 H
Capacitance, C	200 μF
Resistance, R	7 Ω

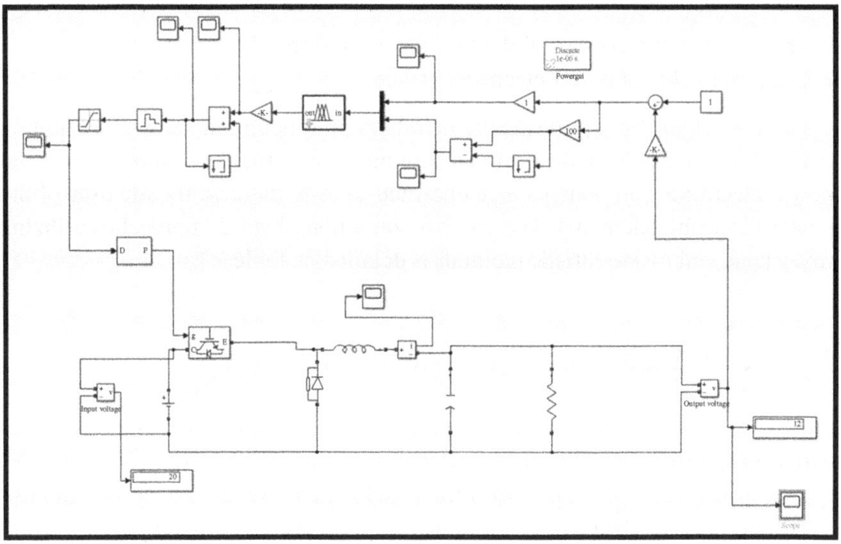

Figure 5.4 Schematic of the fuzzy-logic-controlled buck converter in MATLAB Simulink.

Two Inputs to fuzzy logic controller are:

- **Error Signal:** It is the difference between the reference voltage signal and output voltage.
- **DE:** The change in error.

The output of the fuzzy controller is the duty ratio. This fuzzy-controlled duty ratio is then used to generate the switching pulses for the IGBT.

The schematic of the fuzzy logic-controlled DC–DC converter is shown in Figure 5.4.

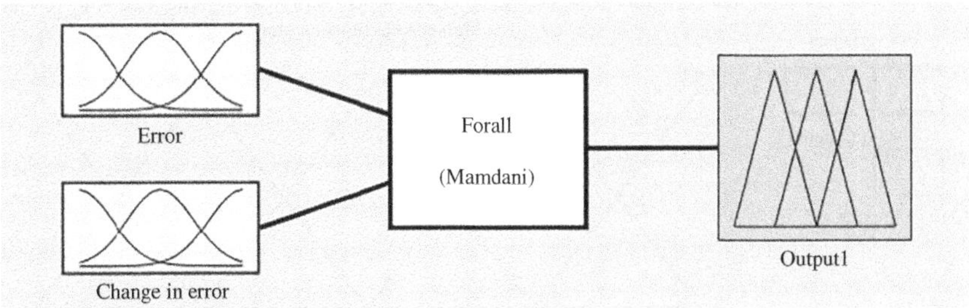

Figure 5.5 Fuzzy logic method.

This fuzzy controller utilizes the Mamdani Inference System which is based on the set of linguistic rules. The fuzzy logic method is shown in Figure 5.5.

For fuzzification, five triangular MF are employed as shown in Figure 5.6.

For the buck converter, duty cycle is given as

$$d = V_{out}/V_{in} \tag{5.1}$$

Thus, for increasing the output, the duty cycle ratio has to be increased. This relation of the buck converter forms the basis for rule base. One of the rules as shown in rule editor is "if error and change in error is NB (Big Negative), then output is PB (Big positive)." This rule implies that the duty ratio should be increased by large value so as to obtain the required output voltage. Rule Editor window is shown in Figure 5.7.

Twenty-five rules for the inputs and output duty cycle are generated. The rules as viewed in Rule Viewer are shown in Figure 5.8

The fuzzy input is converted to the crip value using the centroid defuzzification. The 3D Model for the output of the fuzzy logic controller can be viewed in surface option under view tab, and it is shown in Figure 5.9.

The designed Fis file is imported to workspace and is updated in the Fuzzy logic Controller Block.

Figures 5.10 and 5.11 show the fuzzy logic controller inputs.

The dc voltage at the output is shown in Figure 5.12.

5.5.2 Particle Swarm Optimization-Based MPPT Control Under Different Irradiances

In this case study, the PV array is illuminated under difference conditions of solar irradiations, and the MPP is obtained for each irradiance. The MPPT is achieved with PSO algorithm, and it is implemented in Simulink. The schematic is shown in Figure 5.13.

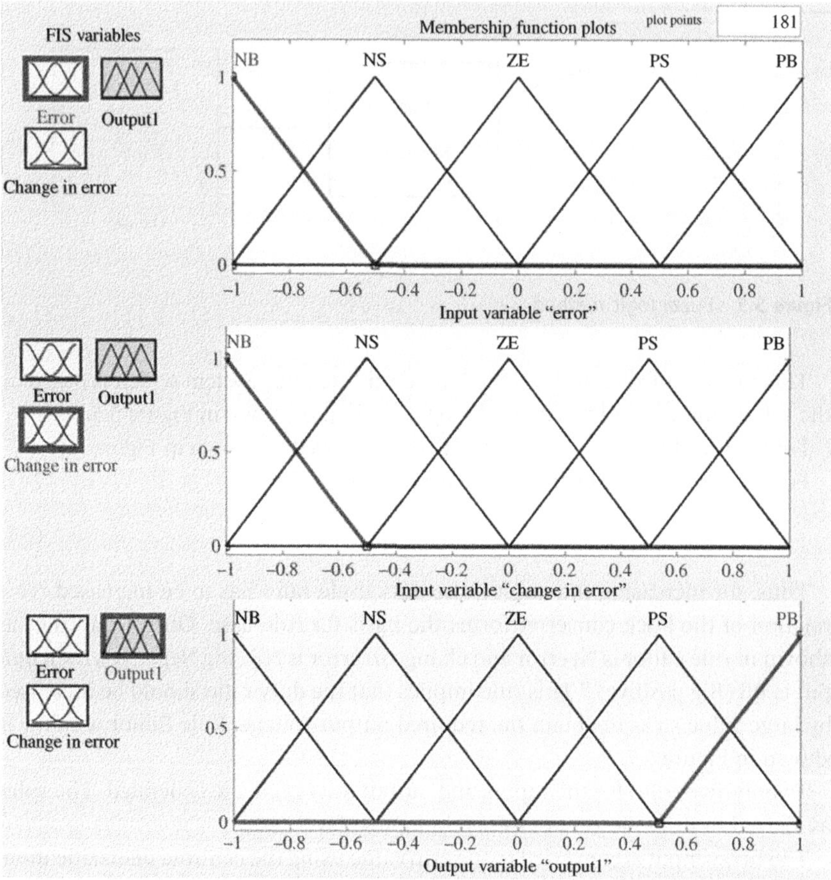

Figure 5.6 Input and output membership functions.

The various elements of the PSO-based solar PV system include:

- **The PV Array:** From the PV Array Block, module Aavid Solar ASMS-220P is selected. The PV array consists of the two parallel PV strings. Each string consists of two series connected modules. The PV array has two inputs, temperature and the solar irradiance. The temperature is set to 25 °C and the irradiance is changed from 700 to 1000 W/m^2 and finally to 12,000 W/m^2.
  The PV array characteristics at 25 °C and under different irradiations is shown in Figure 5.14.
  The outputs from PV array are fed as inputs to the PSO control component.
- **Particle Swarm Optimization Control Block:** PSO control implemented with the MATLAB function block in which the PSO-based algorithm optimizes

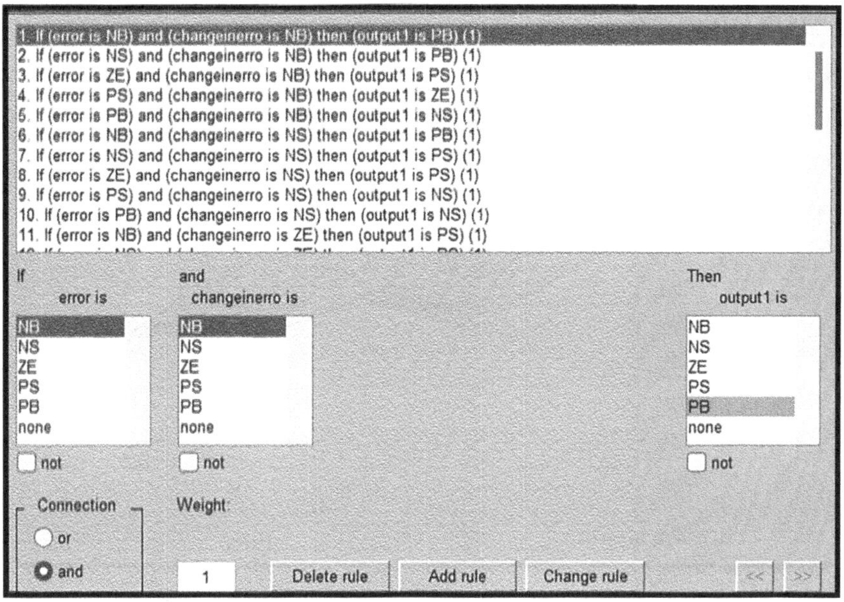

Figure 5.7 Rule Editor depicting the If–Then rules.

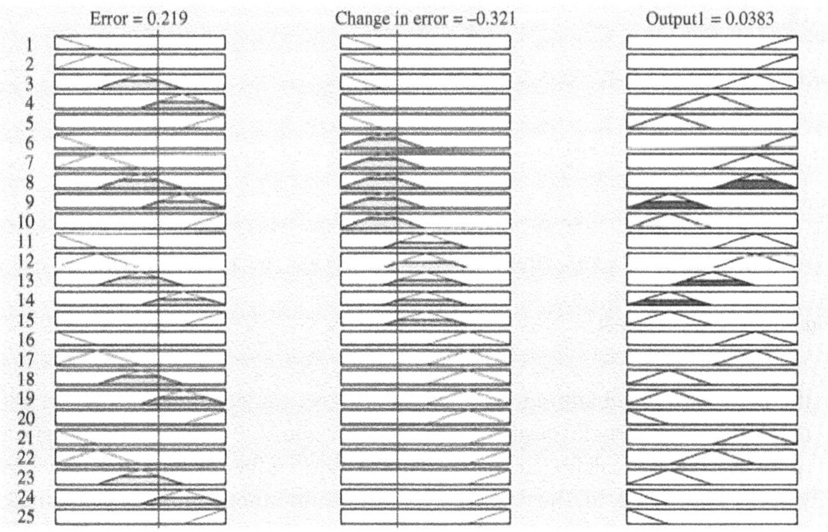

Figure 5.8 Twenty-five rules in the rule view.

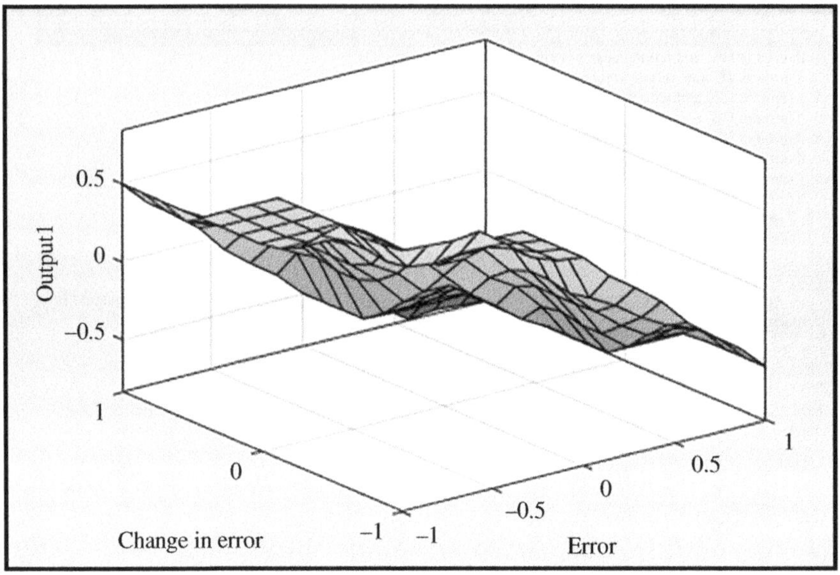

Figure 5.9 3-D model for the inputs and output of the FLC controller.

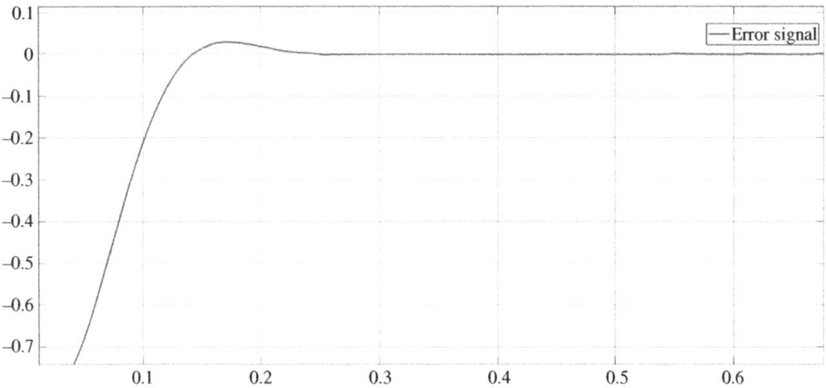

Figure 5.10 Error signal.

the duty ratio for obtaining the maximum power under different solar irradiances. For each particle (designated by the I_PV and V_PV), the fitness function (power, P_PV) is calculated. If the current value is better than the previous value, it is updated, till the particle with maximum fitness function is obtained. Finally, the location (Duty ratio) of the particle that has maximum Power is found.

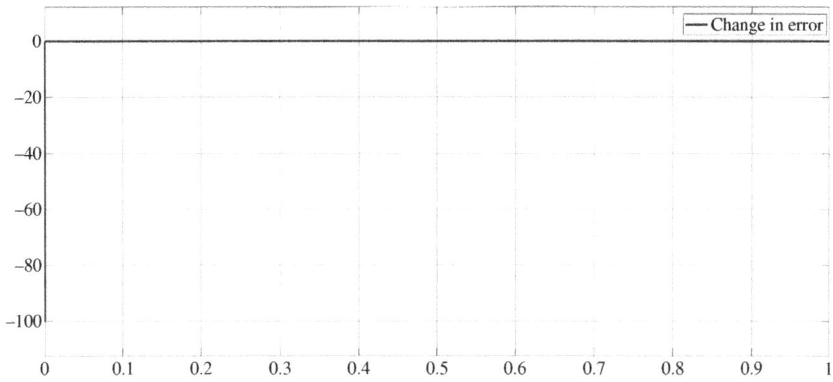

Figure 5.11 Change in error.

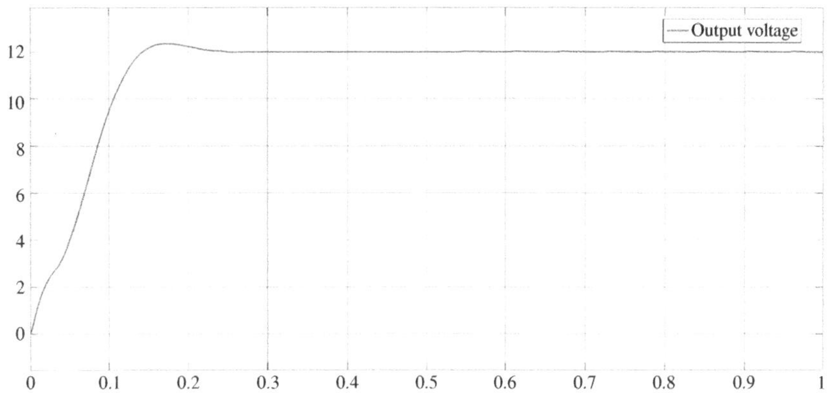

Figure 5.12 Output voltage.

The output Duty ratio is given to the PWM generator. With the switching frequency of 15,000 Hz, the gate pulses for the MOSFETs are generated so as to track the maximum power point.

- **Measurement Block:** The output of the Solar PV array along with the load output is measured at irradiance of 700, 1000, and 12,000 W/m^2.
  Table 5.5 shows the maximum power the specified irradiances along with the voltages and currents at which the MPP is tracked.

The output PV current, voltage, and power at different irradiance is shown in Figure 5.15. The irradiance is changed from 700 to 1000 W/m^2 and finally to 12,000 W/m^2

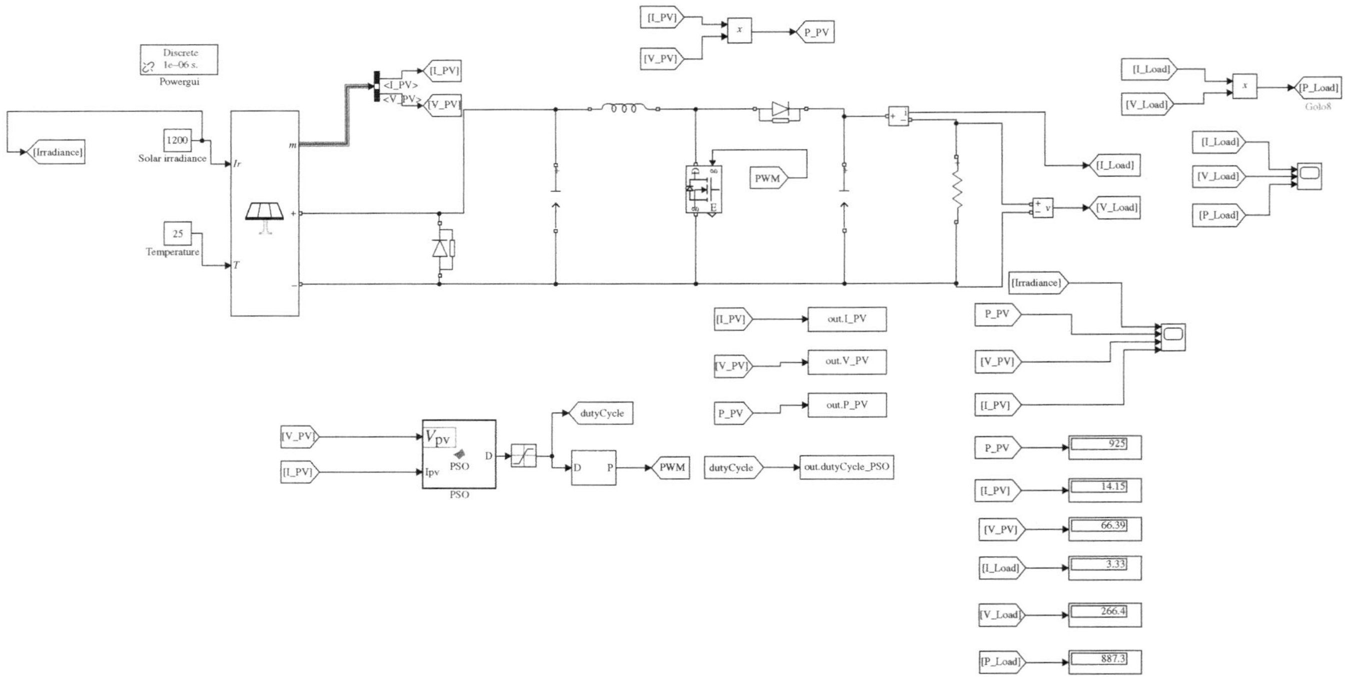

Figure 5.13 Schematic for the PSO-based MPPT control in MATLAB Simulink.

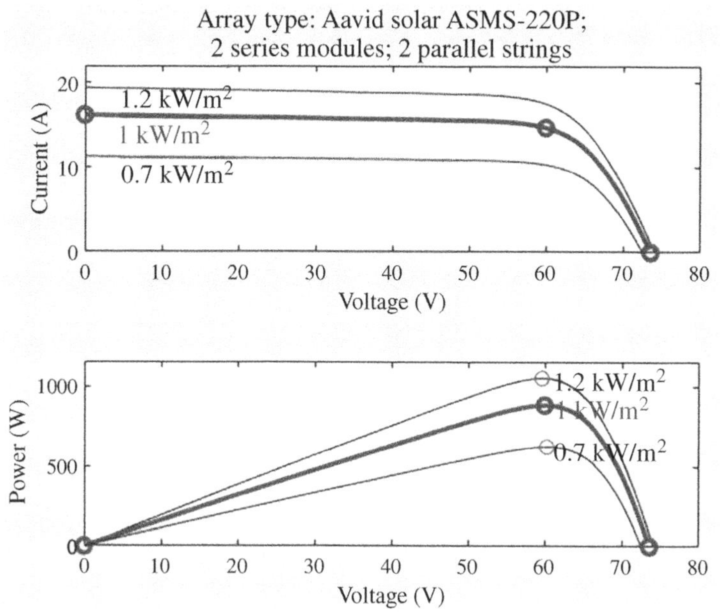

Figure 5.14 The characteristics of the PV array at 25 °C and under different irradiations.

Table 5.5 Maximum Voltage, Current and Power Values Tracked at MPP.

Irradiance (W/m^2)	P_{PV} (W)	I_{PV} (A)	V_{PV} (V)	P_{Load} (W)	I_{Load} (A)	V_{Load} (V)
700	925	14.5	65.39	887.3	3.33	266.4
1000	858.2	13.63	62.96	825	3.21	256.9
1200	551.6	10.8	50.69	536	2.58	207.1

The output voltage, power, and current from the load (80 Ω) is shown in Figure 5.16. The output is obtained at irradiance of 700, 1000, and 1200 W/m^2, respectively.

5.5.3 Fuzzy-Logic-Control-Based MPPT of the PV System

In this example, a fuzzy-logic-based MPPT control of the PV system is simulated in MATLAB Simulink. The schematic of the PV system has been shown in Figure 5.17.

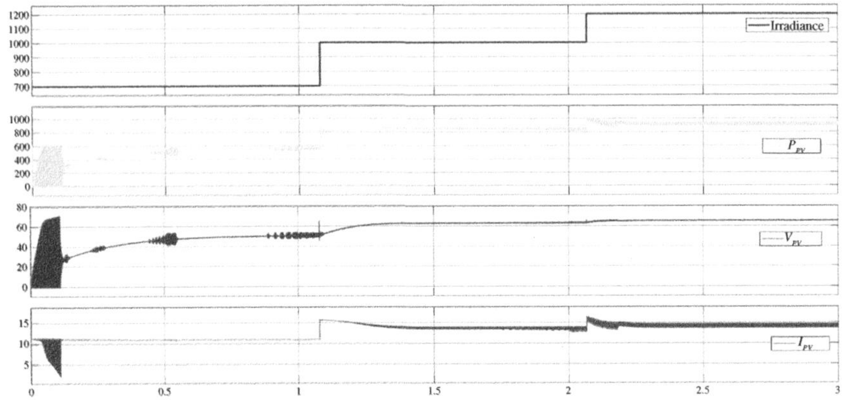

Figure 5.15 Waveform: Irradiance, PV current, PV voltage, and PV power.

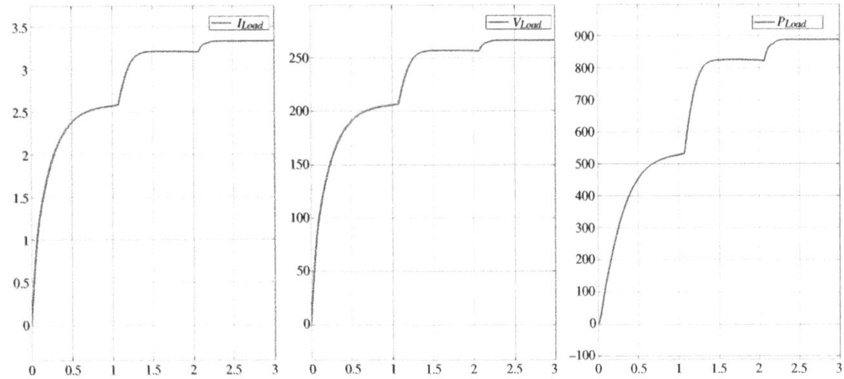

Figure 5.16 Waveform: Output load current, voltage, and power at different irradiance.

The solar array consists of one parallel string having two series-connected modules. Irradiance and temperature are the two inputs to the PV array. PV array is shown in Figure 5.18.

The characteristics of the PV array at 25 °C and under different irradiations shown in Figure 5.19.

As shown in Figure 5.20, the controller block has two inputs: PV voltage and PV current. Fuzzy logic controller (FLC) has two inputs and one output. Fuzzy controller inputs are:

- **Error Signal:** It is the derivative (dP/dV) signal.
- **Change in Error Signal:** DE.

Duty ratio is the output which generates the switching pulses.

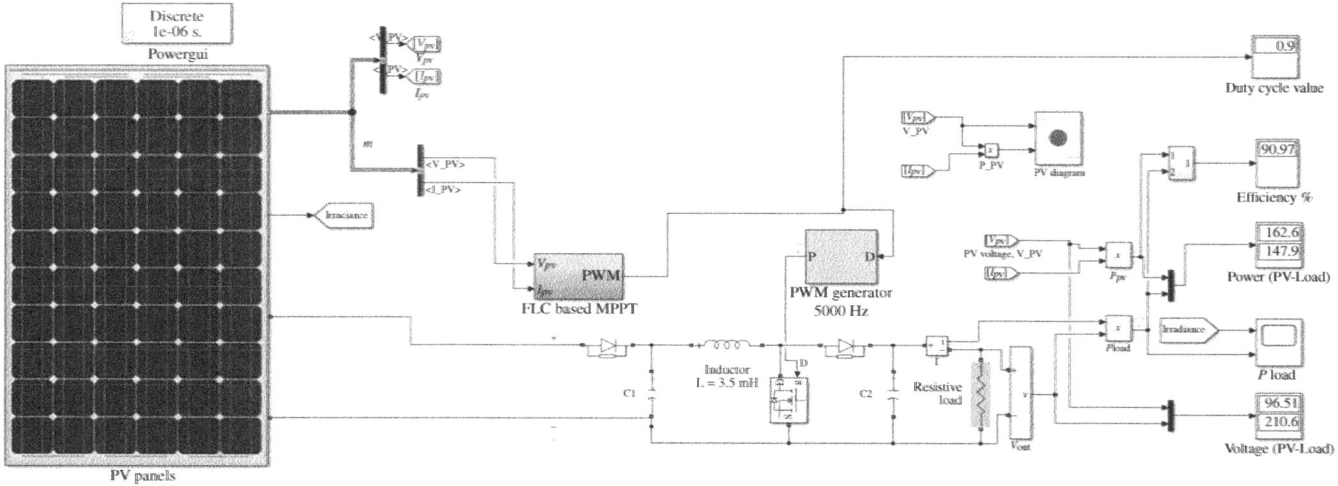

Figure 5.17 The schematic of the Fuzzy-logic-based MPPT of the PV system.

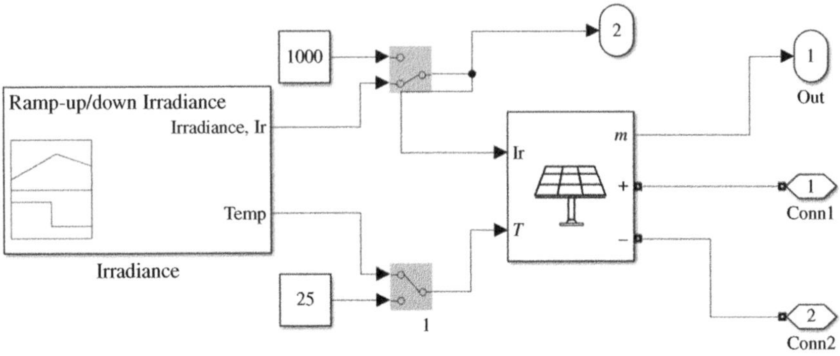

Figure 5.18 Solar photovoltaic array.

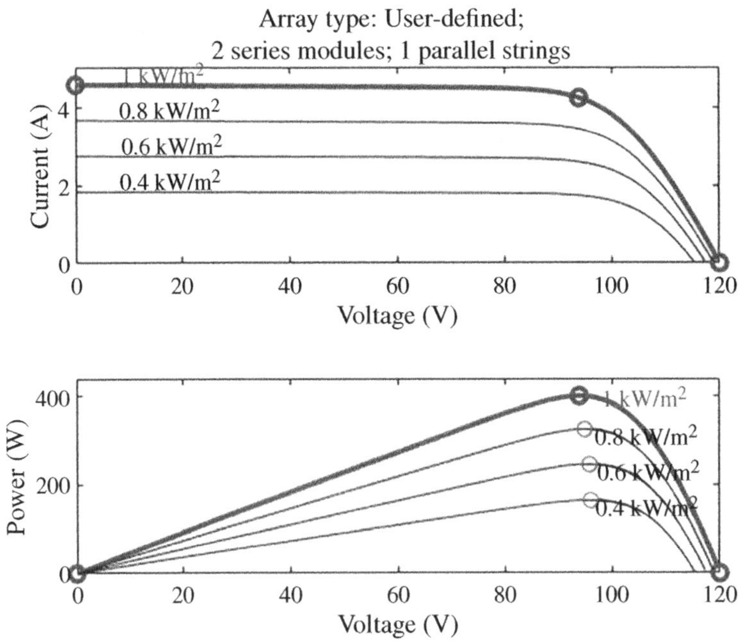

Figure 5.19 The IV and PV characteristics of the PV array at 25 °C and under different irradiations.

The Fuzzy logic controller is shown in Figure 5.21. This fuzzy controller is based on the Mamdani Fuzzy Inference System which is based on the set of linguistic rules. Centroid Defuzzification is implemented to convert the fuzzified values into the crisp values.

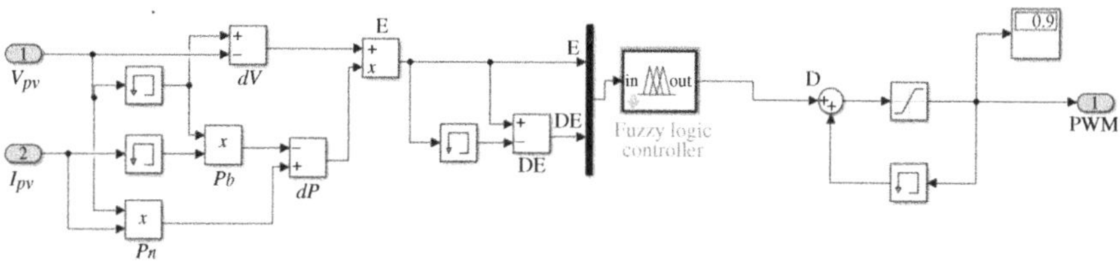

Figure 5.20 Fuzzy logic control block.

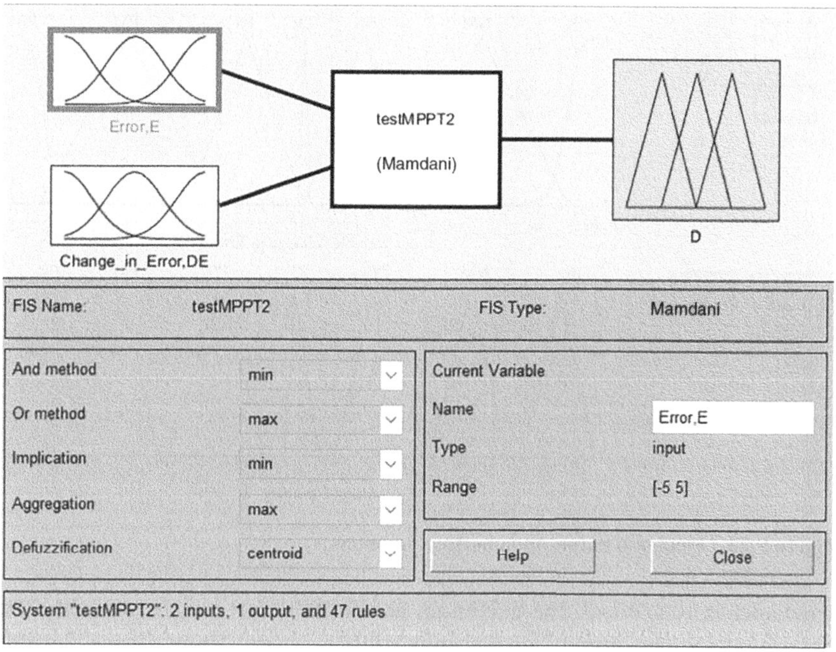

Figure 5.21 Fuzzy inference system.

Each input variable has seven input membership functions. Out of the seven MFs, five are the triangular MFs and two are the generalized bell-shaped membership functions. However, only there are only seven output membership functions.

Figure 5.22 shows the MFs of the input and output variables.

This FLC-Based MPPT control of the PV system has 47 If–Then rules as is shown in Figure 5.23.

The 3D Model for the output of the fuzzy logic controller can be viewed in Surface option under View tab, and it is shown in Figure 5.24.

The designed Fis file is imported to workspace and is updated in the controller block. Thus, with FLC control, the maximum power point is tracked. When the

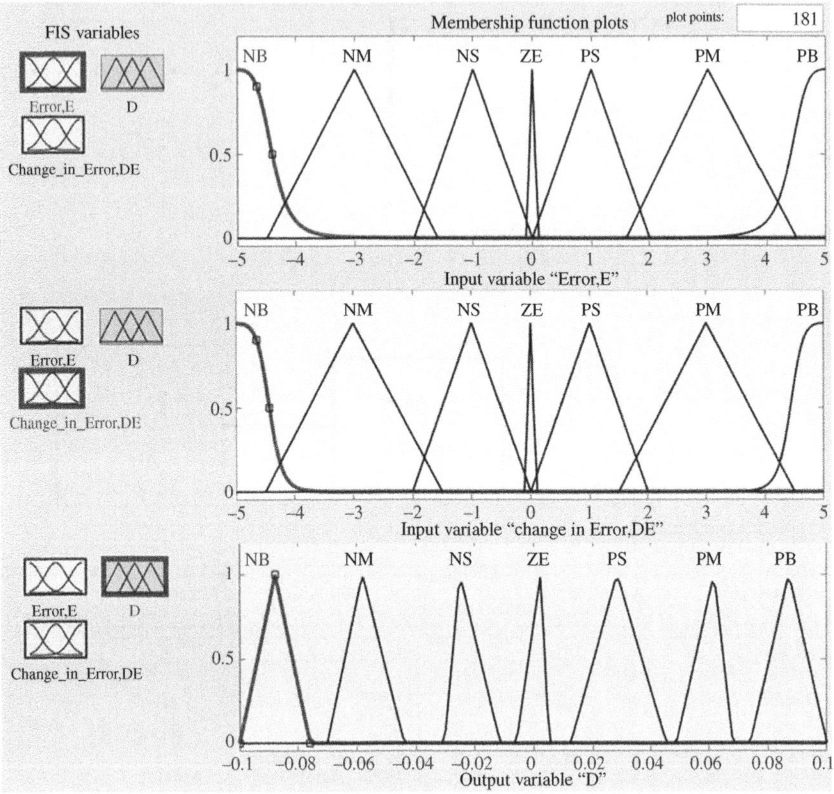

Figure 5.22 Input and output membership functions.

irradiance is 1000 W/m^2, the maximum power (400 W) is tracked when the PV voltage is 95 V. The output load power at MPPT is around 356 W. Hence, the efficiency of the simulated fuzzy logic controller is around 89%.

The output PV curve with the maximum power point at the irradiance of 1000 W/m^2 is shown in Figure 5.25.

The output load voltage at different values of solar irradiance is shown in Figure 5.26.

5.6 Conclusion

This chapter presents a detailed analysis on the applications of the fuzzy logic and metaheuristic methods in power electronics. Fuzzy logic and metaheuristic methods provide powerful tools for improving the performance and reliability of the

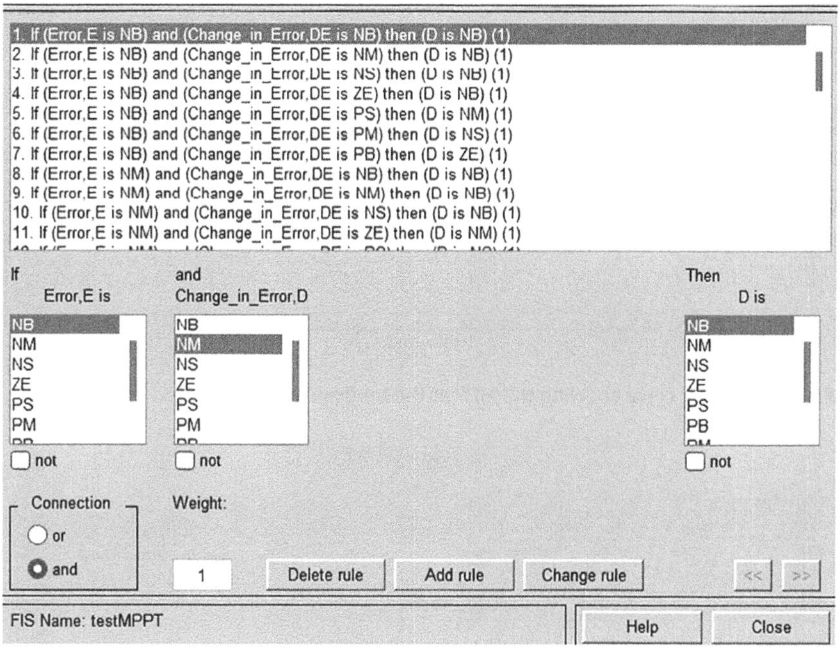

Figure 5.23 Rule Editor window depicting If–Then rules.

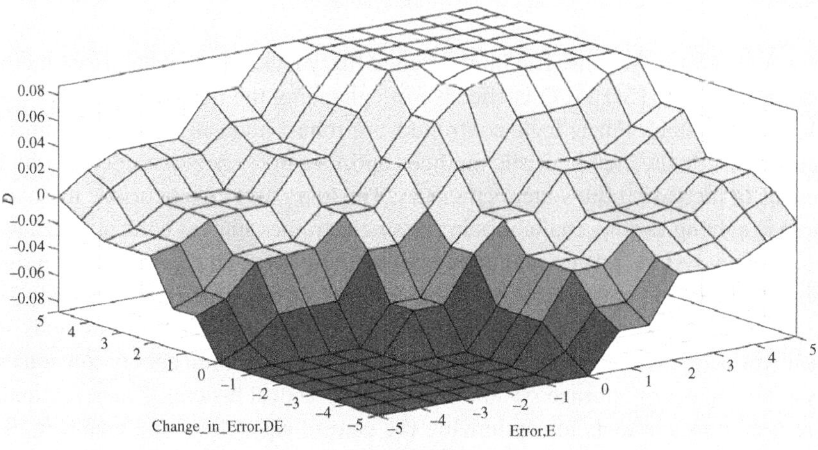

Figure 5.24 3-D model of the Fuzzy logic controller.

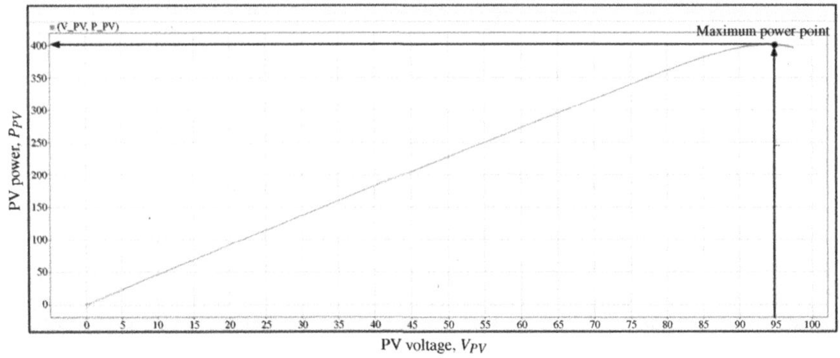

Figure 5.25 PV curve showing the MPP at the irradiance of 1000 W/m^2.

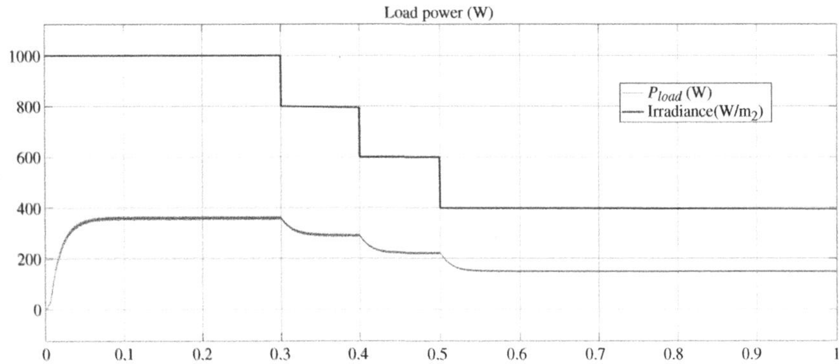

Figure 5.26 Output load voltage at different irradiance.

power electronic systems. The integration of fuzzy logic and metaheuristic methods such as GA, PSO, ACO is effective in enhancing the performance of power electronic system. Fuzzy logic controllers are robust and can handle the uncertainties, while the metaheuristic methods optimize the system parameters which results in increasing the system efficiency. The fuzzy and metaheuristic methods help in solving various challenges in power electronics such as control in renewable energy system, power quality improvement in distributed generation, energy management, and optimal design of the power electronic converters.

The successful application of the fuzzy logic and metaheuristic methods in real-world applications as solar inverter, electric vehicles, wing energy conversion system, and energy storage system depict their practical benefits. These methods are indispensable tools for optimizing the control algorithms and ensuring the system robustness in renewable and sustainable energy systems. As the power electronics system expands in the immensely increasing renewable energy systems and smart grids, the application of the fuzzy and metaheuristic methods is expected to grow. Thus, the continued research for increasing their speed, reducing the computational burden, and integrating with other AI techniques is inevitable. The development and integration into the real-time systems is a key to address the future challenges.

Exercises

5.1 What are the major application areas of the fuzzy logic and metaheuristic methods in power electronics?

5.2 What are the common membership functions used in fuzzy-logic-based MPPT controllers?

5.3 Which type of membership function is most suitable for FLC in Multi-Level Inverters?

5.4 How does the fuzzy-logic-based MPPT technique dynamically adjust the duty cycle to ensure the maximum power extraction from solar panel under varying environmental conditions?

5.5 What the inputs and outputs of the fuzzy logic controller for voltage control of resonant converter?

5.6 A PSO-tuned fuzzy MPPT controller is operated in a buck converter that steps down the voltage form 24 to 12 V. If the power required by the load is 36 W, calculate the current on the output side of the converter.

5.7 How is the fuzzy logic method helpful for the High-Voltage Direct Current (HVDC) Transmission system?

5.8 What type of defuzzification method is commonly used in fuzzy logic MPPT application?

5.9 Which power electronic converter design parameters are commonly optimized with the metaheuristic algorithms?

5.10 In context of a fuzzy controller for the power electronic converters, how can metaheuristic methods improve the performance?

5.11 How is the maximum power point tracked with the ant colony optimization-based FLC method?

5.12 What is the fitness function for the Particle Swarm Optimization-based maximum power point tracking (MPPT) control of the DC–DC converter?

5.13 How does PSO-tuned FLC ensure to enhance the lifetime of power electronic converters?

5.14 A solar panel's power output changes from 40 to 42 W as the voltage changes from 16 to 16.5 V. Calculate the error that will be given as one of the inputs to the fuzzy logic controller.

5.15 In a PSO optimization for an MPPT controller, if the maximum power point is 100 W and the output power achieved by the particle is 90 W, calculate the fitness score for that particle.

5.16 How is the PSO-based control in the PSO control block implemented in Simulink for the solar PV system?

5.17 What is the advantage of the bacterial foraging algorithm in optimizing the parameters for the control of grid-side inverter?

References

1 E. Sayed *et al.*, "Renewable energy and energy storage systems," *Energies*, vol. 16, no. 3, p. 1415, Feb. 2023. 10.3390/en16031415.

2 O. S. Vargas, S. E. De León Aldaco, J. A. Alquicira, L. G. Vela-Valdés, and A. R. L. Núñez, "Adaptive network-based fuzzy inference system (ANFIS) applied to inverters: a survey," *IEEE Trans. Power Electron.*, vol. 39, no. 1, pp. 869–884, Jan. 2024. 10.1109/TPEL.2023.3327014.

3 C. Yadav, M. K. Bhardwaj, S. Patel, S. Yadav, K. K. Bharati, and Y. Shekhar, "Fault detection and location in power system using fuzzy logic controller: a review," in *2024 3rd International Conference on Power Electronics and IoT Applications in Renewable Energy and its Control (PARC)*, IEEE, Feb. 2024, pp. 427–432. 10.1109/PARC59193.2024.10486521.

4 A. R. Ofoli, "Fuzzy-logic applications in electric drives and power electronics," in *Power Electronics Handbook*. Elsevier, 2018, pp. 1221–1243. 10.1016/B978-0-12-811407-0.00040-4.

5 H. R. Chamorro, N. L. Diaz, J. J. Soriano, and H. E. Espitia, "Active and reactive power flow fuzzy controller for VSC HVDC using DBR and DBR type 2," in *2011 Annual Meeting of the North American Fuzzy Information Processing Society*, IEEE, Mar. 2011, pp. 1–6. 10.1109/NAFIPS.2011.5751947.
6 A. R. Ofoli, "Fuzzy-logic applications in electric drives and power electronics," in *Power Electronics Handbook*. Elsevier, 2024, pp. 1233–1260. 10.1016/B978-0-323-99216-9.00003-2.
7 S. M. Almufti, A. Ahmad Shaban, Z. Arif Ali, R. Ismael Ali, and J. A. Dela Fuente, "Overview of metaheuristic algorithms," *Polaris Glob. J. Sch. Res. Trends*, vol. 2, no. 2, pp. 10–32, Apr. 2023. 10.58429/pgjsrt.v2n2a144.
8 P. Kala, P. Joshi, M. Joshi, S. Agarwal, and L. K. Yadav, "Applications of metaheuristics in power electronics," in *Metaheuristic and Evolutionary Computation: Algorithms and Applications*. Springer Nature, 2021, pp. 165–176. 10.1007/978-981-15-7571-6_7.
9 E. Cuevas, D. Zaldívar, and M. Pérez-Cisneros, "Metaheuristic algorithms based on fuzzy logic," in *Studies in Computational Intelligence*. Springer, 2018, pp. 167–218. 10.1007/978-3-319-89309-9_8.
10 J. Kumari and A. K. Singh, "Optimizing solar power generation with bidirectional converter and fuzzy logic for battery management in EVs," in *2024 IEEE International Students' Conference on Electrical, Electronics and Computer Science (SCEECS)*, IEEE, Feb. 2024, pp. 1–6. 10.1109/SCEECS61402.2024.10481954.
11 N. Amulya and R. Dhanalakshmi, "Analysis of fuzzy logic control based resonant DC-DC converter for high voltage gain," in *2019 International Conference on Intelligent Computing and Control Systems (ICCS)*, IEEE, May 2019, pp. 1074–1079. 10.1109/ICCS45141.2019.9065679.
12 C. Buccella, C. Cecati, H. Latafat, and K. Razi, "Comparative transient response analysis of LLC resonant converter controlled by adaptive PID and fuzzy logic controllers," in *IECON 2012 – 38th Annual Conference on IEEE Industrial Electronics Society*, IEEE, Oct. 2012, pp. 4729–4734. 10.1109/IECON.2012.6389483.
13 A. Siddique, Y. Xu, W. Aslam, A. Rehman, and M. K. Aslam, "Performance of fuzzy logic based controller for UPFC on power quality issues in transmission network," in *2018 International Conference on Power System Technology (POWERCON)*, IEEE, Nov. 2018, pp. 2867–2872. 10.1109/POWERCON.2018.8601674.
14 P. Patel, S. Samal, P. K. Mishra, C. Jena, P. K. Barik, and D. K. Gupta, "Power quality analysis using adaptive controller based SAPF under different loading scenario," in *2024 Second International Conference on Smart Technologies for Power and Renewable Energy (SPECon)*, IEEE, Apr. 2024, pp. 1–6. 10.1109/SPECon61254.2024.10537471.

15 S. R. Das *et al.*, "Fuzzy controller designed based multilevel inverter for power quality enhancement," *IEEE Trans. Consum. Electron.*, vol. 70, pp. 4839–4847, Apr. 2024. 10.1109/TCE.2024.3389687.

16 G. Martinez-Teran, A. Garcia-Barrientos, F. R. Trejo-Macotela, F. Coyotl-Mixcoatl, and R. Tapia-Olvera, "Design and implementation of a FLC for CD-CD converters using a microcontroller for PV systems," in *CONIELECOMP 2013, 23rd International Conference on Electronics, Communications and Computing*, IEEE, Mar. 2013, pp. 158–161. 10.1109/CONIELECOMP.2013.6525778.

17 R. K. Dhawan and P. J. Davis, "Fuzzy logic based inductor design program," in *Proceedings of APEC 97 – Applied Power Electronics Conference*, IEEE, 1997, pp. 579–584. 10.1109/APEC.1997.575621.

18 S. Chimplee and S. Khwan-on, "The fault detection method based on fuzzy logic for boost converters," in *2024 12th International Electrical Engineering Congress (iEECON)*, IEEE, Mar. 2024, pp. 1–4. 10.1109/iEECON60677.2024.10537829.

19 V. Singh, A. Yadav, and S. Gupta, "Open-switch fault detection and classification in five-level neutral-point-clamped inverter by using fuzzy interface system," in *2022 IEEE Global Conference on Computing, Power and Communication Technologies (GlobConPT)*, IEEE, Sep. 2022, pp. 1–6. 10.1109/GlobConPT57482.2022.9938287.

20 R. Mahalakshmi, J. Viknesh, M. G. Ramesh, M. R. Vignesh, and K. C. S. Thampatty, "Fuzzy logic based rotor Side converter for constant power control of grid connected DFIG," in *2016 IEEE International Conference on Power Electronics, Drives and Energy Systems (PEDES)*, IEEE, Dec. 2016, pp. 1–6. 10.1109/PEDES.2016.7914302.

21 M. S. Elborlsy, A. E. A. Hussien, and M. A. Ebied, "An intelligent MPPT technique based on fuzzy controller applied to grid-connected PV systems in temperature fluctuations," in *2023 3rd International Conference on Electronic Engineering (ICEEM)*, IEEE, Oct. 2023, pp. 1–6. 10.1109/ICEEM58740.2023.10319534.

22 C. Buccella, C. Cecati, and P. Siano, "A fuzzy-logic-controlled resonant converter for renewable energy sources applications," in *IECON 2011 – 37th Annual Conference of the IEEE Industrial Electronics Society*, IEEE, Nov. 2011, pp. 2450–2455. 10.1109/IECON.2011.6119694.

23 S. Siddula, T. V. Achyut, K. Vigneshwar, and M. V. Uday Teja, "Improvement of power quality using fuzzy based unified power flow controller," in *2020 International Conference on Smart Technologies in Computing, Electrical and Electronics (ICSTCEE)*, IEEE, Oct. 2020, pp. 55–60. 10.1109/ICSTCEE49637.2020.9276496.

24 M. Balasubbareddy, P. Venkata Prasad, and K. Palle, "Power quality conditioner with fuzzy logic controller," in *Information and Communication Technology for Competitive Strategies (ICTCS 2022)*, 2023, pp. 627–636. 10.1007/978-981-19-9304-6_56.

25 A. M. Aboelezz, B. E. Sedhom, M. A. Saeed, A. A. Eladl, and M. M. El-Saadawi, "Fuzzy logic approach based fault Detection in multi terminal HVDC Transmission systems," in *2023 24th International Middle East Power System Conference (MEPCON)*, IEEE, Dec. 2023, pp. 1–6. 10.1109/MEPCON58725.2023.10462447.

26 X. Shen, Y. Zuo, J. Kong, and W. Martinez, "Artificial intelligence applications in high-frequency magnetic components design for power electronics systems: an overview," *IEEE Trans. Power Electron.*, vol. 39, no. 7, pp. 8478–8496, Jul. 2024. 10.1109/TPEL.2024.3381431.

27 S. E. De León-Aldaco, H. Calleja, and J. Aguayo Alquicira, "Metaheuristic optimization methods applied to power converters: a review," *IEEE Trans. Power Electron.*, vol. 30, no. 12, pp. 6791–6803, Dec. 2015. 10.1109/TPEL.2015.2397311.

28 W. Wong and C. I. Ming, "A review on metaheuristic algorithms: recent trends, benchmarking and applications," in *2019 7th International Conference on Smart Computing & Communications (ICSCC)*, IEEE, Jun. 2019, pp. 1–5. 10.1109/ICSCC.2019.8843624.

29 M. Sarbanzadeh, M. A. Hosseinzadeh, A. Salehi, M. Rivera, and P. Wheeler, "Genetic algorithm technique for 7-level cascaded H-bridge multilevel converter THD minimization," in *2019 IEEE 28th International Symposium on Industrial Electronics (ISIE)*, IEEE, Jun. 2019, pp. 2635–2640. 10.1109/ISIE.2019.8781206.

30 I. Alzuguren, A. Garcia-Bediaga, A. Avila, A. Rujas, and M. Vasic, "Genetic algorithm-based optimized modulation for dual active bridge PFC circuit for electric vehicle application," in *2022 IEEE Design Methodologies Conference (DMC)*, IEEE, Sep. 2022, pp. 1–5. 10.1109/DMC55175.2022.9906470.

31 B. Fares, A. Sabrina, B. Achour, and C. Omar, "Power quality improvement by using photovoltaic based PSO as MPPT with space vector modulation (SVM) control strategy of quasi Z-source inverter," in *2023 1st International Conference on Renewable Solutions for Ecosystems: Towards a Sustainable Energy Transition (ICRSEtoSET)*, IEEE, May 2023, pp. 1–6. 10.1109/ICRSEtoSET56772.2023.10525472.

32 S. Choudhury *et al.*, "Energy management and power quality improvement of microgrid system through modified water wave optimization," *Energy Rep.*, vol. 9, pp. 6020–6041, Dec. 2023. 10.1016/j.egyr.2023.05.068.

33 M. M. Shehu, M. Dong, and J. Hu, "Optimization of particle swarm based MPPT under partial shading conditions in photovoltaic systems," in *2021 IEEE*

16th Conference on Industrial Electronics and Applications (ICIEA), IEEE, Aug. 2021, pp. 267–272. 10.1109/ICIEA51954.2021.9516360.

34 M. A. Khazain, N. M. Hidayat, K. Burhanudin, and E. Abdullah, "Boost converter of maximum power point tracking (MPPT) using particle swarm optimization (PSO) method," in *2021 IEEE 12th Control and System Graduate Research Colloquium (ICSGRC)*, IEEE, Aug. 2021, pp. 281–286. 10.1109/ICSGRC53186.2021.9515228.

35 J. Prasanth Ram and N. Rajasekar, "A new robust, mutated and fast tracking LPSO method for solar PV maximum power point tracking under partial shaded conditions," *Appl. Energy*, vol. 201, pp. 45–59, Sep. 2017. 10.1016/j.apenergy.2017.05.102.

36 M. Mallat, O. B. Hadj Brahim Kechiche, K. Ezzeddine, and M. Hamouda, "AN optimized genetic algorithm (GA)-based MPPT for off-grid solar photovoltaic system," in *2023 IEEE International Conference on Artificial Intelligence & Green Energy (ICAIGE)*, IEEE, Oct. 2023, pp. 1–5. 10.1109/ICAIGE58321.2023.10346555.

37 A. G. Agamy, A. H. K. Alaboudy, H. E. Mostafa, and M. Y. Fekry, "Bacterial foraging-based PI controller of inverter-based distributed generators," in *2011 IEEE Trondheim PowerTech*, IEEE, Jun. 2011, pp. 1–7. 10.1109/PTC.2011.6019375.

38 Y. Wang, S. Chakraborty, D. D. Tran, T. Geury, and O. Hegazy, "Design and optimization of the arm inductor for modular multilevel converter," in *2022 International Symposium on Power Electronics, Electrical Drives, Automation and Motion (SPEEDAM)*, IEEE, Jun. 2022, pp. 354–359. 10.1109/SPEEDAM53979.2022.9842277.

39 I. El Hajjami, B. Benhala, and H. Bouyghf, "Comparative study between the genetic Algorithm and the artificial bee colony technique: RF circuits application," in *2020 IEEE 2nd International Conference on Electronics, Control, Optimization and Computer Science (ICECOCS)*, IEEE, Dec. 2020, pp. 1–5. 10.1109/ICECOCS50124.2020.9314521.

40 G. Di Capua, N. Femia, and K. Stoyka, "Differential evolution algorithm-based identification of ferrite core inductors saturation curves," in *2015 IEEE 13th International Conference on Industrial Informatics (INDIN)*, IEEE, Jul. 2015, pp. 1636–1641. 10.1109/INDIN.2015.7281979.

41 D. Izci, S. Ekinci, and H. L. Zeynelgil, "Controlling an automatic voltage regulator using a novel Harris hawks and simulated annealing optimization technique," *Adv. Control Appl.*, vol. 6, no. 2, Jun. 2024. 10.1002/adc2.121.

42 Ö. Akyazı, "GA-based fuzzy logic controller with booster type DC-DC converter," *NWSA Acad. J.*, vol. 13, no. 1, pp. 34–43, Jan. 2018. 10.12739/NWSA.2018.13.1.2A0134.

43 K. Xia, Y. Li, and B. Zhu, "Improved photovoltaic MPPT algorithm-based on ant colony optimization and fuzzy logic under conditions of partial shading," *IEEE Access.*, vol. 12, pp. 44817–44825, 2024. 10.1109/ACCESS.2024.3381345.

44 N. Kumarasabapathy and P. S. Manoharan, "MATLAB simulation of UPQC for power quality mitigation using an ant colony based fuzzy control technique," *Sci. World J.*, vol. 2015, no. 1, p. 304165, Jan. 2015. 10.1155/2015/304165.

45 R. Kumar, "Fuzzy particle swarm optimization control algorithm implementation in photovoltaic integrated shunt active power filter for power quality improvement using hardware-in-the-loop," *Sustain Energy Technol Assess*, vol. 50, p. 101820, Mar. 2022. 10.1016/j.seta.2021.101820.

46 A. Maurya and J. Shukla, "Ant colony optimization tuned fuzzy control for PV battery and super capacitor powered electric vehicle," *Int. J. Intell. Syst. Appl. Eng.*, vol. 12, no. 22s, p. 99, 2024.

47 C.-S. Shieh, "Fuzzy PWM based on genetic algorithm for battery charging," *Appl. Soft Comput.*, vol. 21, pp. 607–616, Aug. 2014. 10.1016/j.asoc.2014.04.009.

48 L. G. da Silva, N. D. de Andrade, R. B. Godoy, M. A. G. de Brito, and E. T. Maddalena, "Differential evolution and fuzzy-logic-based predictive algorithm for V2G charging stations," *Appl. Sci.*, vol. 13, no. 10, p. 5921, May 2023. 10.3390/app13105921.

49 M. A. Ibrahim, M. H. Ibrahim, and S. I. Khather, "Design of fuzzy-ACO based controller for Cuk converter in electric vehicles," *J. Eur. Des Syst. Autom.*, vol. 56, no. 3, pp. 425–430, 2023. 10.18280/jesa.560309.

50 I. Robandi, D. Ajiatmo, and Muhlasin, "Optimization PI-ACO for photovoltaic system battery and supercapacitor on electric vehicle," in *2021 International Seminar on Intelligent Technology and Its Applications (ISITIA)*, IEEE, Jul. 2021, pp. 370–375. 10.1109/ISITIA52817.2021.9502221.

6

Supervised Learning for Power Electronics

Md Zafar Khan and Ahteshamul Haque

Advance Power Electronics and Research Laboratory, Jamia Millia Islamia (A Central University), New Delhi, India

6.1 Introduction

Power electronics (PE) use semiconductor technology to regulate and exchange electrical power. Such systems are used in a variety of fields, including automation in factories, green energy, and mode of transport. PE systems focus heavily on effectiveness and dependability to function effectively. Machine learning (ML) has been used to optimize and control PE systems, improving their performance. These strategies can help lower the computational cost of characterizing DC–DC converters, which is crucial for developing and optimizing PE systems. ML approaches like support vector regression (SVR) and artificial neural networks (ANNs) can effectively forecast DC–DC converter performance. Supervised learning (SL) uses a training dataset of input–output pairs to identify functional correlations and mappings. The inputs and outputs are implied. This function is particularly valuable in PE, where developing system models can be tough. SL mostly focuses on classification and regression. The classification algorithm labels a restricted number of distinct groups based on input–output pairings from the training dataset [1, 2].

SL is a type of ML that trains an algorithm with labeled data to comprehend the link between input and output variables. SL is used in PE management and optimization to predict system behavior based on input variables such as voltage, current, and temperature. PE can benefit from many SL techniques, such as linear regression, support vector machines (SVMs), and neural networks [1, 3].

SL is a ML approach in which a model is trained using feed data (for example, a vector of predictor variables) and an output that is wanted value (also known as a supervisory signal labeled by humans). The training data is analyzed to create a

Artificial Intelligence for Power Electronics, First Edition.
Edited by Ahteshamul Haque, Saad Mekhilef, and Azra Malik.

function that maps fresh data to anticipated outcome values. An optimal situation will enable the programmed to appropriately calculate output values for previously unseen instances. This necessitates that the learning algorithm make "reasonable" generalizations from the training data to previously unknown scenarios. The statistical quality of an algorithm is assessed using the so-called generalization error.

In this chapter, we will look at the use of SL techniques to PE. We will go over regression algorithms for forecasting PE parameters, classification techniques for fault diagnosis, and a case study on predictive maintenance of power converters using ML.

6.2 Types of Supervised Learning

6.2.1 Support Vector Machines (SVMs)

SVM is a sophisticated ML technique that may be used for both nonlinear and linear regression, classification, and identifying outliers. SVMs are useful for a range of applications, including recognizing texts, image classification, recognizing spam, recognizing handwriting, evaluation of gene expression, recognizing faces, and finding anomalies. SVMs are versatile and efficient in a wide range of applications because they can handle high-dimensional data and nonlinear relationships. SVM algorithms are quite effective because they attempt to discover the largest separation hyperplane between the many classes present in the target feature. Figure 6.1 shows the Schematic Diagram of intelligent electronics device (IED)-based controlled electricity management system, where SVM technique is used to get the good output using data line.

The supporting vector machine (SVM) incorporates a classification and regression algorithm that operates on learning data and can be used to forecast new or calculated data. The SVM method is separated into two parts: classification of supporting vectors (SVC) for issues that require classification approaches, and regression of support vectors (SVR) for regression. The method of learning at SVM tries to generate a prediction for the optimum dividing field that not only minimizes the average error in the training data but also has strong generalizability.

Generalization is a hypothesis's capacity to accurately classify or regression data that is not part of the training data set. To ensure this generalization, SVM employs structural risk minimization (SRM). SRM seeks to ensure the upper limit of generalization on test data by limiting the capacity (flexibility) of the learning outcome hypothesis. Figure 6.2 depicts the valuation of margins [4, 5]. Figure 6.3 shows the flow chart of SVM technique.

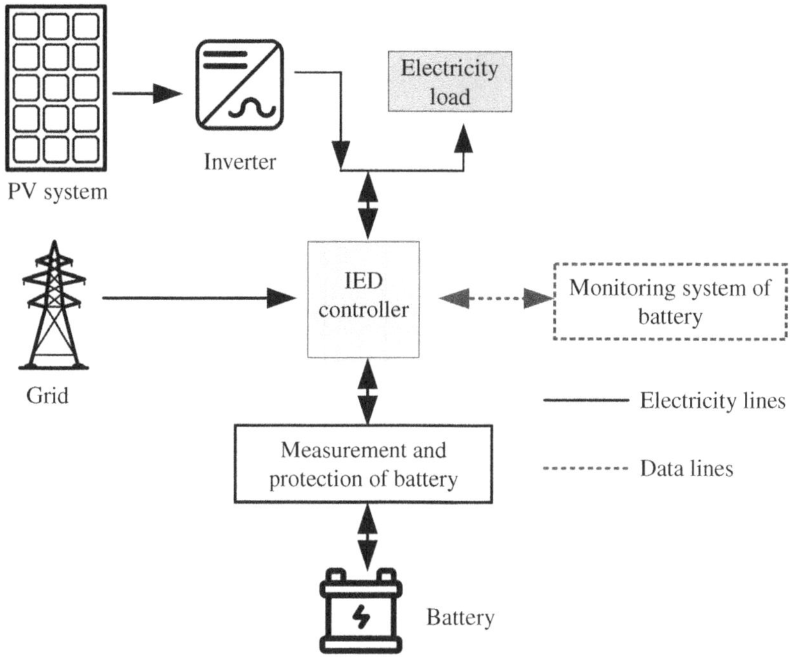

Figure 6.1 Schematic diagram of intelligence-based electricity management system.

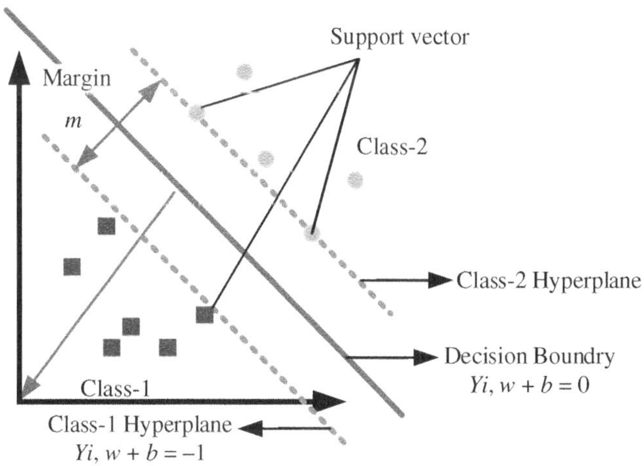

Figure 6.2 The margin valuation.

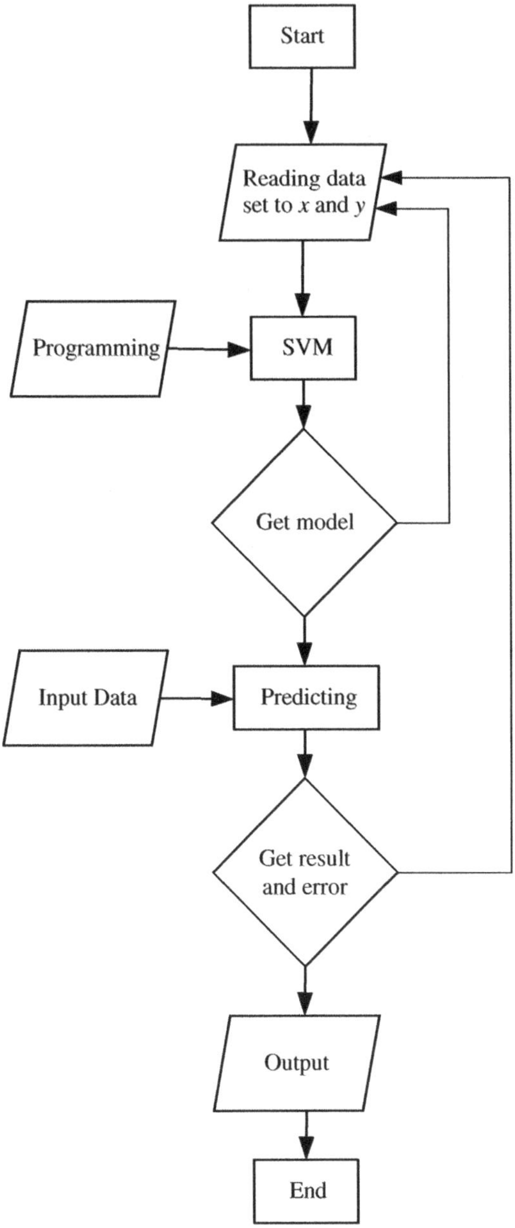

Figure 6.3 Flow chart of SVM technique.

Figure 6.2 shows that the letter "m" indicates the margin used. The margin is the distance between the separating spheres of classes 1 and 2. Data that comes into contact with this boundary region is known as a support vector. Equation (6.1) depicts the equation derived from the splitting of data by the enclosing planes of classes 1 and 2.

$$Y_i \cdot w + b \geq +1 \quad \text{for} \quad Z_i = +1 \tag{6.1}$$

$$Y_i \cdot w + b \leq +1 \quad \text{for} \quad Z_i = -1 \tag{6.2}$$

6.2.2 Neural Networks

Neural networks are a type of artificial intelligence that is designed to mimic the architecture and operations of the human brain. Neural network control, in particular, has received substantial interest due to its ability to improve the profitability of power electronic systems. In this context, neural networks are effective tools for modeling and managing nonlinear and complex systems, particularly where typical linear control tactics fall short. Neural network control techniques are computational models based on natural neural networks. These algorithms learn from behavioral data and can enhance their performance with acquiring or retraining. They can accurately estimate any nonlinear function, making them perfect for dealing with irregularities commonly found in power electronic systems. Neural network (NN) models excel in identifying complicated relationships between variables, coping with data uncertainty (due to redundancy in storage), and predicting trends in real time. NN models have the problem of requiring a large amount of high-quality data, which may not always be available or affordable [6, 7].

A standard function approximator may accurately predict output data when given new input data. A standard function approximator employs a mathematical model of the system, as illustrated in Figure 6.4 (a). When an exact mathematical model of a system is not attainable, model-free approximators are used, as seen in Figure 6.4 (b). Neural networks can simply replace standard function approximators.

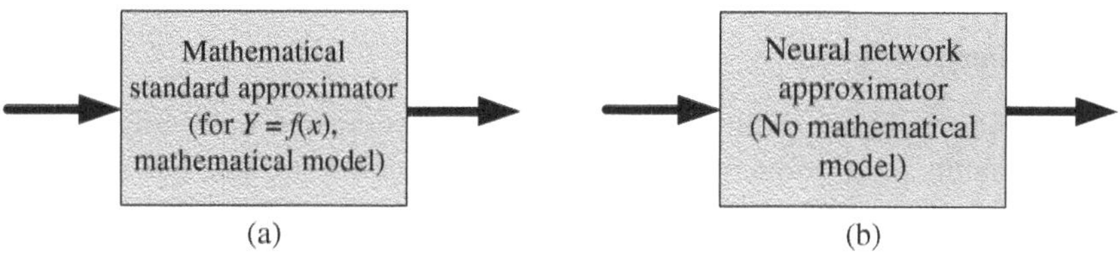

Figure 6.4 (a) Standard function approximator. (b) Neural network approximator.

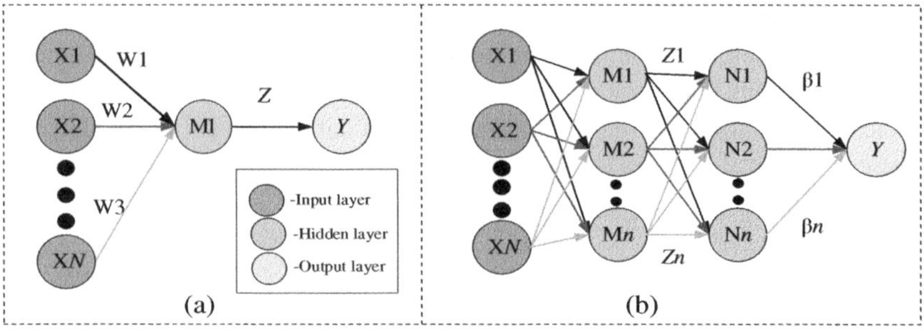

Figure 6.5 (a) One layer and (b) multi-layer neural network model.

An artificial neural network (ANN) learns by training. Error back-propagation is a popular training method. The approach involves providing input along with output data to train to an ANN and configuring its weights. After training, the ANN uses real input data to generate the desired output data. As shown in Figure 6.5 (a) and (b) the One layer and multi-layer neural network model, where inputs output and hidden layers are there. The comparison among various neural networks including advantages and disadvantages is provided in Table 6.1.

In Figure 6.5 (a) the input layer, a neuron (i) produces X_N, while the hidden layer's output, M_1 (Eq. (6.1)), is determined by the activation function, which is a linear combination of the input neurons X_N, adjusted by a threshold W_0. The most common functions for activation are the gradient and exaggeration tangent.

$$Z = \sum_{i=1}^{N} W_i X_i \tag{6.3}$$

While on Figure 6.5 (b) Eq. (6.2) shows the output of the network, the neural network's output, Y, is a straight-line sum of the output of the concealed layer neurons, modified by a further threshold β_j, and taking $Z_0 = 1$. This function is typically the identity in most statistical situations.

$$Y = f(x) = g\left(\sum_{j=1}^{N} \beta_j Z_j\right) \tag{6.4}$$

6.2.3 Decision Trees

A decision-tree-based facilities data link (FDL) was created to classify various inverter faults in an induction motor drive application. The tree-like structure consists of a root at the top that splits into numerous branches below the root. The tree's final decision with respect to the outcome is made at the stage where the branch ceases further separating into more branches. Quinlan's information

Table 6.1 Merits and Demerits of Neural Network in Different Fields.

Ref	Scope	Merits	Demerits
[4]	ANN for motor drive control	• It doesn't require any mathematical model • It can improve the performance after tuned properly • It required less time as compared to standard controller	• It requires large amount of high quality of data • Implementing for real-time control system is challenging
[8]	Recurrent neural network (RNN) impact on control system	• RNN can be trained to detect the fault • RNN can predict the life of the component by daily monitoring its usage • It can improve the performance and efficiency over time	• Training RNNs can be difficult due to vanishing and ballooning gradient issues, which can slow down the learning process for extended sequences • The stability and robustness of RNN is challenging, when it operates in noisy environment
[9]	Convolutional neural network (CNN) for solar PV power	• It processes image efficiently • They are able to identify patterns in photos without requiring feature engineering by hand	• The PV panel works best when the sunlight directly to the panel • It has lack of interpretability • It requires large datasets
[10]	Deep neural network (DNN) for power and energy system	• Reduced computing complexity to resolve optimization-based problems found in power systems • The capacity to handle ill-posed challenges • Smart assistance with fault diagnosis and monitoring in complicated distribution networks	• Applying optimal power flow (OPF) for the distribution system is difficult • Taking data from asynchronous measurement with restricted access to real-world measurements • Modeling of electric vehicle (EV) charging station, House-hold battery can impact on the distribution system
[11]	ANN in pulse width modulation (PWM) control	• It can filter the noise and harmonics from the signal and supply • The ANN is quick implementation, which allows for significantly higher switching frequency	• It requires large data set and consume more time for training the data • Large digital application-specific integrated circuit (ASIC) chip with high resolution is not available due to which parallel operation is not done

Table 6.2 Merits and Demerits of Decision Tree in Different Fields.

Ref	Scope	Merits	Demerits
[12]	Non-intrusive load monitoring	• It is simple and easy to implement • It doesn't require scaling of data • Decision trees are relatively resilient to unusual values, as they partition data based on the most significant traits and avoid unnecessary fluctuations	• It can lead poor generalization of unknown data • It can create only orthogonal decision boundaries • It can't capture the non-linear relationship
[13]	Feasibility of renewable energy	• We can saw its graphical representation clearly • Its help to detect impurity in data • LSTM effectiveness prediction is measured for accuracy	• Its performance is lower in some cases • Decision tree (DT) can't justify the best feature of data and best threshold value we use Gini measurement
[14]	Stability of power system impact due to renewable energy	• The classification tree and regression tree are a real-time data • The suggested approach can recommend optimal corrective steps for any near instability or unstable electromechanical fluctuations, irrespective of prior knowledge of the fluctuating output of renewable source power • The algorithm proposed in the study addresses multi-modal classifications by incorporating a pre-processing phase into the input to the decision tree, which reflects high-dimensional data with only one variable	• Decision trees can be difficult to employ when the factors that predict are multidimensional, as is common in big power systems with numerous oscillation modes • The algorithm proposed in the study addresses multi-modal classifications by including a pre-processing phase in the input to the decision tree, which may enhance the model's complexity • The suggested technique necessitates a substantial quantity of data to construct appropriate trees of classification and regression, which could occasionally not be obtainable in reality
[15]	Fault prediction of machinery	• It is able to work with numeric as well as classified data • To trained data it requires less time • For normalizing or dummy data is not required for missing data	• The data should be raw data so it takes time to extract the data for each attribute • If the broken time data is not much larger than the module will not work well

theory is used to choose roots because it is critical for achieving the best outcomes. The decision tree's input is the time-domain inverter current output acquired from the hall sensors. Relevant characteristics are extracted and fed as input to the decision tree, and these features can represent a variety of decision tree qualities. To design and analyze the right decision tree, thorough training and testing data gathering is required. Table 6.2 shows comparison among various decision-tree-based techniques.

- **Nodes**: Internal nodes play a significant role in decision-making and data training. Features require node attributes. The data is split based on the information obtained at the node level. The tolerance value is used to determine how to split the data, with each node representing a feature. This selection is crucial as it causes data division from hereditary nodes. Figure 6.6 displays an example of a decision tree. The nodes partition the criterion based on the feature's input data. Separated sections display associated decisions, features, and threshold values.
- **Branches**: Branches connect nodes that reflect distinct decisions. This determines the condition of the features. These conditions are assessed in the provided samples. Thus, the decision is made from left to right or from right to left. The branches represent the probable outcomes of each node. The branches facilitate data comprehension by increasing the number of analysis

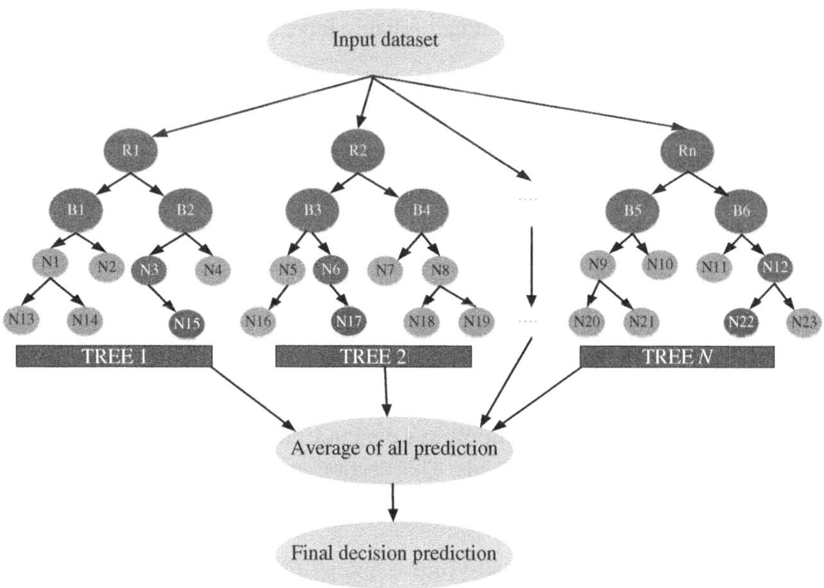

Figure 6.6 The concept of decision tree.

directions. This concept enables us to analyze and complete the decision tree decision-making process in all cases.

- **Leaf Node**: The leaf node is the desired node for the model's final conclusion. When the possibility reaches this node, the method's decision points are completed. The class label and regression values indicate leaf nodes. Leaf nodes continue with the decision tree model's final forecast.

6.3 Applications in Power Electronics

SL applications in PE considerably improve system performance, reliability, and efficiency. Engineers can use advanced ML approaches to optimize different elements of PE, such as problem detection and predictive maintenance, renewable energy forecasting, and control system optimization. These applications not only enhance the functioning and durability of power electronic systems, but they also aid in the transition to more environmentally friendly and effective energy solutions.

6.3.1 Regression Applications

Supervised regression learns the relationship between input characteristics and the target by using labeled training data with a continuous output variable (target). The goal is to create a model that can reliably forecast the target variable given fresh, previously unseen data.

6.3.1.1 Energy Consumption Forecasting

Energy consumption forecasting is an important discipline in PE, as it contributes to effective energy management, grid stability, and planning. This can examine the methodology, applications, and significance of energy consumption forecasting in PE, giving a thorough overview for investigators, technologists, and operators.

Predicting the future energy demand based on historical consumption data. Regression models like linear regression, polynomial regression, and SVR are used to forecast electricity usage. The main application of energy consumption forecasting is as follows.

- Forecasting helps to balance supply and demand, guaranteeing grid stability and avoiding blackouts and overloads. This section emphasizes the importance of accurate predictions in optimizing resource allocation and lowering operational expenses.
- Predictive models help estimate future energy prices and demand, allowing traders to make informed decisions that maximize earnings while minimizing risks.

- Forecasting plays an important role in aligning energy demand with renewable energy production, increasing solar and wind power utilization, and encouraging sustainability.
- Forecasting helps demand response activities by anticipating peak loads and modifying consumption patterns, which lowers energy prices and reduces system strain.
- The incorporation of smart meters and IoT devices for real-time information collecting and monitoring is investigated, demonstrating how such data can improve precision in forecasting and demand-side management.

6.3.1.2 Power Output Prediction

Power output prediction is an essential component of modern PE, allowing for more effective energy management, improved grid stability, and sustainable energy practices. PE can help to make the energy ecosystem more dependable and resilient by utilizing advanced prediction algorithms. This gives a fundamental understanding and practical insights into the methodology and applications of power output prediction, making it an invaluable resource for furthering study and practice in this area.

Power output prediction is an important part of PE, which includes estimating future electricity production and output from various devices and systems. The methodology, applications, and significance of power output prediction in the field of PE make it an invaluable resource for investigators, designers, and industry experts.

For estimating the power output of renewable energy sources such as solar panels or wind turbines, regression models help in predicting the power generation based on weather conditions and historical performance data. The main application of energy consumption forecasting is as follows.

- Predicting the power production of renewable energy sources such as solar panels and wind turbines is critical to integrating renewable energy into the grid. This focuses on how accurate predictions improve the use of renewable energy, reduce dependency on fossil fuels, and promote sustainability.
- Accurate power output projections help to balance supply and demand, guaranteeing grid stability and avoiding overloads. This section emphasizes the importance of forecasting in optimizing resource allocation and lowering operating expenses.
- Predicting the power output of energy storage systems, such as batteries, is critical for optimizing charging and discharging cycles. This section explains how correct predictions increase energy storage efficiency and battery life.
- Power output prediction is critical for managing microgrids, especially in remote or off-grid locations. This way precise predictions increase microgrid reliability and enable more effective energy management.

6.3.1.3 Converter Control

Converters are important in PE because they convert electrical power from one form to another, such as AC to DC (rectifiers), DC to AC (inverters), DC to DC (choppers), and AC to AC (cycloconverters). Effective control of these converters is required to achieve desired performance characteristics such as voltage regulation, power factor correction, harmonic reduction, and efficiency optimization [16].

Linear controllers, such as cascaded voltage and current loops, are extensively used in power converter systems to control power flow and maintain stability. However, because these controllers were built for linear systems, they may not perform best in nonlinear systems. As a result, for the best performance, linear control loop settings should be modified to fit changing operating conditions in nonlinear situations. In this scenario, AI-based methods can be used to change the parameters of linear controllers in power converters [17].

In addition to ANN applications, in [18], local model networks (also known as neuro-fuzzy systems) were utilized to identify the model of a DC–DC buck converter. This model can be used to create a local linear controller to regulate the converter's output voltage. Liu et al. improved the transient response performance of DC–DC converters by combining two fuzzy-logic controllers with a proportional–integral–derivative controller (PID) compensator for coefficient and error changes.

Figure 6.7 displays the online and offline ANN-based design diagrams for a linear controller. The picture depicts a DC–AC voltage source inverter (VSI) supplying an ac load through an output LC filter. However, ANN can also be used in other topologies, such as an AC (micro)grid with an LCL filter. Offline design differs from online design in that it does not require rigorous internet training. Independent offline learning often necessitates a large number of offline simulations to gather training data. Liu et al. proposed employing ANNs to change PID coefficients and controller structure for DC–DC converters. Maruta et al. developed an ANN-based predictor to improve the transient response of a PID controller by modifying its output voltage references. This research used offline ANN training to achieve great accuracy in predicting the output voltage of a DC–DC converter.

6.3.2 Classification Applications

Classification is an important use of ML in PE, since it provides sophisticated solutions for defect identification, condition monitoring, and system diagnosis. This investigates the approaches, applications, and significance of classification techniques for improving the reliability and efficiency of power electronic systems.

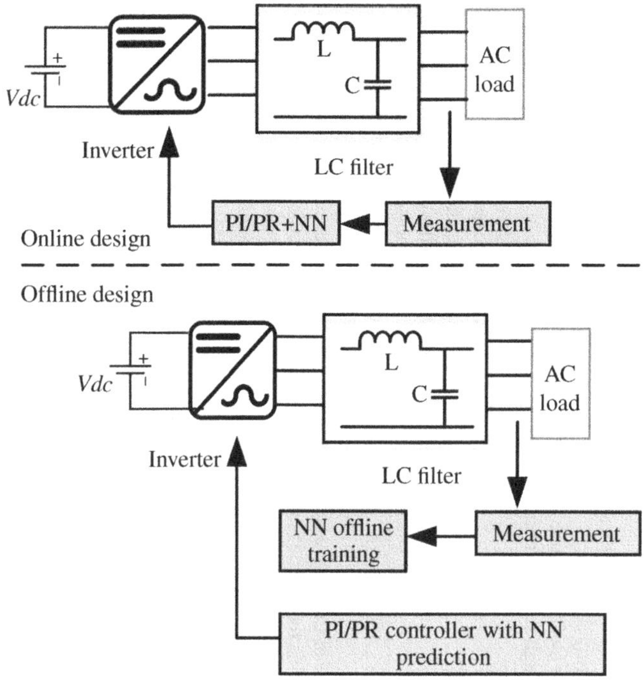

Figure 6.7 The online/offline ANN Linear controller.

Classification entails dividing data into preset classes or categories based on input characteristics. In PE, classification techniques aid in defining operating states, detecting problems, and anticipating failures, hence assuring system dependability and safety.

6.3.2.1 Fault Location and Type Detection

PE defects can arise for a variety of reasons, including component failures, ambient conditions, and operational stressors. Comparison among various machine-learning-based fault detection techniques is illustrated in Table 6.3. Accurate detection and localization of these defects is critical for reducing downtime, mitigating damage, and guaranteeing smooth operation.

The fault detection in [23] is as follows.

- The long short-term memory (LSTM) technique accurately localizes malfunctioning smote algorithms (SMs) within 10 ms without the need for training data.
- The hybrid ML approach doesn't need correct mathematical models or additional sensors.
- The quantum particle swarm optimization (QPSO) technique was used to create a more effective LSTM neural network for detecting SM open-circuit (OC) faults.

Table 6.3 Method, Application, Fault Type and Their Merits and Demerits.

Ref	Method	Fault Type	Application in PE	Merits	Demerits
[19]	Multi-layer perceptron (MLP)	Rotor fault	Converter fed induction motor and changeable rotor	• Capable of implementing complex nonlinear problems and datasets • Fast post-training performance and cost savings during the testing phase • Ideal and accurate performance on tiny datasets • The basic structure allows for straightforward network design	• In this strategy, the efficiency of independent variables versus dependent variables is unknown • The computations for this method are difficult and time-consuming • Model performance and test outcomes are heavily reliant on the quality of the training procedure • Overfitting is a common issue when processing high-dimensional data • Does not have the capacity to model time-series data and find correlations between the input and output variables
[20]	Backpropagation in neural network (BPNN)	Fault in thyristor	Three phase full bridge rectifier	• One of the most essential aspects of this technique is its ability to correct trajectories using gradient descent • Due to the elimination of a very basic network structure • It allows for quick and easy programming No prior knowledge of networks is necessary • Its training method is independent of the function's characteristics • Allows for efficient calculation of gradients in each layer	• Most issues require certain inputs to be solved effectively • It is particularly sensitive to complicated and turbulent information • Cannot model time-series data Over-fitting is an issue when processing huge amounts of data with this method • Not able to obtain attributes from the input data

[21]	Self organizing map (SOM)	Driver power is under rated and overrated	DC–DC converters for multi-level motor	• Network training allows for rapid testing and evaluation of fresh data • The SOM's network structure is incredibly simple, avoiding the complexity of computing • It can also be utilized in an unsupervised manner • It can be used as a method to reduce the dimensions of large datasets	• The method of training a network to handle high-dimensional data is time-consuming • Despite processing massive amounts of data, it is unable to model time-series data • The level of accuracy of input data has a significant impact on the network's training and test performance
[22]	Radial basis function (RBF)	Rotor fault	Converter fed induction motor and changeable rotor	• It quickly determines network parameters and moves on to the training step • RBF excels at approximating functions for data and surfaces with predictable peaks and valleys • This technique's most notable features are its excellent performance and resilience to noisy data • In the test stage, performance improves quickly following training	• Having a large number of neurons complicates the network's ability to analyze and correlate input and output information • The RBF network training approach is unable to handle and model robust and complicated nonlinear systems • Traditional neural networks struggle with time-series data modeling and high-dimensional data

- An updated K-means approach with an evaluation function lowers reliance on manual settings, allowing for easier data clustering.
- A hybrid ML model integrates both supervised and unsupervised learning.

6.3.2.2 Islanding Detection

There are two categories of island detecting techniques: distant and local. Remote islanding detection approaches rely on costly communication methods. Local islanding detection approaches might be passive, active, or hybrid. Passive islanding detection involves measuring local characteristics like voltage, phase angle, and frequency, and comparing them to grid-connected mode values. If the discrepancy exceeds a specific threshold, islanding is identified. However, this strategy results in a huge non-detection zone (NDZ). The NDZ refers to the operating points when islanding detection techniques are ineffective. The active islanding technique involves injecting disturbance signals externally. Passive island detection strategies match system parameters to threshold values. If the quantity exceeds the threshold, islanding is identified. Passive approaches can easily implement. The downsides of these approaches are their high NDZ and ineffectiveness in multi-inverter systems [3].

6.3.3 Clustering Applications

Supervised clustering approaches use labeled data to improve the clustering process. These strategies can make clusters more interpretable and relevant in real applications. The primary goal is to use the labels to drive cluster creation in a way that maximizes both within-cluster homogeneity and between-cluster heterogeneity in relation to the labeled information.

6.3.3.1 Load Profiling

Load profiling is a critical approach in PE that entails analyzing the consumption patterns of electrical loads over time. Using clustering methods, these loads can be classified into distinct groups depending on their profiles. This chapter delves into the methodology, applications, and significance of load profiling for clustering in PE, emphasizing its advantages in optimizing energy management and increasing system efficiency. Load profiling entails gathering and analyzing information on the power usage of various loads. Clustering is a ML technique that combines data points with similar qualities together. In PE, integrating load profiling with clustering aids in discovering trends and categorizing loads based on consumption behavior. Some of the clustering techniques for load profiling are as follows.

- ***K*-Means Clustering:** J.B. MacQueen proposed the K-Means algorithm, which is based on division. This unsupervised technique is commonly used for data mining and pattern recognition. This approach is based on the square-error and error criterion, with the goal of minimizing the cluster performance index. To achieve the best possible conclusion, this method attempts to discover K divisions that satisfy a specific requirement. Firstly, select some dots to represent the initial cluster focal points (usually, we select the first K sample dots of income to represent the initial cluster focal point); secondly, gather the remaining sample dots to their focal points in accordance with the criterion of minimum distance, then we will get the initial classification, and if the classification is unreasonable, we will modify it, iterate repetitively until we get reasonable classification.

The K-Means algorithm, which is based on division, is a type of cluster algorithm that has the advantages of being quick, efficient, and timely. However, this method is heavily reliant on beginning dots and the differences in selecting initial samples, which always results in varied conclusions. Furthermore, this technique based on target functional ways employs the gradient approach to obtain the extremum. The gradient method's search orientation is always in the direction of decreasing energy, which means that if the starting cluster focal point is incorrect, the entire algorithm will simply sink into a local minimum point [24].

- **DBSCAN:** DBSCAN (Density-Based Spatial Clustering for Applications with Noise) is a density-based clustering technique that assigns data points to clusters based on density. Unlike K-Means or hierarchical clustering, DBSCAN does not require a predetermined number of clusters and can detect clusters with any form. The method operates by detecting dense regions of data points and expanding clusters from them.

  DBSCAN is a highly cited density-based clustering technique that accurately identifies random groups of varying shapes and sizes in huge, noisy databases. The DBSCAN algorithm eliminates the need to predetermine the number of clusters for datasets, which is a significant advantage. The DBSCAN algorithm effectively handles noise points, making it useful for identifying both noise-surrounded groups and other groups. Additionally, it is considered insensitive to noise and outliers. The DBSCAN method requires two crucial parameters: Eps and MinPts. Eps is the radius of a data point's neighboring neighborhood, while MinPts is the minimal number of data points in the region. Users can determine Eps and MinPts for datasets without prior knowledge of the number of clusters. Manually predicting an Eps value might be challenging due to the wide range of differences between datasets. The DBSCAN algorithm's outputs are heavily influenced by Eps, which can lead to substantial variations depending on the value. DBSCAN's performance suffers significantly when identifying clusters with varying data density due to the global fixed Eps value. To detect clusters with various densities, change the Eps and MinPts values based on the data distribution [25].

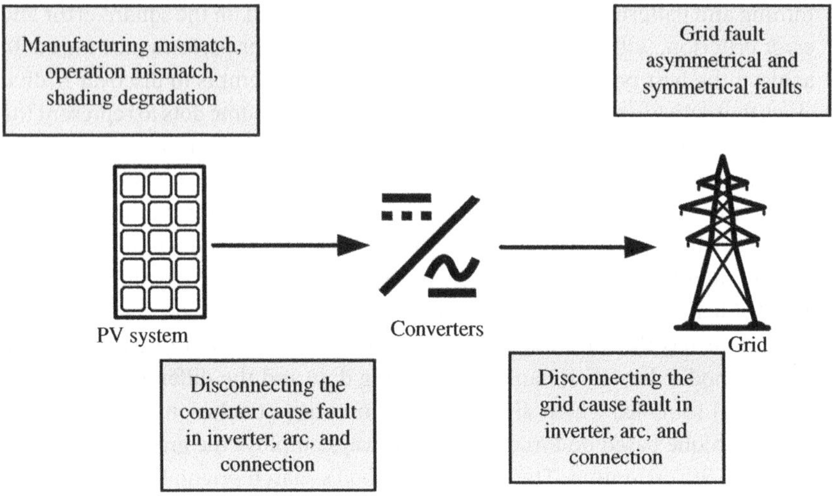

Figure 6.8 Various problems cause anomaly's fault in PV system.

6.3.3.2 Anomaly Detection

Anomalies in photovoltaic (PV) systems can significantly impact the efficiency of electrical energy generation, which is vital for an enterprise's financial return. Deviations in PV module grouping and grid integration, as seen in Figure 6.8, can be caused by many factors. PV plant difficulties are often reflected in module behavior, either directly or indirectly. This simplifies the use of detection algorithms. A failure in a converter might result in modules entering a short-circuit (SC) or OC state. Anomalies in PV projects can be caused by a variety of circumstances, including a designer's failure to prioritize critical aspects. One example of this is a lack of understanding about how to prevent abnormalities caused by environmental conditions or using low-quality components. Anomalies are a normal part of any installation. This is due to inevitable factors like grid outages or system component degradation.

6.4 Case Study: Predicting Power Consumption in an Electric Motor Using Support Vector Regression (SVR) in MATLAB

To predict a Predicting Power Consumption in an Electric Motor Using SVR in MATLAB which is based on input voltage, input current, speed, Temperature, and load, we'll implement by using SVM. Generate data, model training, define

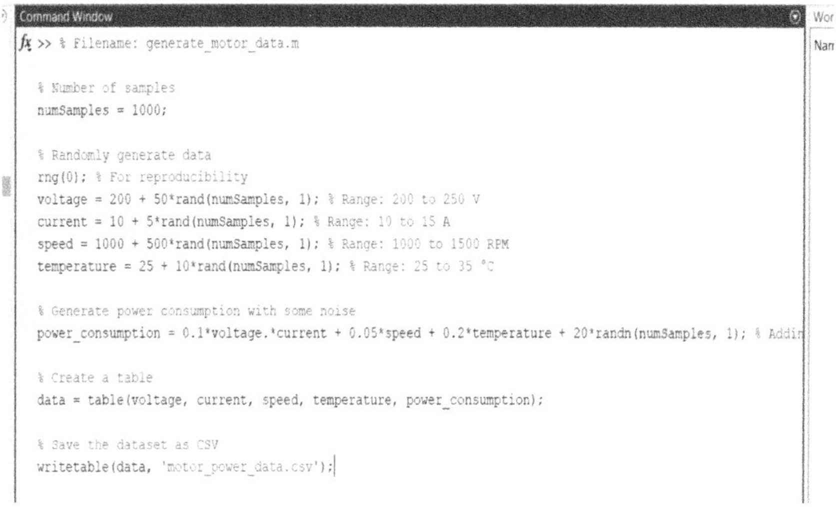

Figure 6.9 The command to generate motor data.

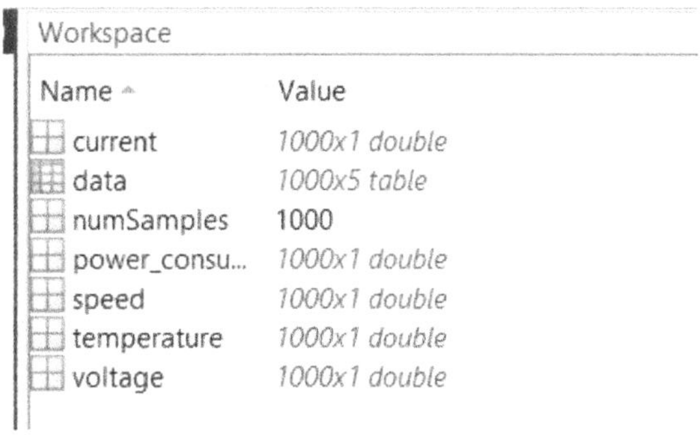

Figure 6.10 Work space table created.

prediction function, and testing are all steps in the workflow as shown in Figure 6.9.

Step 1: The above program automatically creates a table at the workspace after it is completed, as seen in Figure 6.10.
The demonstrated data is now found in a CSV file identified as converter_data.csv.

Step 2: Load, Preprocess Data, and Train Model: To load, preprocess data, and train model in MATLAB: We write a program in the command window that is shown in Figure 6.11.

```
% Save the trained model
save('motor_svr_model.mat', 'svrModel');

% Predict using the test set
predictions = predict(svrModel, XTest);

% Calculate and display the performance metrics
rmse = sqrt(mean((predictions - yTest).^2));
fprintf('Root Mean Squared Error (RMSE): %.2f\n', rmse);

% Plot actual vs predicted values
figure;
scatter(yTest, predictions);
xlabel('Actual Power Consumption (W)');
ylabel('Predicted Power Consumption (W)');
title('Actual vs Predicted Power Consumption');
grid on;

% Plot residuals
residuals = yTest - predictions;
figure;
scatter(predictions, residuals);
xlabel('Predicted Power Consumption (W)');
ylabel('Residuals (W)');
title('Residuals of Predictions');
grid on;
```

Figure 6.11 Load the data to the model.

The above program automatically train the Support Vector Mechanism a table with column title Input Voltage, Temperature, Speed, Current, and root mean square error (RMSE) at the command window after it is finished (run), as seen in Figure 6.12.

Step 3: Define the Prediction Function.

To extract relevant features and the target variable, we develop a program in the editor window that is shown in Figure 6.13.

After the completion of the above program, we successfully extracted relevant features and the target variable in the Command window and workspace, respectively, which is shown in Figure 6.14.

Step 4: Test the Prediction Function.

A key phase in the ML process is model training, which involves teaching an algorithm to make predictions or perform a specific task. Here we

Workspace

Name	Value
current	1000x1 double
cv	1x1 cvpartition
data	1000x5 table
numSamples	1000
power_consu...	1000x1 double
predictions	300x1 double
residuals	300x1 double
rmse	23.5335
speed	1000x1 double
svrModel	1x1 RegressionSV...
temperature	1000x1 double
voltage	1000x1 double
X	1000x4 double
XTest	300x4 double
XTrain	700x4 double
y	1000x1 double
yTest	300x1 double
yTrain	700x1 double

Figure 6.12 Data loaded at the work book.

predict_motor_power.m

```
% Filename: predict_motor_power.m

function power = predict_motor_power(voltage, current, speed, temperature)
    % Load the trained model
    persistent svrModel
    if isempty(svrModel)
        load('motor_svr_model.mat', 'svrModel');
    end

    % Prepare the data point for prediction
    dataPoint = [voltage, current, speed, temperature];

    % Predict the power consumption
    power = predict(svrModel, dataPoint);
end
```

Figure 6.13 Prediction of motor power.

write a program to Split the data into training and testing sets as shown in Figure 6.15.

After the completion of the above program, we successfully Split the data into training and testing sets in the workspace, which is shown in Figure 6.16.

```
Command Window
>> predict_motor_power(voltage, current, speed, temperature)

ans =

   385.4516         339.0662         351.2309
   367.5180         391.0443         371.3754
   379.2626         350.4563         422.1555
   344.8715         327.4872         341.7148
   370.5464         330.2912         305.9053
   346.9750         381.4612         320.1143
   302.6593         347.2433         329.0521
   335.2200         378.8111         320.3211
   328.6592         388.9138         348.3363
   351.9992         305.3423         305.0459
   282.2328         369.6281         361.2356
   420.6884         293.3039         396.2228
   371.0837         413.3167         350.3615
   378.4919         374.2678         310.6273
   423.7032         323.2571         411.7307
   324.6715         422.9343         353.5324
   289.6602         269.3048         401.3946
   417.0285         359.3192         407.8641
   336.4436         387.9591         322.5983
   415.5341         368.8677         340.0480
   319.6277         386.4195         345.6975
   316.8002         401.8883         357.9915
   339.1094         357.6074         421.7831
   407.3669         339.5477         402.5822
   323.8287         307.6411         324.7920
   331.9886         321.1594         364.2036
   371.2926         310.5652         314.6983
   345.7402         331.4440         365.9103
   400.4910         381.9716         378.2309
                    410.0534         371.2571
                    307.4156
                    325.4875
fx 303.6273         373.8796         369.8029
```

Figure 6.14 Predicted data.

```
Command Window
>> % Filename: test_motor_prediction.m
% Example usage
voltage = 220;
current = 12;
speed = 1300;
temperature = 30;
power = predict_motor_power(voltage, current, speed, temperature);
fprintf('Predicted Power Consumption: %.2f W\n', power);
% Load the dataset (for plotting purposes)
data = readtable('motor_power_data.csv');
X = data{:, {'voltage', 'current', 'speed', 'temperature'}};
y = data.power_consumption;
% Make predictions for the entire dataset
all_predictions = zeros(size(y));
for i = 1:length(y)
    all_predictions(i) = predict_motor_power(X(i,1), X(i,2), X(i,3), X(i,4));
end
% Plot actual vs predicted values
figure;
scatter(y, all_predictions);
xlabel('Actual Power Consumption (W)');
ylabel('Predicted Power Consumption (W)');
title('Actual vs Predicted Power Consumption');
grid on;
% Plot residuals
residuals = y - all_predictions;
figure;
scatter(all_predictions, residuals);
xlabel('Predicted Power Consumption (W)');
ylabel('Residuals (W)');
title('Residuals of Predictions');
grid on;
```

Figure 6.15 Command to compare the predicted data with actual data.

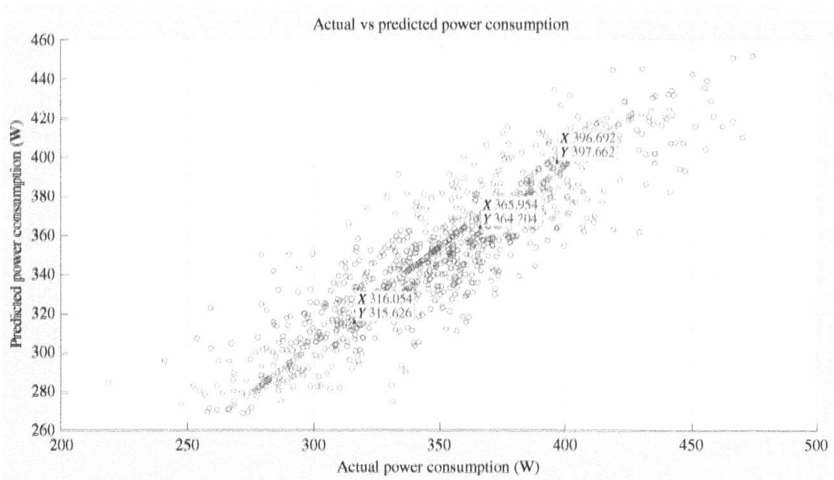

Figure 6.16 Relation graph between actual and predicted power.

6.5 Challenges and Future Prospects

There are several key challenges to overcome when incorporating SL techniques into PE systems. The quantity and quality of data with labels needed for efficient model training is one of the main issues. It is necessary to gather extensive and precise datasets covering a wide range of operational situations. The time-consuming and costly nature of this data collection procedure may make SL models less applicable in real-world scenarios. Furthermore, PE systems function in dynamic and complex conditions, where several factors might impact nonlinear relationships between variables. It is a significant technological problem to develop models that can effectively capture these intricate relationships and generalize to new data. Black-box models might hinder adoption in safety-critical applications, which actually require understandable, and explainable data, since they offer no visibility into their decision-making processes.

Despite these challenges, the future of SL in PE seems bright. The acquisition of richer and more complete datasets will be made easier by advances in data collection technology, such as IoT devices and high-resolution sensors, which will improve the accuracy and resilience of models. PE systems are dynamic and complex; therefore, developing hybrid models that incorporate SL with other ML methods could be a solution. The implementation of SL in PE is anticipated to grow as ML advances, propelling advancements in fault diagnostics, predictive maintenance, and system optimization. These developments will support the growing need for resilient and sustainable energy solutions by helping to create PE systems that are more intelligent, dependable, and efficient.

6.6 Conclusion

In conclusion, the incorporation of SL into PE is an unprecedented approach for improving these systems' efficiency, reliability, and performance. SL approaches may effectively anticipate system behavior by employing labeled datasets, which fulfils the crucial need for efficient system modeling and optimization in the field of PE. This chapter has illustrated the significance and adaptability of SL in this domain by thoroughly examining regression method for parameter forecasting. SL will surely play a more important role as the need for more resilient, and efficient PE systems grows across a range of sectors, propelling more developments and breakthroughs in the field of PE technology.

Exercises

6.1 How does ML improve the performance and optimization of DC–DC converters in PE?

6.2 What are the key differences between classification and regression in SL for PE?

6.3 How does the SVM technique work in identifying anomalies in power systems?

6.4 What is the significance of the margin in the SVM algorithm, as shown in the margin valuation diagram?

6.5 Why are neural networks effective in managing nonlinear and complex systems in PE?

6.6 What are the benefits and limitations of using ANNs for controlling motor drives in power systems?

6.7 How do decision trees work in fault diagnosis for inverter applications, and what are their key features like nodes and branches?

6.8 What role do regression models play in predicting energy consumption and power output in PE systems?

6.9 How can clustering techniques, such as K-Means, improve load profiling in PE?

6.10 What is the advantage of using density-based clustering techniques like DBSCAN in PE applications?

6.11 How is anomaly detection applied to photovoltaic (PV) systems, and what common causes lead to anomalies in these systems?

6.12 In SL applications, what are the main challenges associated with acquiring labeled data for training ML models in PE?

6.13 How does the integration of SL techniques with PE improve system reliability and efficiency?

6.14 What are the potential future directions and prospects for using SL in PE, particularly in areas like predictive maintenance and fault diagnostics?

References

1 M. Novak and T. Dragicevic, "Supervised imitation learning of finite-set model predictive control systems for power electronics," *IEEE Trans. Ind. Electron.*, vol. 68, no. 2, pp. 1717–1723, 2021. 10.1109/TIE.2020.2969116.

2 Z. Khashroum, H. Rahimighazvini, and M. Bahrami, "Applications of machine learning in power electronics: A specialization on convolutional neural networks," *Eng. Trans.*, vol. 4, no. 1, pp. 1–5, 2023. 10.61186/engt.4.1.2866.

3 A. Mishra and P. Jena, "Superimposed positive sequence voltage and current based islanding detection techniques for microgrid," in *2017 6th International Conference on Computer Applications In Electrical Engineering-Recent Advances (CERA)*, IEEE, Oct. 2017, pp. 282–285. 10.1109/CERA.2017.8343341.

4 B. Karanayil and M. F. Rahman, *Artificial Neural Network Applications in Power Electronics and Electrical Drives*, Third Edition. Elsevier Inc., 2010. 10.1016/B978-0-12-382036-5.00038-0.

5 A. Zamhuri Fuadi, I. N. Haq, and E. Leksono, "Support vector machine to predict electricity consumption in the energy management laboratory," *J. RESTI (Rekayasa Sist. Teknol. Informasi)*, vol. 5, no. 3, pp. 466–473, 2021. 10.29207/resti.v5i3.2947.

6 A. Malik, A. Haque, V. S. B. Kurukuru, M. A. Khan, and F. Blaabjerg, "Overview of fault detection approaches for grid connected photovoltaic inverters," *e-Prime - Adv. Electr. Eng. Electron. Energy*, vol. 2, no. November 2021, p. 100035, 2022. 10.1016/j.prime.2022.100035.

7 A. Malik, A. Haque, V. S. B. Kurukuru, M. A. Khan, and F. Blaabjerg, "Overview of fault detection approaches for grid connected photovoltaic inverters," *e-Prime Adv. Electr. Eng. Electron. Energy*, vol. 2, p. 100035, 2022. 10.1016/j.prime.2022.100035.

8 P. Dev, S. Jain, P. Kumar Arora, and H. Kumar, "Machine learning and its impact on control systems: A review," *Mater. Today Proc.*, vol. 47, no. 13, pp. 3744–3749, 2021. 10.1016/j.matpr.2021.02.281.

9 A. M. Alam, Nahid-Al-Masood, M. Iqbal Asif Razee, and M. Zunaed, "Solar PV power forecasting using traditional methods and machine learning techniques," *2021 IEEE Kansas Power Energy Conf. KPEC 2021*, pp. 3–7, 2021. 10.1109/KPEC51835.2021.9446199.

10 M. Marković, M. Bossart, and B. M. Hodge, "Machine learning for modern power distribution systems: Progress and perspectives," *J. Renew. Sustain. Energy*, vol. 15, no. 3, pp. 1–26, 2023. 10.1063/5.0147592.

11 B. Karanayil and M. F. Rahman, "Artificial neural network applications in power electronics and electric drives," in *Power Electronics Handbook*, Fourth Edition. Butterworth-Heinemann, 2017, pp. 1245–1260. 10.1016/B978-0-12-811407-0.00041-6.

12 T.-T.-H. Le, H. Kang, and H. Kim, "Household appliance classification using lower odd-numbered harmonics and the bagging decision tree," *IEEE Access.*, vol. 8, pp. 55937–55952, 2020. 10.1109/ACCESS.2020.2981969.

13 F. F. Fadoul, A. A. Hassan, and R. Çağlar, "Assessing the feasibility of integrating renewable energy: Decision tree analysis for parameter evaluation and LSTM forecasting for solar and wind power generation in a campus microgrid," *IEEE Access.*, vol. 11, pp. 124690–124708, 2023. 10.1109/ACCESS.2023.3328336.

14 T. Wang, T. Bi, H. Wang, and J. Liu, "Decision tree based online stability assessment scheme for power systems with renewable generations," *CSEE J. Power Energy Syst.*, vol. 1, no. 2, pp. 53–61, Jun. 2015. 10.17775/CSEEJPES.2015.00019.

15 T. Q. Vinh, P. M. Hoang, and T. Q. Hoa, "Machinery monitoring and maintenance system using decision tree model for fault prediction," in *2021 International Symposium on Electrical and Electronics Engineering (ISEE)*, IEEE, Apr. 2021, pp. 36–41. 10.1109/ISEE51682.2021.9418694.

16 M. Z. Khan, A. Haque, A. Malik, M. Amir, F. S. Zahgeer, and H. M. Khalid, "A critical review on control techniques for parallel operated inverters in grid connected and standalone mode," in *2024 International Conference on Green Energy, Computing and Sustainable Technology (GECOST)*, IEEE, Jan. 2024, pp. 66–71. 10.1109/GECOST60902.2024.10475094.

17 A. Haque, K. V. S. Bharath, M. Amir, and Z. Khan, "Role and applications of power electronics, renewable energy and IoT in smart cities," in *Smart Cities: Power Electronics, Renewable Energy, and Internet of Things*. Boca Raton, FL: CRC Press, 2023, pp. 66–95. 10.1201/9781032669809-4.

18 K. Rouzbehi, A. Miranian, E. Rakhshani, A. Luna, and P. Rodriguez, "Identification and local linear control of a DC-DC buck converter using local model networks," in *2013 International Conference on Renewable Energy Research and Applications (ICRERA)*, IEEE, Oct. 2013, pp. 745–750. 10.1109/ICRERA.2013.6749852.

19 M. Kaminski and T. Orlowska-kowalska, "Application of radial basis neural networks for the rotor fault detection of the induction motor," no. 1.

20 R. Wang, Y. Zhan, and H. Zhou, "Application of S transform in fault diagnosis of power electronics circuits," *Sci. Iran.*, vol. 19, no. 3, pp. 721–726, 2012. 10.1016/j.scient.2011.06.013.

21 P. Xiao, G. K. Venayagamoorthy, K. A. Corzine, and R. Woodley, "Self-healing control with multifunctional gate drive circuits for power converters," in *2007 IEEE Industry Applications Annual Meeting*, pp. 1852–1858, 2007.

22 B. X. Cui Li, Z. Liu, Y. Zhang, and L. Chai, "Diagnosis and location of the open-circuit fault in modular multilevel converters: An improved machine learning method," *Neurocomputing*, vol. 331, pp. 58–66, 2018. 10.1016/j.neucom.2018.09.041.

23 Y. An, X. Sun, B. Ren, and X. Zhang, "Open-circuit fault diagnosis for a modular multilevel converter based on hybrid machine learning," *IEEE Access.*, vol. 12, pp. 61529–61541, 2024. 10.1109/ACCESS.2024.3395121.

24 Y. Li and H. Wu, "A clustering method based on K-means algorithm," *Phys. Procedia*, vol. 25, pp. 1104–1109, 2012. 10.1016/j.phpro.2012.03.206.

25 M. M. R. Khan, M. A. B. Siddique, R. B. Arif, and M. R. Oishe, "ADBSCAN: Adaptive density-based spatial clustering of applications with noise for identifying clusters with varying densities," *4th International Conference of Electrical Engineering Information Communication Technology iCEEiCT* 2018, pp. 107–111, 2018. 10.1109/CEEICT.2018.8628138.

7

Unsupervised Learning for Anomaly Detection

Ahteshamul Haque[1] and Mohammed Ali Khan[2]

[1]*Advance Power Electronics and Research Laboratory, Jamia Millia Islamia (A Central University), New Delhi, India*
[2]*Centre of Industrial Electronics, University of Southern Denmark, Sønderborg, Denmark*

7.1 Introduction

The concerns related to the climate changes have made renewable energy essential investment for the future. The increase in renewable energy installation has also enabled the distributed power generation, further reducing the cost of energy and dependence on the grid [1]. Even different governments over the years have been planning and promoting different strategies and initiatives for the promotion of renewable energy. A sustainability development goal has been established by the United Nations for promoting sustainable energy projects [2]. Even the European Union climate directives have aimed at attaining carbon neutrality by 2050.

As renewable energy has emerged as an alternative for the conventional power generation, the interaction between the power generation unit and the load is regulated by multistages of power converters depending on the nature of the current being generated [3]. As the power generation unit is mostly unregulated and depends on the various environmental parameters, the power electronics converter ensures that the system operates in a stable mode [4]. Further, renewable energy sources have enabled distributed generation which further makes the power system more vulnerable to failure.

As the power electronics circuit operates at a high switching frequency, the possibility of the system failure is high [5]. There are a range of failures that may take place, from component degradation to overall frequency mismatch. The abnormality in the power system can result in a widespread system failure which can lead to a major blackout [6]. There have been multiple studies focusing on the system failure concerning the PV-based distributed generation. In case of a grid-based failure, the PV system is disconnected from the grid and starts to

Artificial Intelligence for Power Electronics, First Edition.
Edited by Ahteshamul Haque, Saad Mekhilef, and Azra Malik.

operate in islanded mode of operation [7]. However, before the islanded mode of operation [8] can be enabled the DG's check if the recovery of the grid is possible via fault ride-through (FRT) mechanism; in case the system is not recovered the DG is disconnected form the network and starts to operate in stand-alone mode [9–11]. The main goal of the DGs is to ensure the operation and satisfaction of load even when the grid is not available.

For ensuring the optimum operation of DGs, it is necessary to identify the different operating conditions well in time and attain the controllability on the converters. The controller needs to be able to comprehend for the failures that are occurring in the converter and if the fault is identified in time, then the controller can take up the appropriate action [12]. During the accidental trip, an improper grounding can cause issues related to significant overvoltage. The converters can be monitored with techniques such as thermal imaging and converter output monitoring for fault identification [13]. The detection time was one of the major concerns for such fault detection algorithm [14]. However, the drawbacks can be negated by implementation of intelligent error detection methods which happen to improve the response time significantly. Techniques such as fuzzy logic controller have been widely used for such fault identification [15]. The dependence on set of rules can lead to misclassification of the operating condition. The operation of the system is determined by the data logging and data acquiring devices, and the data is then used to estimate for the health of the individual components and further evaluate the performance and economics potential of the power system.

In case of the malfunction, the principle of FRT mechanism [16] can be explored. During the FRT mechanism the operation control switches to a reactive power injection mode with an aim to lead the value of current to that of the voltage so that the voltage dips can be taken care off. In a grid-connected system, the FRT is determined by the grid code [17] which is being followed. The operating parameters such as voltage, frequency, and current at the point of common coupling (PCC) are responsible for the performance of such controllers [18].

The understanding of system under the operation can be associated with a wide spectrum of faults that are taking place at different instance. The localization of fault is necessary to operate smoothly as it makes the fault clearance faster and easier. In this chapter a brief about the fault in grid connected power system is discussed, and an analysis of different unsupervised learning algorithm has been accounted for to understand the fault classification.

7.2 Faults in Power Electronics

Power converter is the primary building block of the modern electrical system which ensures efficient power conversion and controllability. The system is

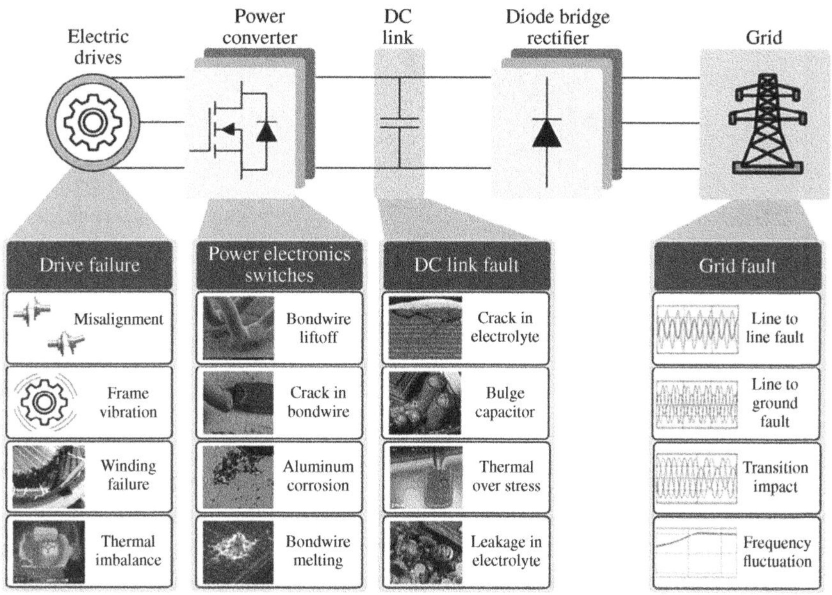

Figure 7.1 Different fault possibilities in a power electronics drive.

integrated for a wide range of industrial to renewable energy application [19]. However, the operating condition and the complexity make it susceptible to the faults. The identification of such fault is essential for ensuring the safe, reliable, and optimal performance. In Figure 7.1, an electric power drive [20, 21] is depicted along with few of the common faults that can be encountered during the operation.

7.2.1 Open Circuit Faults

An open circuit fault occurs when the electrical interconnection in the circuit is broken and the current stops to flow, it can result because of the broken wire, loose connection, or a faulty component. The environmental wear and tear may be responsible for the physical damage and corrosion [22]. Poor solder joint can also cause for the system to break away. The open circuit fault can prevent electrical circuit to function and can further result in reduced efficiency and interruption in power delivery for the device. If the location on the fault can be identified, then the failure can be mitigated by securing the connection and repairing the failure [23]. In case of the component failure [24], the damage is calculated and based on the assessment that either the component is repaired or replaced completely. The regular maintenance of the circuit [25] can determine such issues prior to the failure. One of the common examples of open circuit failure is the IGBT liftoff illustrated in Figure 7.2.

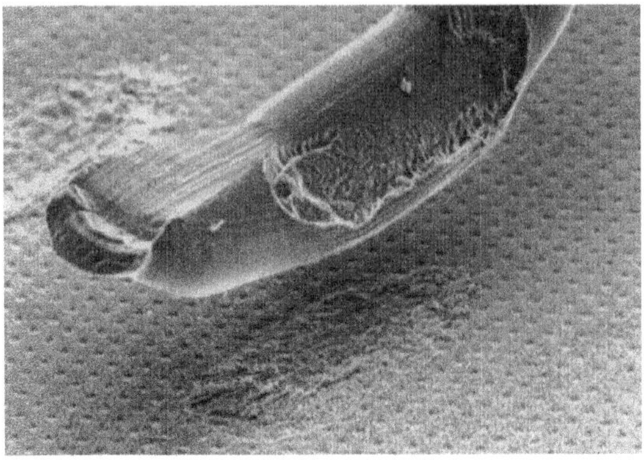

Figure 7.2 Bond wire liftoff causing open circuit failure. Source: Marcel Held.

7.2.2 Short Circuit Faults

The short circuit fault occurs when there is a low resistance interconnection between two points in an electrical circuit which are not required to be interconnected but may end up getting entangled. This can result in a significant increase in current flow, which can cause various issues, including equipment damage, power outages, fire hazard, and electric shock [26]. Few of the common causes for the short circuit are faulty wiring, overloading of the circuit, and environmental factors such as dust and moisture. Few of the most comment occurrences of such faults are line to line, line to ground, or three-phase faults [27]. Such fault can be taken care by integrating circuit breaker and fuses in the circuit that may insure that in case of the overcurrent the system is disconnected. It is necessary to identify such a fault at the earliest so that the affected section of the circuit can be disconnected from the power supply; the other solution to the issue could be replacement of the damaged fault part with the repaired or new unit [28]. The short circuit failure can lead to a burned power electronics module as illustrated in Figure 7.3.

7.2.3 Gate Drive Faults

Gate failure refers to the malfunction of the gate driver circuit, a crucial component in the operation of power electronics switches like MOSFETs and IGBTs [29]. The gate driver circuit is responsible for enabling the high-frequency switching operation of these power switches, which is essential for efficient power management and conversion in various electronic devices and systems. When the gate driver circuit fails, it disrupts the control of the power switches, potentially leading to the malfunction or failure of the entire power electronic system [30]. This can result in reduced performance, system instability, or even complete shutdown, highlighting the importance of the gate driver circuit in maintaining the reliability

Figure 7.3 Power electronics switch burnout because of short circuit failure.

and functionality of power electronics. The gate driver fault is usually caused due to incorrect gate resistance, noise interference, and a faulty gate driver. It can further lead to improper switching and increase in the switching losses along with a potential of damaging the device.

7.2.4 Thermal Faults

Thermal faults are of a significant concern in electrical systems, often caused due to excessive heat generated caused because of the poor connections, overloaded circuits, or faulty components. The increase in electrical resistance on a particular node, such as loose connections or corroded terminals, can further cause localized heating [31]. This raised temperature can damage insulation materials, potentially directing to short circuits or complete system failures. Additionally, thermal faults can be aggravated by environmental factors, such as insufficient ventilation or high ambient temperatures, further preventing effective dissipation of heat [32]. Identifying and deal with thermal faults is crucial, as its persistent exposure to high temperatures can impact system integrity, decrease the lifespan of electrical components, and pose safety hazards, including the risk of fire [33]. Methods such as infrared thermography and temperature monitoring are usually employed to identify hotspots and prevent the escalation of thermal faults, ensuring reliable and safe operation of electrical systems.

7.2.5 Component Aging and Degradation

Component aging and degradation imply the steady deterioration in the performance and reliability of a component over time owing to various factors. These factors can include chemical reactions, physical degradation, environmental conditions, and operational stress [34]. Aging can reveal in numerous ways, such as material fatigue, corrosion, and thermal degradation. Electronic components could encounter alterations in electrical properties caused because of the continued exposure to high temperatures and electrical stress [35].

Degradation cannot be identified uniformly as it may be affected by the given conditions in which a component operates. Components subjected to harsh environments, such as high temperatures, humidity, tend to degrade at a faster rate. Furthermore, the way a component is used may also influence its degradation rate. Predicting the lifespan of a component requires knowledge of variables and often needs detailed modeling and testing.

Maintenance and frequent inspections are key strategies to mitigate the influences of aging and degradation. By detecting early indicators of wear and addressing them proactively, the risk of unexpected failures can be diminished. In some cases, materials and designs can be selected precisely for their resistance to aging and degradation, extending the operational life of the component. Identification the mechanisms of aging and degradation is necessary for improving the reliability and security of systems that depend on these components.

7.2.6 Electromagnetic Interference (EMI)

Electromagnetic Interference (EMI) implies the disruption caused by external sources that influences an electrical circuit through electromagnetic induction. These interferences can interrupt the performance of electronic devices, indicating to data loss, communication breakdowns, and even equipment failure [36]. EMI can initiate from various sources such as power lines, industrial machinery, and even household appliances. The effect of EMI is particularly significant in susceptible environments like hospitals, where interference can influence medical devices, and in the aerospace and defense sectors, where it can compromise the operation of critical systems [37]. To mitigate EMI, engineers employ shielding, filtering, and grounding techniques to protect electronic equipment. Controlling standards established by the Federal Communications Commission (FCC) and the International Electrotechnical Commission (IEC) are crucial in during devices design. It minimizes EMI and can operate without causing further interference. Understanding EMI is essential for maintaining the reliability and efficiency of modern electronic systems.

7.3 Unsupervised Learning

Various machine learning algorithms can be characterized as in Figure 7.4. Unsupervised learning deals with data with no labeled responses, making it distinct from supervised learning, where the data are trained on a labeled dataset [38]. The objective in unsupervised learning is to expose hidden patterns within the data.

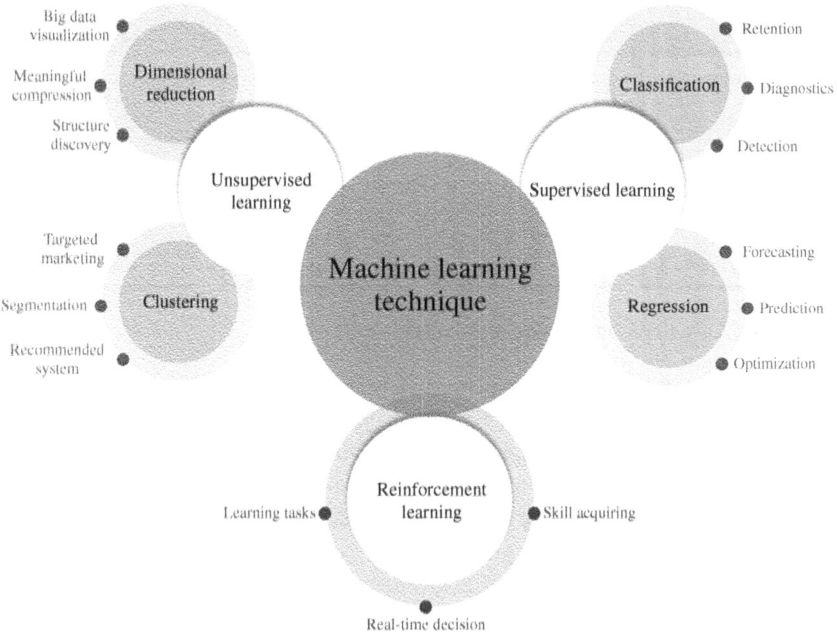

Figure 7.4 Different types of machine learning algorithm.

Techniques such as clustering and dimensionality reduction are commonly used. Clustering algorithms, such as K-means and hierarchical clustering, group data points into clusters depending on the resemblances, further aiding to identify natural groupings in the data. Dimensionality reduction methods, such as Principal Component Analysis (PCA) and t-Distributed Stochastic Neighbor Embedding (t-SNE), reduce the number of variables under consideration, reducing the data and providing a visualized interpretation ensuring the information lost is not significant. Unsupervised learning is particularly useful in experimental data analysis, anomaly detection, and preprocessing steps for supervised learning tasks [39]. Some common approaches and techniques used in unsupervised anomaly detection as illustrated in Figure 7.5 are discussed in this section.

7.3.1 Statistical Methods

Statistical methods of unsupervised learning are predicated assuming that normal data points are in high likelihood regions of a stochastic model; however, anomalies are discovered in low-probability regions [40]. In principle, these methods employ the probability distribution of the data to distinguish between typical and atypical occurrences. Modeling the data's distribution can identify regions

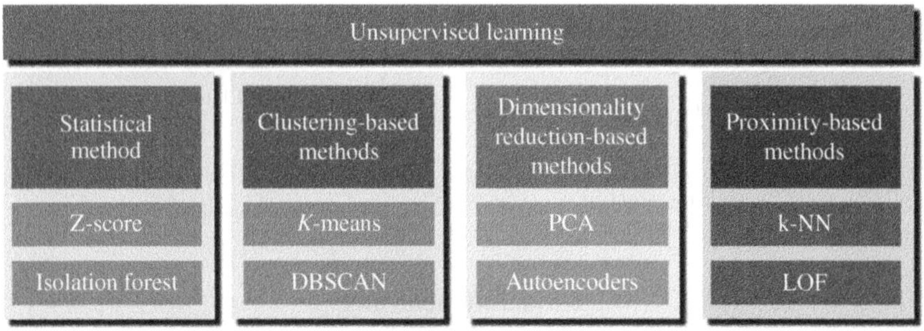

Figure 7.5 Unsupervised learning techniques.

where data points are most likely to occur, which are considered normal. Similarly, data points that fall outside these high-probability regions are considered anomalies due to their rarity. This approach permits for the identification of outliers or unusual patterns without the need for labeled training data, making it particularly useful in scenarios where labeling is impractical or impossible. The efficacy of these methods relies heavily on the accuracy of the stochastic model in capturing the true distribution of the data, as any deviation could lead to misclassification of normal data points as anomalies.

7.3.1.1 Z-Score

The Z-score is a statistical measurement that explains a relationship between the mean of a group of values, expressed in terms of standard deviations. Z-scores can be a valuable tool for anomaly detection and data normalization [41]. By transforming data into Z-scores, the dataset is standardized, making it easier for algorithms to detect outliers and anomalous behavior, as these will have Z-scores significantly higher or lower than the rest of the data. Furthermore, Z-scores aids the identification of unusual patterns by emphasizing data points that vary significantly from the norm, thus enabling more effective unsupervised learning processes.

7.3.1.2 Isolation Forest

Isolation Forest is an unsupervised learning algorithm consistent for anomaly detection. Distinct from other anomaly detection techniques that depend on density measures, Isolation Forest [42] influences the principle of isolating anomalies. The algorithm operates on the hypothesis that anomalies are few and diverse, which makes them simpler to isolate.

The Isolation Forest algorithm makes a set of decision trees where data points are partitioned. The process includes randomly selecting of feature and afterward splitting the choosing value randomly between the minimum and maximum values of the selected feature [43]. This random partitioning creates paths to isolate observations. Anomalies tend to have shorter isolation path compared to normal points. Hence, the average path length from the root to a leaf node can be used as a scoring mechanism.

One of the key advantages of Isolation Forest is its linear time complexity, requiring it be scalable to large datasets. In addition, it is less affected by the dimensionality of the data compared to other anomaly detection methods, and it does not require the estimate of data distribution, which is complex.

Isolation Forest has efficacy in identifying outliers without prior labeling which makes it a powerful tool in the unsupervised learning. However, like all algorithms, it has limitations; its performance can degrade if the dataset contains clusters of anomalies, and the random nature of tree building can sometimes lead to variability in results.

7.3.2 Clustering-Based Methods

Clustering-based methods are used for identifying patterns and structures within datasets, extensively. These methods group data points into clusters based on resemblance, intending to maximize intra-cluster similarity and minimize inter-cluster resemblance [44]. Common clustering algorithms include *K*-means, hierarchical clustering, and DBSCAN (Density-Based Spatial Clustering of Applications with Noise). *K*-means is extensively used for its simplicity and efficiency, where data points are designated to clusters by minimizing the variance within each cluster. Hierarchical clustering forms nested clusters by either merging or splitting them iteratively, producing a dendrogram that represents the data's hierarchical structure. Further, DBSCAN is good at handling noise and discovering clusters of arbitrary shapes by grouping densely packed points and identifying outliers. Despite their utility, these methods often require thorough consideration of parameters and assumptions, such as the number of clusters in *K*-means or the density threshold in DBSCAN, to ensure significant and accurate clustering outcomes.

7.3.2.1 *k*-Means

K-means is a widely used clustering algorithm, with its aims to divide a set of data points into *k*-clusters, where each data point is appropriate to the cluster with the nearest mean [45]. The algorithm operates iteratively, starting with an initial set of *k* randomly chosen centroids. Each data point is allocated to the nearest centroid, forming clusters. The centroids are recalculated as the mean of all points in each cluster. This process of allocation and centroid update is recurring until no significant changes are noticed in the centroids which further implies to convergence. The efficiency of *K*-means makes it popular, though it requires prior knowledge of *k* [46], and it converges to a local minimum, which may lead to a suboptimal clustering result. Regardless of these constraints, K-means is efficient for large datasets and presents a straightforward approach to detecting essential patterns in data.

7.3.2.2 DBSCAN

DBSCAN is a fundamental clustering algorithm recognized for its capability to detect clusters of shifting shapes and sizes in spatial data. Unlike traditional clustering algorithms like K-means, DBSCAN does not necessitate the number of clusters to be stated in advance, making it mostly useful in scenarios where the data distribution is not well defined.

One of the key concepts in DBSCAN is regarding the notion of density reachability and connectivity [47]. Points are classified into three categories: core, border, and noise. Core points are those that have enough points within a restricted epsilon to be deemed part of a cluster. Border points are within the epsilon radius of a core point; however, it does not have enough neighbors to be core points themselves. Noise points do not belong to any cluster since they are neither core nor border points.

The algorithm advances by first recognizing core points and developing clusters by recursively adding core-reachable points to them. This allows DBSCAN to determine the number of clusters and handle unevenly shaped clusters effectively [48]. However, the performance of DBSCAN can be sensitive to the choice of parameters, particularly epsilon and the minimum number of points required to form a cluster.

DBSCAN has several advantages over other clustering algorithms. It can manage noise and outliers effectively by defining them as noise points, which is crucial for data where not all points may belong to an individual cluster. Moreover, DBSCAN can detect clusters of arbitrary shapes and sizes, making it versatile for various applications such as anomaly detection. Despite its advantages, DBSCAN may represent issues with datasets having varying densities. Additionally, the parameter tuning process can be challenging and requires consideration based on the specific characteristics of the dataset.

7.3.3 Dimensionality Reduction-Based Methods

Dimensionality reduction-based methods are essential for data analysis and machine learning, predominantly aimed at streamlining datasets by reducing the number of random variables under consideration. These methods assistant in mitigating dimensionality, which is an issue when dealing with high-dimensional datasets. High dimensionality can further lead to numerous issue, including risen computational cost, overfitting, and complications in visualizing data. By dimensions reductions, the performance and efficiency of machine learning algorithms, regarding identify patterns and insights from complex data, can be made effortless.

7.3.3.1 Principal Component Analysis (PCA)

PCA is a method in unsupervised learning used for dimensionality reduction and empirical data analysis. By transforming data into a set of linearly uncorrelated variables called principal components, PCA simplifies the complexity of high-dimensional data while preserving its necessary patterns and structures. Through PCA [49], the variance in the data is acquired in fewer dimensions, assisting simpler visualization, analysis, and modeling.

The objective of PCA is to recognize the directions along which the variance of the data is maximized. These components are orthogonal to each other, ensuring that each captures distinctive information about the data's structure [50]. The first principal component relations for the largest possible variance, with each subsequent component capturing the remaining variance in decreasing order. This ordered variance representation helps in aiming on the most revealing aspects of the data, often leading to significant understandings that might not be evident in the original high-dimensional space.

Further, by projecting data onto a lower-dimensional subspace, PCA can filter out fewer significant components that often correspond to noise, thereby improving the signal-to-noise ratio. This is specifically advantageous when dealing with noisy datasets, as it allows for apparent pattern recognition and more robust machine learning tasks. Furthermore, by reducing the dimensionality, PCA can better mitigate the curse of dimensionality, which can adversely affect the performance of many machine learning algorithms.

It assumes that the principal components with the highest variance are the most important, which may not always be true, especially in cases where the data has underlying non-linear relationships. PCA is also susceptible to the scale of the data, requiring normalization of variables to ensure meaningful results. Moreover, interpreting the principal components can sometimes be challenging, as they are linear combinations of the original variables and may not always have clear physical or practical significance.

7.3.3.2 Autoencoders

Autoencoders have developed as a robust tool for unsupervised learning due to its ability to learn efficient representations of data. Autoencoders are neural networks constructed to compress input data into a lower-dimensional space and then rebuild the data from this compacted representation. The process involves are encoder and decoding [51]. During the encoding, the input data is transformed into a lower-dimensional code, and the decoder, which attempts to reconstruct the original data from this code. The training process goals is to reduce the reconstruction error, thus ensuring that the latent representation captures the most salient features of the input data.

One of the significant advantages of autoencoders is its ability to perform dimensionality reduction, like PCA, but with the extra advantage of being able to capture non-linear relationships, making autoencoders remarkably useful for dealing with complex datasets where linear methods fall short. The non-linear nature of autoencoders allows them to denote data more compactly and effectively, which is critical for tasks involving high-dimensional data.

Autoencoders also play an essential role in generative modeling. Variants such as Variational Autoencoders (VAEs) [52] extend the basic autoencoder framework to allow for probabilistic analysis of the latent space. VAEs are trained to confirm that the latent space follows a predefined distribution, typically Gaussian, which facilitates the generation of new data samples by sampling from this distribution. This expertise is highly beneficial in scenarios where synthetic data generation is needed, such as in creating and enhancing data.

Additionally, the concept of self-supervised learning, closely related to unsupervised learning, has secured traction with the start of autoencoders. In self-supervised learning, the model is trained on automatically generated labels rather than manually noted ones. Autoencoders fit well into this concept as they generate their individual supervisory signals through the reconstruction task. This approach has proved remarkable success in pretraining models on large amounts of unlabeled data, which can then be fine-tuned for a smaller labeled dataset.

7.3.4 Proximity-Based Methods

Proximity-based approaches for unsupervised learning depend on the norm that similar data points are nearby to each other in the feature space. These methods use the concept of similarity to group data points without predefined labels, compelling them predominantly useful in exploratory data analysis and pattern. One of the fundamental approaches in this category is clustering, the goal is to separate a dataset into clusters based on the intrinsic similarities between the data points.

7.3.4.1 *k*-Nearest Neighbors (*k*-NNs)

The *k*-Nearest Neighbors (*k*-NNs) can be adapted for several unsupervised learning tasks, such as, clustering and anomaly detection. For clustering, *k*-NN can be used to detect sets of similar data points within a dataset without pre-labeled classes. This involves estimating the distances between data points and recognizing clusters based on the closeness of these points to each other [53]. By exploring the *k*-NNs of each point, the algorithm can gather the natural group within the data. This method, regularly linked with other clustering techniques like *k*-means or hierarchical clustering, aids to define more precise clusters and enhance the overall estimation of the dataset's structure.

In anomaly detection, k-NN is utilized for recognize outliers data points that do not fit to the expected pattern. By calculating the distance between each point and its k-NNs, the algorithm can determine which points are substantially different from most of the data [54]. Points that have a high average distance to their nearest neighbors are identified as anomalies. This approach is useful in applications such as fraud detection, network security, and quality control.

The flexibility of k-NN in unsupervised learning is mostly due to its nonparametric nature and minimalism in implementation. Unsupervised learning with k-NN depend on inherent structure of the data. This allows for a versatile application across different domains. However, there are limitations to using k-NN in an unsupervised environment [55]. The select of k is critical and can substantially impact the results. Furthermore, k-NN can be computationally exhaustive for large datasets due to the need to calculate distances between all pairs of points, and it can be subtle to dimensionality, where the distance measures become less sensitive in high-dimensional spaces.

7.3.4.2 Local Outlier Factor (LOF)

The Local Outlier Factor (LOF) is particularly effective in recognizing outliers in datasets where the perception of an outlier may vary depending on local data density. Unlike global methods that utilize a uniform threshold for outlier detection across the entire dataset, LOF estimates the local density deviation of a given data point concerning its neighbors [56]. This local approach allows LOF to reveal anomalies that might be forgotten by global techniques, making it a flexible tool for various applications.

LOF functions by relating the density of a data point to the densities of its neighbors. Explicitly calculating the Local Reachability Density (LRD) [57] of each point, which is fundamentally an inverse measure of the average distance from the point to its neighbors. The LOF score is then verified by evaluating the LRD of a point to the LRDs of its neighbors. If a point has a substantially lower density than its neighbors, it is considered an outlier. This method safeguards that outliers are identified based on the characteristics of their local environment rather than a global standard, permitting more effective detection.

The advantage of LOF is its ability to manage data with variable density regions. In real-world datasets, data points often cluster in regions of varying densities. Conventional outlier detection methods might struggle in such scenarios, by avoiding outliers in sparse regions. LOF refers this challenge by vigorously adapting to the local density, providing a more accurate and situation-aware assessment of an outlier.

However, the application of LOF has its own challenges. The technique requires the selection of several parameters, such as the number of neighbors (k) to consider while calculating the local density. Selecting the optimal value for k is essential, as it guides the sensitivity and accuracy of the outlier detection. A small k could make the algorithm too sensitive to noise, while a large k could smooth out local variations and weaken the detection of some outliers. Additionally, LOF can be computationally intensive, mainly with large datasets, as it requires the calculation of distances connecting data points.

Despite these challenges, the LOF remains a popular choice for unsupervised outlier detection due to its adaptability and effectiveness in varying density scenarios. Its ability to discern local density variations makes it particularly suited for complex, real-world datasets where outliers are context dependent.

7.4 Modeling System for the Case Study

A two-stage grid-connected PV system is considered for the analysis. The DC–DC converter boosts the DC voltage using a low-current rating switch. For MW-scale PV systems, incorporating a DC–DC converter in a two-stage configuration increases costs and reduces efficiency [58]. Therefore, two-stage systems are suitable only for power ratings below 100 kW. Due to these factors, a two-stage single-phase grid-connected PV system is implemented for the case study [59]. The overall system layout and control parameters are shown in Figure 7.6.

The proposed configuration offers an advantage under low solar irradiance conditions, as the boost converter enhances the flexibility of active power control. To manage the boost converter, a perturb and observation (P&O)-based MPPT control technique is employed. This technique involves introducing a small disturbance in V_{pv} and observing its effect on P_{pv}. The gathered information is then used to track the maximum power point [60]. For inverter control, a dual-loop cascaded controller is implemented. The outer loop controller regulates the DC link voltage v_{dc}, while the inner loop functions as a current controller. A Proportional-Integral (PI) controller is used in the outer loop, incorporating feedforward power from the panels (P_{pv}), and PQ theory is applied in the inner loop to generate the i_g^*.

A Proportional-Resonant (PR) controller is utilized for current control because it can track a sinusoidal reference signal without any steady-state error when operating at the resonant frequency. This is due to the high gain of the PR controller around the resonant frequency [61]. Considering the delays introduced by the PWM generator and sampling, the transfer function for the inverter is derived [16]. The grid filter transfer function receives the difference between the inverter output voltage and the grid voltage, providing grid feedback current as an input to the

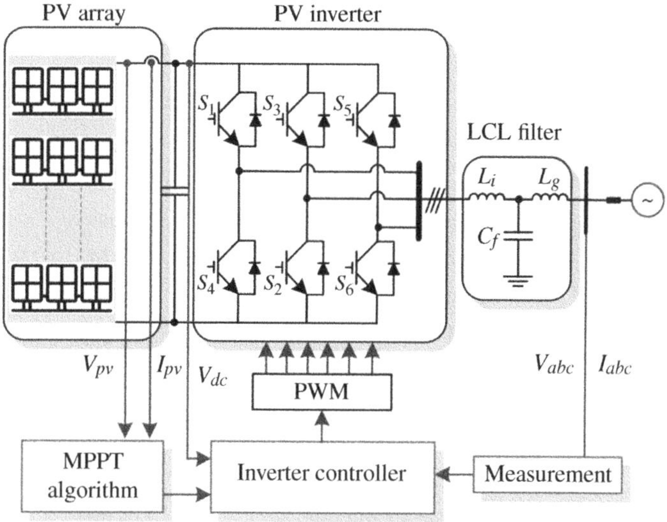

Figure 7.6 Schematic for a single-phase grid connected photovoltaic system.

PR controller. In this research, a combination of a PR controller and a repetitive controller is used to achieve harmonic compensation.

The grid-connected PV system was simulated. Various sections were designed through mathematical analysis, as shown in Table 7.1.

A balanced V_{rms} 230 V AC is taken as the nominal voltage at V_{pcc}. The robustness of the developed algorithm and the proposed one is tested for the grid connected system, with design parameters outlined in Table 7.1. Further, the system was modulated with different faults such as line-to-line and line-to-ground fault. Even short circuit and open circuit fault was exported. The different signals related to different modes of operation are illustrated in Figure 7.7. Once the data is collected for different operating conditions, the data preprocessing is performed to ensure that there is no noise and data redundancy. The initial step in this process involves pre-processing the input data and transforming the raw data by normalizing it to achieve better results. During normalization, the inputs are scaled to a range of [−1, +1]. This pre-processing stage occurs before the data is propagated through the network for learning. The technique employed is min–max normalization, which distributes the input values across a chosen normalized interval.

The classification process begins by configuring 2 input nodes, 12 hidden layers, and 2 output nodes. Features extracted using discrete wavelet transform are compiled into a matrix of size 8 × 1000, which serves as the input data. Various operating conditions are categorized based on the similarity in operating pattern, represented as a binary data matrix (000001b to 100000b) of size 1 × 4013. Of the

Table 7.1 Overall system description for 1 kW solar inverter.

System		Parameter
PV array rating		4 kWp
Boost converter	Inductance	1 mH
	Time sample	100 μs
	Capacitance	1000 uF
	Operation time	200 μs
	MPPT algorithm	P&O algorithm
Inverter (Full Bridge Topology)	DC link voltage	400 V
	Power switch (IGBT) Rating	600 V, 30 A
	Diode rating	400 – 600 V, 15 A
Filter (LCL)	Inductance (converter side)	3.8 mH
	Inductance (grid side)	2.02 mH
	Capacitance	3.01 uF
Operating frequency (System)		50 Hz
Output voltage (System)		230 V AC

total input samples, 70% (700 samples) are allocated for training, 15% (150 samples) for validation, and the remaining 15% (150 samples) for testing. After partitioning the data, the Scaled Conjugate Gradient algorithm is employed for network training due to its efficiency and performance with large datasets.

The analysis was performed for the isolated forest, *k*-mean, and *k*-NN algorithm. The analysis is assisted by confusion plot and Receiver Operating Characteristic (ROC) curve. A confusion matrix is a tool used to evaluate the performance of a classification algorithm. It summarizes the results of a classification by showing the number of correct and incorrect predictions broken down by each class. Whereas the ROC curve is used to evaluate the performance of a binary classifier system, it is plotted by varying the cut-off points and graphically representing the true-positive rate (sensitivity) on the *y*-axis and the false-positive rate (1-specificity) on the *x*-axis. For this specific analysis, the ROC curves are constructed using 100 different specificity values across multiple cut-off points, where each curve represents the sensitivity for a specific sample.

In Figure 7.8, the confusion plot and the ROC for the training and testing of isolated forest method are presented. It can be observed that in the training set there is certain amount of misclassification; however, in the testing data set no

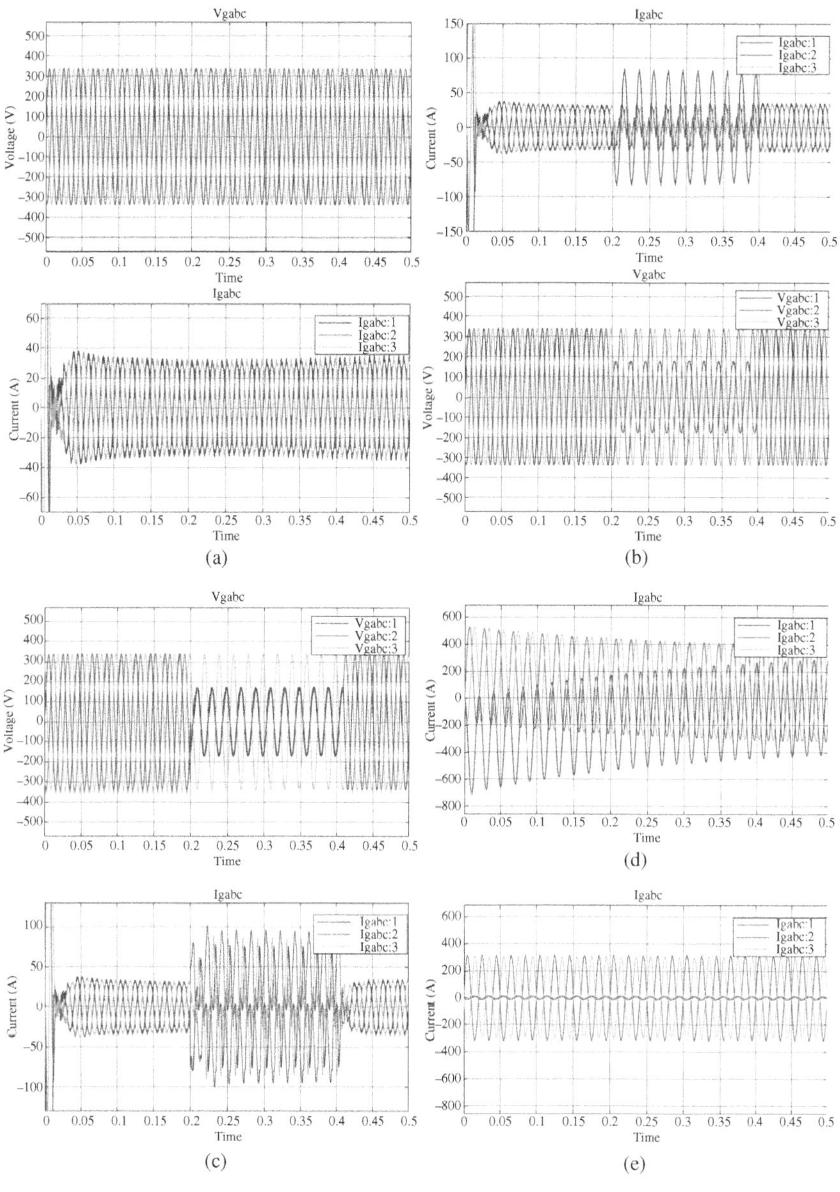

Figure 7.7 Different operating conditions for the database creations. (a) Normal operation, (b) LG fault, (c) LL fault, (d) short circuit, and (e) open circuit.

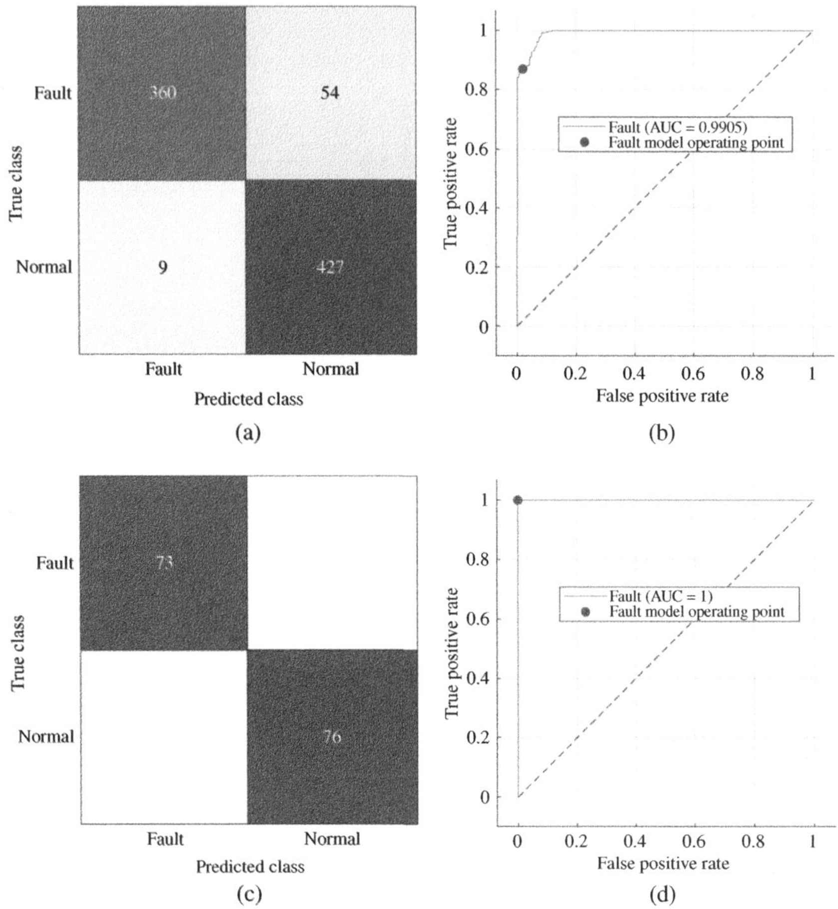

Figure 7.8 Performance parameters for isolated forest. (a) Training confusion plot, (b) training ROC curve, (c) testing confusion plot, and (d) testing ROC curve.

data is being misclassified. The area under curve for the ROC is in the positive class of the curve, with training data presenting 0.9905 sensitivity, whereas testing data present 1 as sensitivity.

In Figure 7.9, the confusion plot and the ROC for the training and testing of *k*-mean method is presented. It can be observed that in the training and testing set that there is certain amount of misclassification. The area under curve for the ROC is in the positive class of the curve, with training data presenting 0.968 sensitivity, whereas testing data present 0.966 as sensitivity.

In Figure 7.10, the confusion plot and the ROC for the training and testing of *k*-NN are presented. It can be observed that in the training set there is certain

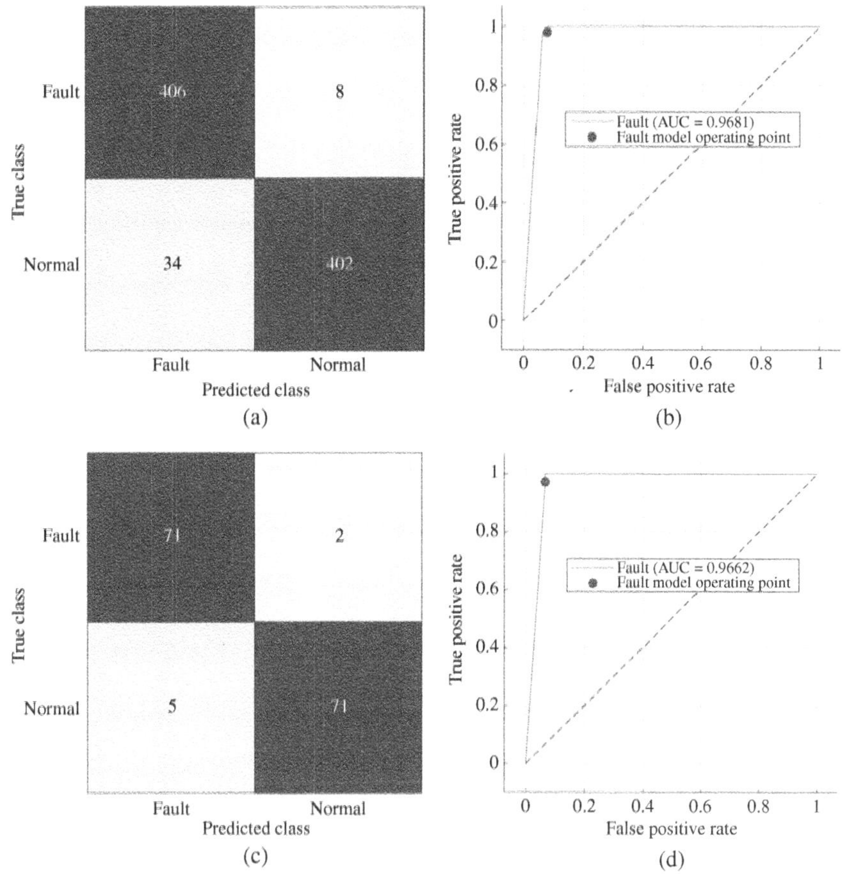

Figure 7.9 Performance parameters for *k*-Mean. (a) Training confusion plot, (b) training ROC curve, (c) testing confusion plot, and (d) testing ROC curve.

amount of misclassification; however, in the testing data set no data is being misclassified. The training and testing confusion plot of the classifier is presented in Figure 7.10; it can realize that the misclassification is significantly lower in case of *k*-NN as 408 fault and 431 normal condition operation are classified correctly in training and the accuracy of testing data is 100%. The area under curve for the ROC is in the positive class of the curve, with training data presenting 0.9979 sensitivity, whereas testing data present 1 as sensitivity.

Further the different algorithm operational parameters are compared in Table 7.2, and it can be observed that the *k*-NN is one of the most accurate and fast training algorithms with a low training time and high prediction speed.

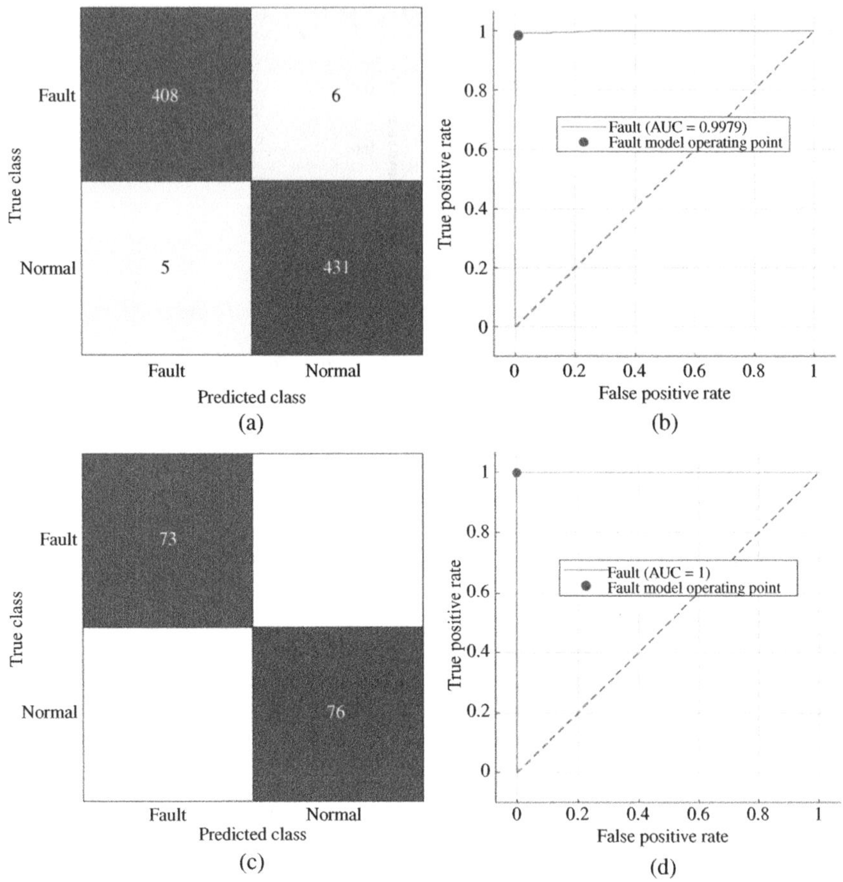

Figure 7.10 Performance parameters for *k*-NN. (a) Training confusion plot, (b) training ROC curve, (c) testing confusion plot, and (d) testing ROC curve.

Table 7.2 Comparison of different machine learning techniques.

	Techniques		
Parameters	**Isolation Forest**	***k*-Mean**	***k*-NN**
Accuracy	92.6%	95.1%	98.7%
Total cost	63	42	11
Prediction speed	4100 obs/s	1900 obs/s	4500 obs/s
Training time	13.02 s	10.852 s	3.28 s

7.5 Conclusion

This chapter presents an unsupervised learning approach for power electronics converters in a grid-connected environment. Initially, a real-time model-driven estimation for a three-phase inverter is established to capture the complex operating dynamics under various faults and conditions affecting different components and sensors. By comparing the measured and estimated outputs of the converter, residual error vectors are generated for all operating states. These residual vectors are then compiled into a fault signature library and trained using an isolation forest, k-mean, and k-nearest neighbor classifier. The fault classification mechanism is tested on a three-phase grid-connected inverter dataset. When a fault occurs, the residual error reflects the fault's magnitude and signature, and classifies it as fault. The k-NN classification process achieves an accuracy of 98.7% with a training time of 3.28 seconds.

Exercises

7.1 Elaborate the principle of fault ride through (FRT).

7.2 Explain the causes of open-circuit faults.

7.3 Explain the potential causes of short-circuit faults.

7.4 What is the function of the gate driver circuit in power electronic switches like MOSFETs and IGBTs?

7.5 What are the consequences of a gate driver circuit failure in terms of impacting device performance?

7.6 How do thermal faults arise in electrical systems, and what are the potential consequences if they are not addressed promptly?

7.7 What factors contribute to the aging and degradation of electronic components, and why is it important to predict a component's lifespan?

7.8 What techniques are commonly used to mitigate EMI in electronic devices?

7.9 How do statistical methods in unsupervised learning help identify anomalies, and why are anomalies found in low-probability regions of a stochastic model?

7.10 How does the Z-score assist in anomaly detection and data normalization in machine learning applications?

7.11 What is the primary mechanism by which the Isolation Forest algorithm identifies anomalies, and why is it less affected by the dimensionality of data?

7.12 What role does dimensionality reduction play in improving the efficiency of machine learning algorithms, and how does PCA achieve this?

7.13 How do autoencoders perform dimensionality reduction?

7.14 What makes the autoencoders more effective than linear methods like PCA in capturing non-linear relationships in data?

7.15 What are the different metrics to evaluate the machine learning model performance?

References

1 E. Bassam, "Renewable energy for rural communities," *Renew. Energy*, vol. 24, no. 3–4, pp. 401–408, 2001.

2 European Commission, "Progress Report 2023 Climate Action," 2023.

3 E. Villanueva, P. Correa, J. Rodríguez, and M. Pacas, "Control of a single-phase cascaded H-bridge multilevel inverter for grid-connected photovoltaic systems," *IEEE Trans. Ind. Electron.*, vol. 56, no. 11, pp. 4399–4406, 2009.

4 M. A. Khan, "Overview and impact of faults in grid-connected photovoltaic systems," in *Fault Analysis and its Impact on Grid-connected Photovoltaic Systems Performance*. Wiley-IEEE Press, 2022, pp. 1–42. https://doi.org/10.1002/9781119873785.ch1.

5 V. S. Bharath Kurukuru, A. Haque, R. Kumar, M. A. Khan, and A. K. Tripathy, "Machine learning based fault classification approach for power electronic converters," in *9th IEEE International Conference on Power Electronics, Drives and Energy Systems, PEDES 2020*, 2020. doi: 10.1109/PEDES49360.2020.9379365.

6 A. Malik, A. Haque, V. S. B. Kurukuru, M. A. Khan, and F. Blaabjerg, "Overview of fault detection approaches for grid connected photovoltaic inverters," *e-Prime Adv. Electr. Eng. Electron. Energy*, vol. 2, p. 100035, 2022. doi: 10.1016/j.prime.2022.100035.

7 M. A. Khan, V. S. Bharath Kurukuru, A. Haque, and S. Mekhilef, "Islanding classification mechanism for grid-connected photovoltaic systems," *IEEE J. Emerg Sel. Top. Power Electron.*, vol. 9, no. 2, 2021. doi: 10.1109/JESTPE.2020.2986262.

8 M. A. Khan, A. Haque, V. S. B. Kurukuru, and M. Saad, "Islanding detection techniques for grid-connected photovoltaic systems – A review," *Renew. Sust. Energ. Rev.*, vol. 154, p. 111854, 2022. doi: 10.1016/j.rser.2021.111854.

9 M. A. Khan, A. Haque, and V. S. B. Kurukuru, "Performance assessment of stand-alone transformerless inverters," *Int. Trans. Electr. Energy Syst.*, vol. 30, no. 1, p. e12156, 2020. doi: 10.1002/2050-7038.12156.

10 M. A. Khan, A. Haque, V. S. B. Kurukuru, H. Wang, and F. Blaabjerg, "Stand-alone operation of distributed generation systems with improved harmonic elimination scheme," *IEEE J. Emerg Sel. Top. Power Electron.*, vol. 9, no. 6, pp. 6924–6934, 2021. doi: 10.1109/JESTPE.2021.3084737.

11 M. A. Khan, A. Haque, F. Blaabjerg, V. S. B. Kurukuru, and H. Wang, "Intelligent transition control between grid-connected and standalone modes of three-phase grid-integrated distributed generation systems," *Energies.*, vol. 14, no. 13, p. 3979, 2021. doi: 10.3390/en14133979.

12 M. S. Arani and M. A. Hejazi, "The comprehensive study of electrical faults in PV arrays," *J. Electr. Comput. Eng.*, vol. 2016, p. 8712960, 2016.

13 V. S. Bharath Kurukuru, F. Blaabjerg, M. A. Khan, and A. Haque, "A novel fault classification approach for photovoltaic systems," *Energies.*, vol. 13, no. 2, p. 308, 2020. doi: 10.3390/en13020308.

14 K. Dadhich, V. S. B. Kurukuru, M. A. Khan, and A. Haque, "Fault identification algorithm for grid connected photovoltaic systems using machine learning techniques," in *2019 International Conference on Power Electronics, Control and Automation, ICPECA 2019 – Proceedings*, 2019. doi: 10.1109/ICPECA47973.2019.8975397.

15 A. Saedian, H. Zarabadipour, M. A. Shoorehdeli, and F. Farivar, "Inverted pendulum fault tolerant control based on fuzzy backstepping design and anti-control of chaos," *IFAC Proc. Vol.*, vol. 45, no. 12, pp. 243 248, 2012. doi: 10.3182/20120620-3-MX-3012.00013.

16 M. A. Khan, A. Haque, and V. S. Bharath Kurukuru, "Enhancement of Fault ride through strategy for single-phase grid-connected photovoltaic systems," in *2019 IEEE Industry Applications Society Annual Meeting, IAS 2019*, 2019. doi: 10.1109/IAS.2019.8911895.

17 M. A. Khan, A. Haque, and V. S. Bharath Kurukuru, "Reliability analysis of a solar inverter during reactive power injection," in *9th IEEE International Conference on Power Electronics, Drives and Energy Systems, PEDES 2020*, 2020. doi: 10.1109/PEDES49360.2020.9379776.

18 M. Khan, A. Haque, and V. S. B. Kurukuru, "Dynamic voltage support for low-voltage ride-through operation in single-phase grid-connected photovoltaic systems," *IEEE Trans. Power Electron.*, vol. 36, no. 10, pp. 12102–12111, 2021. doi: 10.1109/TPEL.2021.3073589.

19 F. Blaabjerg, F. Iov, R. Teodorescu, and Z. Chen, "Power electronics in renewable energy systems," in *2006 12th International Power Electronics and Motion Control Conference*, IEEE, Aug. 2006, pp. 1–17. doi: 10.1109/EPEPEMC.2006.4778368.

20 A. Wang, Y. Qi, and K. Ma, "Adaptive DC voltage control for optimal junction temperature redistribution under stall condition of electric machine drive inverter," *IEEE Trans. Power Electron.*, vol. 38, no. 4, pp. 4229–4234, Apr. 2023. doi: 10.1109/TPEL.2022.3231867.

21 V. S. B. Kurukuru, M. A. Khan, and R. Singh, "Health monitoring framework for electric vehicle drive train in digital twin," in *2023 25th European Conference on Power Electronics and Applications (EPE'23 ECCE Europe)*, IEEE, Sep. 2023, pp. 1–10. doi: 10.23919/EPE23ECCEEurope58414.2023.10264291.

22 U.-M. Choi, J.-S. Lee, F. Blaabjerg, and K.-B. Lee, "Open-circuit fault diagnosis and fault-tolerant control for a grid-connected NPC inverter," *IEEE Trans. Power Electron.*, vol. 31, no. 10, pp. 7234–7247, 2015. doi: 10.1109/TPEL.2015.2510224.

23 A. Haque, V. S. B. Kurukuru, M. A. Khan, A. Malik, and F. Fayaz, "Centralized intelligent fault localization approach for renewable energy-based islanded microgrid systems," in *Applications of AI and IOT in Renewable Energy*. Academic Press, 2022, pp. 129–149. doi: 10.1016/B978-0-323-91699-8.00007-3.

24 T. Balachandran, A.-K. Manoharan, V. Aravinthan, and C. Singh, "Component-level reliability evaluation model for cyber power devices," *IEEE Trans. Smart Grid.*, vol. 12, no. 1, pp. 692–703, Jan. 2021. doi: 10.1109/TSG.2020.3016524.

25 M. Alam, A. Haque, M. A. Khan, N. M. Sobahi, I. M. Mehedi, and A. I. Khan, "Condition monitoring and maintenance management with grid-connected renewable energy systems," *Comput. Mater. Cont.*, vol. 72, no. 2, pp. 3999–4017, 2022. doi: 10.32604/cmc.2022.026353.

26 Y. Chen, W. Li, F. Iannuzzo, H. Luo, X. He, and F. Blaabjerg, "Investigation and classification of short-circuit failure modes based on three-dimensional safe operating area for high-power IGBT modules," *IEEE Trans. Power Electron.*, vol. 33, no. 2, pp. 1075–1086, Feb. 2018. doi: 10.1109/TPEL.2017.2682114.

27 M. A. Khan, V. S. B. Kurukuru, N. Bayati, and T. Ebel, "Digital twin paradigm for fault ride through in grid-integrated distributed generation," in *2023 8th IEEE Workshop on the Electronic Grid (eGRID)*, IEEE, Oct. 2023, pp. 1–8. doi: 10.1109/eGrid58358.2023.10380947.

28 D.-P. Sadik *et al.*, "Short-circuit protection circuits for silicon-carbide power transistors," *IEEE Trans. Ind. Electron.*, vol. 63, no. 4, pp. 1995–2004, Apr. 2016. doi: 10.1109/TIE.2015.2506628.

29 H. Guzman, F. Barrero, and M. J. Duran, "IGBT-gating failure effect on a fault-tolerant predictive current-controlled five-phase induction motor drive," *IEEE Trans. Ind. Electron.*, vol. 62, no. 1, pp. 15–20, Jan. 2015. doi: 10.1109/TIE.2014.2331019.

30 M. A. Rodriguez-Blanco *et al.*, "A failure-detection strategy for IGBT based on gate-voltage behavior applied to a motor drive system," *IEEE Trans. Ind. Electron.*, vol. 58, no. 5, pp. 1625–1633, May 2011. doi: 10.1109/TIE.2010.2098355.

31 Irshad, Z. A. Jaffery, A. Haque, A. K. Dubey, M. A. Khan, and V. S. B. Kurukuru, "Thermography based real time intelligent condition monitoring system for solar power inverter," in *2019 International Conference on Power Electronics, Control and Automation, ICPECA 2019 – Proceedings*, 2019. doi: 10.1109/ICPECA47973.2019.8975472.

32 C. Yan, J. Chen, H. Liu, and H. Lu, "Model-based fault tolerant control for the thermal management of PEMFC systems," *IEEE Trans. Ind. Electron.*, vol. 67, no. 4, pp. 2875–2884, Apr. 2020. doi: 10.1109/TIE.2019.2912772.

33 P. E. Sutherland, "Mechanical forces and thermal effects in substation equipment due to high fault currents," in *Principles of Electrical Safety*. Wiley, 2014, pp. 299–324. doi: 10.1002/9781118886397.ch13.

34 M. Xu, K. Ma, and Q. Zhong, "Degradation diagnosis of power module based on frequency characteristics in heat flow," *IEEE Trans. Power Electron.*, vol. 38, no. 8, pp. 10292–10301, Aug. 2023. doi: 10.1109/TPEL.2023.3274756.

35 H. Liu, J. Wang, and Y. Ji, "A novel high-step-up coupled-inductor DC–DC converter with reduced power device voltage stress," *IEEE J. Emerg Sel. Top. Power Electron.*, vol. 7, no. 3, pp. 1941–1948, Sep. 2019. doi: 10.1109/JESTPE.2018.2868769.

36 L. Devaraj, A. R. Ruddle, and A. P. Duffy, "Electromagnetic risk analysis for EMI impact on functional safety with probabilistic graphical models and fuzzy logic," *IEEE Lett. Electromagn. Compat. Pract. Appl.*, vol. 2, no. 4, pp. 96–100, Dec. 2020. doi: 10.1109/LEMCPA.2020.3017483.

37 J. E. Timperley, U. Z. Ahmad, and J. M. Vallejo, "Electric generation condition assessment with electromagnetic interference analysis," in *2019 IEEE Petroleum and Chemical Industry Committee Conference (PCIC)*, IEEE, Sep. 2019, pp. 179–186. doi: 10.1109/PCIC30934.2019.9074512.

38 A. Arif, K. Imran, Q. Cui, and Y. Weng, "Islanding detection for inverter-based distributed generation using unsupervised anomaly detection," *IEEE Access.*, vol. 9, pp. 90947–90963, 2021. doi: 10.1109/ACCESS.2021.3091293.

39 V. S. B. Kurukuru, A. Haque, M. A. Khan, and A. K. Tripathy, "Fault classification with robust knowledge transfer for single phase grid connected photovoltaic systems," in *2019 International Conference on Power Electronics, Control and Automation, ICPECA 2019 – Proceedings*, 2019. doi: 10.1109/ICPECA47973.2019.8975554.

40 S. Zhao, F. Blaabjerg, and H. Wang, "An overview of artificial intelligence applications for power electronics," *IEEE Trans. Power Electron.*, vol. 36, no. 4, pp. 4633–4658, Apr. 2021. doi: 10.1109/TPEL.2020.3024914.

41 O. A. Gashteroodkhani, M. Majidi, M. S. Fadali, M. Etezadi-Amoli, and E. Maali-Amiri, "A protection scheme for microgrids using time-time matrix z-score vector," *Int. J. Electr. Power Energy Syst.*, vol. 110, pp. 400–410, Sep. 2019. doi: 10.1016/j.ijepes.2019.03.040.

42 Y. Chen, Z. Zhao, H. Wu, X. Chen, Q. Xiao, and Y. Yu, "Fault anomaly detection of synchronous machine winding based on isolation forest and impulse frequency response analysis," *Measurement*, vol. 188, p. 110531, Jan. 2022. doi: 10.1016/j.measurement.2021.110531.

43 V. S. B. Kurukuru, A. Haque, and M. A. Khan, "Fault detection in single-phase inverters using wavelet transform-based feature extraction and classification techniques," in *Applications of Computing, Automation and Wireless Systems in Electrical Engineering*, vol. 553. Springer, 2019, pp. 649–661. doi: 10.1007/978-981-13-6772-4_56.

44 S. M. Miraftabzadeh, C. G. Colombo, M. Longo, and F. Foiadelli, "K-means and alternative clustering methods in modern power systems," *IEEE Access.*, vol. 11, pp. 119596–119633, 2023. doi: 10.1109/ACCESS.2023.3327640.

45 H. Jang, H. Lee, H. Lee, C.-K. Kim, and H. Chae, "Sensitivity enhancement of dielectric plasma etching endpoint detection by optical emission spectra with modified K-means cluster analysis," *IEEE Trans. Semicond. Manuf.*, vol. 30, no. 1, pp. 17–22, Feb. 2017. doi: 10.1109/TSM.2017.2653062.

46 A. K. Gangwar, O. P. Mahela, B. Rathore, B. Khan, H. H. Alhelou, and P. Siano, "A novel k-means clustering and weighted k-NN-regression-based fast transmission line protection," *IEEE Trans. Indus. Inform.*, vol. 17, no. 9, pp. 6034–6043, Sep. 2021. doi: 10.1109/TII.2020.3037869.

47 K. Choksi, A. B. Mirza, S. Salehi Vala, and F. Luo, "Comprehensive evaluation and detection of partial discharge in WBG motor drive using DBSCAN based feature extraction," in *2023 IEEE Energy Conversion Congress and Exposition (ECCE)*, IEEE, Oct. 2023, pp. 3503–3509. doi: 10.1109/ECCE53617.2023.10362669.

48 F. Jiao *et al.*, "Topology identification method for low-voltage distribution node networks based on density clustering using smart meter real-time measurement data," *IEEE Access.*, vol. 12, pp. 83600–83610, 2024. doi: 10.1109/ACCESS.2024.3407827.

49 I. Jaffel, O. Taouali, E. Elaissi, and H. Messaoud, "A new online fault detection method based on PCA technique," *IMA J. Math. Control. Inf.*, vol. 31, no. 4, pp. 487–499, Dec. 2014. doi: 10.1093/imamci/dnt025.

50 M. Rezamand, M. Kordestani, R. Carriveau, D. S.-K. Ting, and M. Saif, "A new hybrid fault detection method for wind turbine blades using recursive PCA and wavelet-based PDF," *IEEE Sensors J.*, vol. 20, no. 4, pp. 2023–2033, Feb. 2020. doi: 10.1109/JSEN.2019.2948997.

51 Y. Lin and J. Wang, "Probabilistic deep autoencoder for power system measurement outlier detection and reconstruction," *IEEE Trans. Smart Grid.*, vol. 11, no. 2, pp. 1796–1798, Mar. 2020. doi: 10.1109/TSG.2019.2937043.

52 S. R. Khazeiynasab, J. Zhao, I. Batarseh, and B. Tan, "Power plant model parameter calibration using conditional variational autoencoder," *IEEE Trans. Power Syst.*, vol. 37, no. 2, pp. 1642–1652, Mar. 2022. doi: 10.1109/TPWRS.2021.3107515.

53 L. Cai, N. F. Thornhill, S. Kuenzel, and B. C. Pal, "Real-time detection of power system disturbances based on *k*-nearest neighbor analysis," *IEEE Access.*, vol. 5, pp. 5631–5639, 2017. doi: 10.1109/ACCESS.2017.2679006.

54 L. Cai, N. F. Thornhill, S. Kuenzel, and B. C. Pal, "Wide-area monitoring of power systems using principal component analysis and *k*-nearest neighbor analysis," *IEEE Trans. Power Syst.*, vol. 33, no. 5, pp. 4913–4923, Sep. 2018. doi: 10.1109/TPWRS.2017.2783242.

55 S. Jo *et al.*, "Hybrid genetic algorithm with k-nearest neighbors for radial distribution network reconfiguration," *IEEE Trans. Smart Grid.*, vol. 15, no. 3, pp. 2614–2624, May 2024. doi: 10.1109/TSG.2023.3324328.

56 S. Liu *et al.*, "Data-driven event detection of power systems based on unequal-interval reduction of PMU data and local outlier factor," *IEEE Trans. Smart Grid.*, vol. 11, no. 2, pp. 1630–1643, Mar. 2020. doi: 10.1109/TSG.2019.2941565.

57 Z. Zhu, L. Jin, E. Song, and C.-C. Hung, "Quaternion switching vector median filter based on local reachability density," *IEEE Signal Process Lett.*, vol. 25, no. 6, pp. 843–847, Jun. 2018. doi: 10.1109/LSP.2018.2808343.

58 Y. Xue, B. Ge, and F. Z. Peng, "Reliability, efficiency, and cost comparisons of MW-scale photovoltaic inverters," in *2012 IEEE Energy Conversion Congress and Exposition (ECCE)*, IEEE, Sep. 2012, pp. 1627–1634. doi: 10.1109/ECCE.2012.6342618.

59 N. A. Rahim, R. Saidur, K. H. Solangi, M. Othman, and N. Amin, "Survey of grid-connected photovoltaic inverters and related systems," *Clean Techn. Environ. Policy*, vol. 14, no. 4, pp. 521–533, Aug. 2012. doi: 10.1007/s10098-011-0411-z.

60 Y. Yang and F. Blaabjerg, "Overview of single-phase grid-connected photovoltaic systems," *Electric Power Compon. Syst.*, vol. 43, no. 12, pp. 1352–1363, Jul. 2015. doi: 10.1080/15325008.2015.1031296.

61 Y. Yang, F. Blaabjerg, H. Wang, and M. G. Simões, "Power control flexibilities for grid-connected multi-functional photovoltaic inverters," *IET Renewable Power Gener.*, vol. 10, no. 4, pp. 504–513, Apr. 2016. doi: 10.1049/iet-rpg.2015.0133.

8

Reinforcement Learning and Control

Azra Malik, Suwaiba Mateen, and Ahteshamul Haque

Advance Power Electronics and Research Laboratory, Jamia Millia Islamia (A Central University), New Delhi, India

8.1 Introduction

Power electronics is witnessing a notable surge in the application of Reinforcement Learning (RL), a subfield of machine learning, which offers innovative approaches to complex control and optimization issues. Power electronics, which employs semiconductor devices to convert and control electrical power efficiently, is an essential technology for many applications such as industrial automation, electric vehicles (EVs), and renewable energy systems [1]. Although they work well in many situations, traditional control techniques frequently struggle to keep up with the complexity and dynamic nature of today's power electronics systems. A viable substitute is provided by RL, which enables systems to adjust and learn by interactions with their surroundings, eventually improving performance [2]. RL is especially well suited for real-time control and adaptive management because it focuses on training an agent to execute actions that maximize cumulative rewards, in contrast to supervised learning, which depends on a static dataset [3]. This capability is essential to handle issues like changing load situations, variable renewable energy inputs, and the requirement for high reliability and efficiency in power electronics systems.

The procedure as given in Figure 8.1, begins with the agent, the primary actor, engaging with the environment and trying to accomplish a multi-step objective there. The agent identifies and analyzes a state that defines the environment [4]. However, as a result of the agent's various actions, the state of the environment varies in line with these changes. At this point, the agent eventually receives reward signals as they come closer to their objective. These reward signals can be employed by the agent to identify successful and unsuccessful activities. The

Artificial Intelligence for Power Electronics, First Edition.
Edited by Ahteshamul Haque, Saad Mekhilef, and Azra Malik.

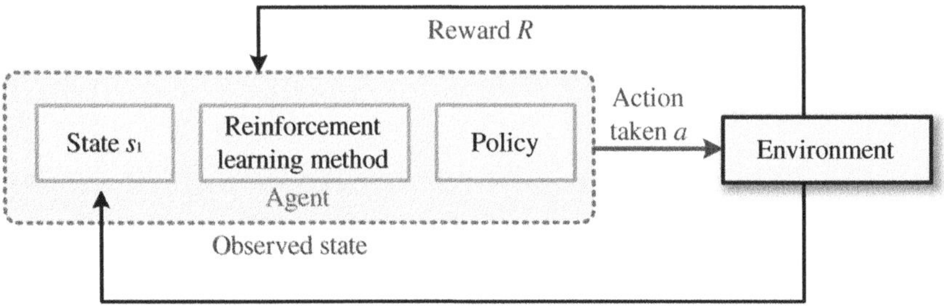

Figure 8.1 General RL framework.

reward is then repeated and the state action keeps happening until the agent masters the trial-and-error idea of working efficiently in the environment [5]. Unlike supervised learning, the target label cannot be recognized in RL due to the lack of a dataset. Similarly, a Policy Network is a network that transforms input frames into actions. Using the Policy Gradient is one of the simple methods for training a Policy Network. Additionally, the network's output will include two numbers: the likelihood of each choice and what to do during training. The goal of RL is to provide the agent the freedom to learn everything on its own. The agent's notable goal is to maximize its policy in order to maximize the quantity of incentives it receives.

Over the years RL has observed tremendous developments as illustrated in Figure 8.2. Even while RL has had some success previously, earlier methods were essentially restricted to low-dimensional issues and did not have scalability. RL methods face complexity issues similar to those in other algorithms, such as sample, computational, and memory complexity [6]. However, recent advancements in deep learning have introduced methods to address these issues by harnessing the powerful function approximation and representation learning abilities of deep

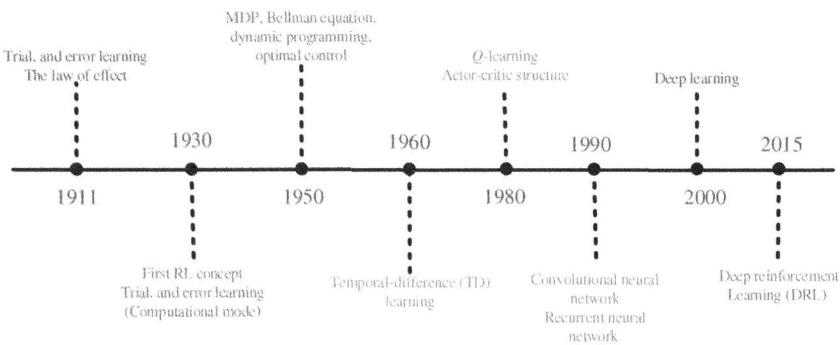

Figure 8.2 Emergence of RL over the years.

neural networks (DNNs). The adoption of deep learning has led to significant breakthroughs across various machine learning tasks. A core advantage of deep learning is its ability to automatically extract compact, low-dimensional features from high-dimensional inputs, such as text, audio, and images. The curse of dimensionality has been effectively addressed by machine learning specialists by incorporating inductive biases into neural network architectures, particularly those with layered structures. The topic of "deep reinforcement learning" (DRL) is characterized by the application of deep learning methods in RL, which has similarly accelerated advancements in RL [7]. Deep learning enables RL to tackle complex decision-making problems that were earlier unsolvable, especially those concerning high-dimensional spaces.

In this chapter, the significance of RL in power electronics applications is explored, focusing over its potential to transform and revolutionize system design and functionality. We will look at several RL-based approaches to power electronics control, talk about how they are better than traditional techniques, and provide case studies. This study provides an overview of how DRL, an advanced machine learning approach, might drive future advancements in power electronics by delving into its principles and real-world applications.

8.2 Basics of Reinforcement Learning (RL)

8.2.1 Markov Decision Process (MDP)

The Markov decision process (MDP) is a robust and dynamic framework in the discipline of RL. In scenarios where outcomes are partially controlled by an agent and partially random, an MDP is a mathematical framework for simulating decision-making [8]. The agent continuously learns the best course of action in each state through ongoing interactions with the surroundings. Because it enables an agent to make successive decisions with the aim of maximizing cumulative rewards over time, it is crucial to the concept of RL. The RL method typically includes the following elements:

- **State S:** It shows the agent's present position in relation to the environment. Depending on the task, it might be either continuous or discrete. It is important to ensure that state is properly descriptive and has the desired information to define the system conditions. It is further depicted as

$$S = [s_1, s_2, s_3, \dots, s_n] \tag{8.1}$$

- **Action A:** It outlines a collection of methods because of which the agent interacts with the surroundings. Action space is represented as either continuous or discrete as given below

$$A = [a_1, a_2, a_3, \dots, a_n] \tag{8.2}$$

- **Transition Function *P*:** When taking action from the action space, this function specifies the likelihood that state S will change to the subsequent state.
- **Reward Function *R*:** It provides an immediate reward for taking an action *A* in the provided state (state space *S*). The major aim of an RL agent is to achieve the maximum reward. It can be defined as

$$R = [r_1, r_2, r_3, \dots, r_n] \tag{8.3}$$

The description above specifies that the agent observes the states $s_t \in S$ and acquires the $r_t \in R$ from the environment at each time step t. In order to interact with the environment, it simultaneously selects an action $s_t \in S$ considering the policy $\pi(s)$. The agent repeats the aforementioned actions till the termination at the subsequent time step. An RL agent's primary aim is to determine the optimal policy $\pi^*(s)$ that maximizes the long-term expected reward that is defined and discounted as follows:

$$G_t = \sum_{n=t+1}^{T} \gamma^{n-t-1} r_t \tag{8.4}$$

where the discount factor is denoted by γ, and the time step is denoted by T, which terminates the interaction, prevent long-term rewards from expanding in an unbound form (for the circumstances of some environments, which do not possess terminal state).

8.2.2 Bellman Equation

A recursive equation known as the Bellman equation is employed in dynamic programming and RL to compute the value function, which quantifies the predicted cumulative reward of possessing a given state and conforming to a certain policy. When evaluating a state's or an action's quality within the MDP framework, it plays a crucial role. An action value function $Q(s, a)$, also referred to as the *Q*-value function, is assessed in order to assess the outcome of action a_t in state s_t. The Bellman equation is employed to update the action value function $Q(s, a)$ at every step of time. It is defined as

$$Q_{t+1}(s.a) = E\left[r + \gamma \max_{a_{t+!}} Q_t(s_{t+1}, a_{t+1}) | (s, a) |\right] \tag{8.5}$$

The *Q*-learning is a value-based algorithm that represents the traditional RL method. *Q*-learning determines the best course of action $\pi^*(s)$ through the following

$$Q_{t+1}(s.a) \leftarrow Q(s_t, a_t) + \alpha(r_{t+1} + \gamma \max_{a} Q_t(s_{t+1}), a) - Q(s_t, a_t) \tag{8.6}$$

where the learning rate that can be selected to prevent overfitting is denoted by α. When the observation space is discrete and finite, the traditional Q learning technique works well. However, the exponential growth of information in large-scale issues like power systems poses a significant challenge to standard Q learning, making it more time consuming and even divergent. However, it may also be difficult for the traditional RL algorithm to identify the pattern behind the state, reward, and action due to the signal noise in the real environment [9].

8.2.3 On-policy and Off-policy Methods

The primary distinction between on-policy and off-policy training methods revolves around whether the behavior policy and the target policy are identical. The policy that interacts with the environment to provide training data is known as the behavior policy. The policy we want the agent to learn is referred to as the target policy. In one on-policy technique, data for the subsequent policy optimization round is generated straight from the target policy [10]. After the subsequent round of policy optimization is complete, the data are removed. The behavior policy and the target policy are therefore identical. When engaging with the environment, an off-policy approach buffers samples produced by behavior policies. Samples from the buffer are utilized to modify the target policy during training. The target policy and the behavior policy are different because the training data may be from outdated policies. The ability to directly optimize the policy is a major advantage of the on-policy approaches. On the other hand, the off-policy approaches are more data efficient.

8.2.4 Dynamic Programming Methods

The Bellman equations can be resolved using dynamic programming techniques after the environment model is well understood. Using dynamic programming, difficult problems may be solved by breaking them down into smaller, more manageable issues. Policy iteration and value iteration are the two basic algorithms based on dynamic planning [11]. Policy iteration is a process that alternates between assessing the current policy and improving it until it reaches the optimal solution. The assessment phase uses the policy to compute the value function, while the improvement phase updates the policy based on these values. In contrast, value iteration calculates Q-values for all possible actions in each state, selecting the optimal one to represent the state's value. Unlike policy iteration, where value function convergence is required, value iteration allows both the value function and policy to converge simultaneously, streamlining the process once the optimal values are achieved.

8.2.5 Monte Carlo Method

The Monte Carlo (MC) method is capable of handling larger state-space problems as it does not require knowledge of the environment's dynamics, unlike dynamic programming techniques. This approach determines the value function by running multiple episodes and averaging the returns for individual state or state–action pair. It is model-free, meaning no transition probabilities are needed. However, convergence relies on two key conditions: (i) a large number of episodes, and (ii) frequent visitation of all states and actions. MC methods can be either on-policy or off-policy. On-policy methods use the same policy for evaluation and exploration, requiring a stochastic policy, while off-policy methods allow for a deterministic policy. Off-policy methods may be preferred owing to their simplicity, though on-policy approaches are more stable in continuous state spaces and when paired with function approximators like neural networks.

8.2.6 Temporal Difference Method

Like MC, the temporal difference (TD) approach (model-free method) also learns from experiences. However, the TD learning, in contrast to MC, updates before the next episode closes. By using the Bellman equation, it updates each step in the episode and may therefore offer faster convergence. The TD method is shown in one step in the equation below

$$V'(s_t) \leftarrow \alpha V(s_t) + (1 - \alpha)(R_{t+1} + \gamma V(s_{t+1})) \tag{8.7}$$

where α is the step-size and $0 < \alpha < 1$. The current values V' are updated with the help of the preceding values V. This is particularly referred to as the bootstrapping approach. In general, the bootstrapping approach learns more quickly than non-bootstrapping ones. Additionally, the TD learning is separated into two groups: (i) State–Action–Reward–State–Action or SARSA (on-policy TD control) and (ii) Q-learning (off-policy TD control) [12]. In reality, MC and TD learning frequently save the value function of each state or each state–action combination using a table memory structure (tabular technique). When it comes to solving difficult problems with a high number of states, their lack of memory renders them inefficient. The solution to this is actor–critic method, which will be discussed later in this chapter.

8.2.7 Policy Gradient Method

A probability distribution or probability density function of all actions is typically used to create the policy $\pi : s \rightarrow a$, which is a mapping between states to actions that represents the likelihood of selecting a particular action in each state. It can be represented as

$$\pi_\theta(a|s) = P(a|s, \theta) \tag{8.8}$$

where policy parameter is denoted by θ. The aforementioned dynamic programming, MC, and TD methods, all referred to as value-based methods, need to determine the optimal Q values prior to applying Bellman equation to determine the best course of action. The policy gradient methods are a different class of techniques that have the ability to directly optimize the policy. The performance metric for policy π_θ can be defined through expected return as

$$J(\theta) = E_{\pi_\theta(s)}\left[\sum_a Q(s, a)\pi_\theta(a|s)\right] \tag{8.9}$$

Now, the policy optimization can be achieved through policy gradient method by differentiating with respect to θ.

$$\nabla_\theta J(\theta) = E_{\pi_\theta(s)}\left[\sum_a Q(s, a)\nabla_\theta \pi_\theta(a|s)\right] \tag{8.10}$$

$$\theta = \theta + \alpha\nabla_\theta J \tag{8.11}$$

The action spaces need to be discretized before applying value-based approaches to problems with continuous action spaces. Nevertheless, the curse of dimensionality inevitably affects discretization. Even worse, an intolerable lack of smoothness in control results may be a consequence of an excessively large step size for discretization. A probability distribution or probability density function on actions, which might be continuous or discrete, is the ideal policy that policy gradient approaches produce, avoiding the aforementioned drawbacks.

8.3 RL Methods

RL can determine optimal control strategies by consistently interacting with the environment and using the gradient derived from the feedback provided by the reward signal. The key RL algorithms and their connections are illustrated in the Figure 8.3. Conventional RL techniques have various drawbacks. They struggle with the "curse of dimensionality" in high-dimensional and continuous state–action spaces and often necessitate hand-crafted state representations. DNNs can mitigate these issues by approximating the state–action function using neural network parameters. Merging the DNNs with RL provides two key benefits: (i) DNNs' robust feature extraction eliminates the need for manual feature design, enabling end-to-end learning, and (ii) DNNs assist RL in generalizing

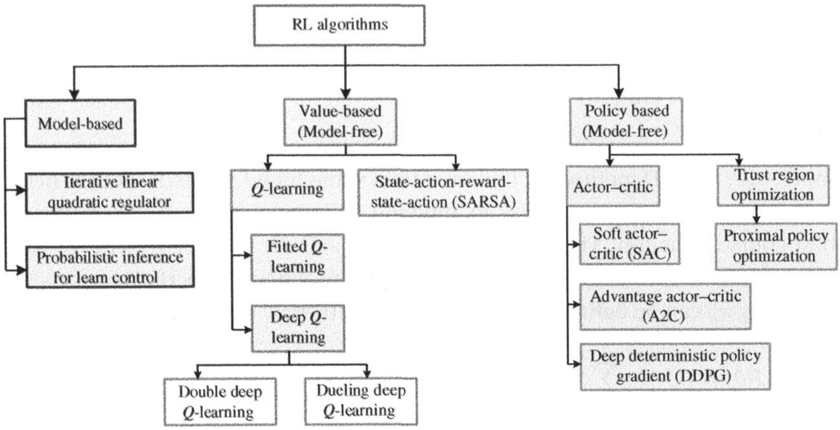

Figure 8.3 Classification of RL-based methods.

across large state spaces. However, the independence assumption is broken since RL algorithms are typically used to tackle sequential-control issues, which results in highly correlated training data. Furthermore, the training data may not be uniformly distributed as the agent's exploration of the environment may cause the distribution to become non-stationary. The training process may diverge due to the highly correlated and non-stationary training data. Many approaches have been developed to mitigate these issues, falling into two categories: policy gradient and value-based algorithms.

8.3.1 Value-based DRL

One of the important breakthroughs in DRL is considered to be the value-based technique. Finding the optimal value function and describing the value function are the main goals of value-based approaches.

8.3.1.1 Deep *Q*-learning

The most traditional value-based method is the *Q*-learning algorithm, which was previously discussed. For issues with vast state spaces and action spaces, conventional *Q*-learning is impractical since it updates *Q* values using the TD technique and saves them in the *Q* table. The *Q*-learning and DNNs can be integrated together to define Deep *Q*-Network (DQN), which was proposed by Mnih et al., and it outperforms human players in the majority of Atari 2600 games. Three characteristics of improvement over traditional *Q*-learning are as follows:

- **Deep *Q*-network (DQN):** Without any further domain expertise, a basic DNN known as Deep *Q* Network is utilized to estimate the action–value function and extract low-level data from initial images of Atari games on the premise of employing Q-learning as the RL method. Each action's Q-value is the output layer's outcome. At time step t, the estimated value of DQN is

$$y = R + \gamma \max_{a} Q(s_{t+1}, a; \theta) \tag{8.12}$$

where θ denotes the DQN parameters, and these parameters are refined by reducing the mean squared error (MSE) between the predicted Q value and the true Q value. To improve exploration capabilities, DQN employs the ε-greedy strategy, where random actions are taken with a probability of ε, while optimal actions are executed with a probability of $1 - \varepsilon$, effectively introducing a level of noise into the policy.

- **Experience Replay:** In the time domain, there is some relationship between the interaction-derived trajectories. There may be an increasing discrepancy between the estimated and predicted values if these trajectories are used directly for training. By randomly and evenly selecting training samples from a buffer containing past transitions, experience replay is suggested as a way to remove correlations among training samples. Additionally, experience replay makes it possible for DQN to train in an off-policy way, which greatly increases data efficiency.
- **Target Network:** Because of parameter updating, which can lead to issues like unstable output, the same network with different parameters obtains the values of adjacent time steps used to update DQN. Target network provides the solution to this issue and has a similar structure as that of the primary network. Both networks start training with the identical set of parameters. The primary network is in charge of interacting with the environment and obtaining training samples during the whole training process. The primary network is updated at each training step using the goal value from the target network and the primary network's predicted Q value. The parameters of this network are synchronized to the target network each time the training process reaches an appropriate amount of steps. The stability of DQN is improved as the target network maintains the target value constant for a while. The target network assists in converting the estimated value to

$$y = R + \gamma \max_{a} Q^{-}(s_{t+1}, a; \theta^{-}) \tag{8.13}$$

where Q^{-} denotes the target network with required set of parameters.

8.3.1.2 Double Deep Q-learning

The issue of overestimated Q-values in the Q-learning process, attributing it to limited function approximation capability, was discovered during the early 1993. van Hasselt et al. [13] linked this problem to environmental noise. Later, it was demonstrated that any form of error, whether from function approximation, environmental noise, or instability, could cause Q-value overestimation. The solution to this discussed issue is double Q-learning, which, when combined with DQN, forms the Double Deep Q Network (DDQN). DDQN separates action selection and value estimation into two networks: the primary network selects the optimal action, while the target network estimates the value function. In the double deep Q-learning framework, the target network proves to be a proper second value function excluding the necessity of adding an extra network. Assuming the primary network parameters as θ, the target network can be explained through the following expression, where the parameter update criterion is like that of DQN.

$$y_{j+1} = R_{j+1} + \gamma \max_{a} Q(s_{j+1}, \operatorname{argmax}_{a} Q(s_{j+1}, a; \theta); \theta^{-}) \tag{8.14}$$

8.3.1.3 Dueling Deep Q-learning

Gradually, the major focus shifted from refining control strategies and RL algorithms to innovating neural network architectures. Wang et al. [14] introduced the dueling architecture, which builds on the advantage updating algorithm [15]. In this approach, the advantage function $A(s, a)$ represents the relative benefit of taking an action in a specific state, and is calculated as

$$A(s, a) = Q(s, a) - V(s, a) \tag{8.15}$$

The architecture estimates both the advantage $A(s, a)$, and value $V(s, a)$ combining them to get $Q(s, a)$. This approach improves policy evaluation, especially when several actions have similar values, enhancing performance. Unlike the original DQN, the dueling architecture splits features into two streams: one for the advantage function $A(s)$, and another for the state value function $V(s)$. These are then combined to derive the Q-value.

$$Q(s, a; \theta, \alpha, \beta) = V(s; \theta, \beta) + A(s, a; \theta, \alpha) \tag{8.16}$$

where α and β are the fully connected layer parameters from the advantage and state value streams. It is necessary to restrict the advantage function, since for a particular Q-value, it results in infinite state value and advantage function combinations. Ultimately, an optimal method for estimating the Q-value was adopted as given below

$$Q(s, a; \theta, \alpha, \beta) = V(s; \theta, \beta) + A(s, a; \theta, \alpha) - \frac{1}{|A| \sum_{a'} A(s, a'; \theta, \alpha)} \tag{8.17}$$

8.3.1.4 Fitted Q-learning

A variant of the traditional Q-learning method, fitted Q-learning is intended to perform better in high-dimensional state and action spaces, particularly where function approximation is necessary. It is especially helpful in situations when tabular approaches are difficult due to a continuous or vast state–action space.

Fitted Q-learning enables the agent to generalize across states and actions more efficiently by approximating the Q-value function using regression techniques. Fitted Q-learning modifies the classic Q-learning process by estimating the Q-values using function approximators, usually regression models like decision trees or neural networks. It fits a model to estimate the Q-value function throughout the whole state–action space based on accumulated experience, rather than updating the Q-values directly using the tabular technique. Fitted Q-learning's main concept is to repeatedly fit a regression model in order to estimate the Q-value function by minimizing the TD error – the difference between the target Q-values and the current estimate of Q-values. A batch of transitions kept in a replay buffer is utilized for this. For each sample, target value is computed as

$$y_t = R_t + \gamma \max_a Q(s_{t+1}, a'; \theta) \tag{8.18}$$

Q-value approximator is updated by minimizing the loss function expressed below

$$L(\theta) = \frac{1}{N \sum_{i=1}^{N} \left(y_i - Q\left(s_i, a_i; \theta\right)^2 \right)} \tag{8.19}$$

Fitted Q-learning improves stability and generalizes over high-dimensional state–action spaces by using function approximators and batch learning. However, because it uses regression techniques and requires careful hyperparameter adjustment, it has problems including instability and overfitting.

8.3.2 Policy-based DRL

In general, the actor–critic technique refers to learning a policy and a value function at the same time, with the policy being evaluated using the value function. The actor is in charge of choosing actions, creating policies, and dealing with the environment. The gradients computed from the policy gradient methods as discussed before, are used to update the actor. At every time step, the critic assesses the actor's policy value function. The actor's policy may be assessed using a variety of metrics, including advantage function $A(s)$, state-value function $V(s)$, and action–value function $Q(s, a)$. Policy-based DRL methods refer to the extension of these actor critic techniques, and policy gradient algorithms, found to be highly suitable for the high-dimensional state–action spaces.

8.3.2.1 Trust-region Optimization (TRO)

In the policy optimization problems, α regulates the update step size. A well-chosen step size can accelerate policy convergence, whereas an improper one can lead to instability or degradation of the policy. In practice, the step size should ensure that the new policy performs at least as well as the previous one, ideally improving it incrementally. Schulman et al. [16] introduced the trust-region optimization (TRO) algorithm, demonstrating that the expected advantage of one policy compared to another can be mathematically defined as follows

$$\eta(\pi') - \eta(\pi) = E_{s_0,a_0,s_1,a_1\ldots\ldots}\left[\sum_{t=0}^{\infty}\gamma^t A_\pi(s_t, a_t)\right] \tag{8.20}$$

where π is the old policy and π' is the new policy. The estimated value of policy is denoted through η. As long as the difference $\eta(\pi') - \eta(\pi)$ remains positive, the policy will improve incrementally. A trust-region constraint was proposed [16] to make sure that neither old nor new policy diverge after going through various updates. This further helps in ensuring stability in training and continuous enhancement. Moreover, an improved TRO algorithm was developed [17], referred to as proximal policy optimization (PPO), which utilizes a truncated objective function simplifying the TRO. The PPO objective function can be expressed as

$$L_t = \frac{1}{N}\sum_{t=1}^{N}[\min(r(\theta)A_\pi(s_t, a_t)\, clip(r(\theta), 1 - \varepsilon, 1 + \varepsilon)A_\pi(s_t, a_t))] \tag{8.21}$$

where r represents the ratio of new and old policy. To prevent unstable policy updates resulting from an excessively large update step size or policy ratio, PPO incorporates two safeguards into its objective function. The first safeguard limits the ratio within the range $(1 - \varepsilon,\ 1 + \varepsilon)$ ensuring that updates remain controlled and do not fluctuate drastically. The second safeguard involves a minimum function that selects the lesser of two calculated values, further reducing the risk of instability during the update process.

8.3.2.2 Soft Actor–Critic (SAC)

Soft Actor–Critic (SAC) is an algorithm designed to maximize the balance between exploration and exploitation by incorporating the concept of entropy into the learning process. By incorporating an entropy factor in the objective function, SAC promotes exploration in contrast to typical RL techniques that only concentrate on maximizing expected reward. This helps to retain a variety of actions even as the optimal policy is being learnt. An entropy-augmented reward objective is utilized by SAC. Entropy reduces premature convergence to suboptimal policies by increasing the amount of randomness introduced into action selection. The policy objective function for SAC can be expressed as

$$J_\pi = \sum(R(s_t, a_t) + \alpha H(\pi(\cdot|s_t))) \tag{8.22}$$

where R denotes the reward function and α is the temperature parameter controlling the trade-off between maximizing reward and entropy $\mathcal{H}$. SAC learns the stochastic policy instead of a deterministic policy, allowing the agent to explore different actions more effectively, reducing the risk of getting stuck in local optima.

8.3.2.3 Advantage Actor–Critic (AAC)

Traditional policy gradient algorithms often rely on on-policy updates, leading to slow convergence and inefficient data use. To address this, Mnih et al. [18] introduced the asynchronous advantage actor–critic (A3C) method, which collects data asynchronously across multiple parallel threads. Each thread interacts with its environment independently, resulting in varied interaction trajectories, thus accelerating sample collection. After gathering data, each thread trains independently and asynchronous global model parameters updates are observed. Additionally, the advantage function helps balance bias and variance, while the use of the N-step return technique and entropy maximization improves value model updates and exploration, respectively. Further, to improve the model exploration capability, policy entropy is provided in the objective function, and therefore, the policy gradient expression can be written as

$$\nabla_\theta J(\theta) = \frac{1}{T}\sum_{t}^{T}\nabla_\theta \log \pi(\theta)\left(\sum_{i=1}^{n}\gamma^{i-1} r_{t+1} + v(s_{t+n}) - v(s_t)\right) + \beta\nabla_\theta \mathcal{H}(\pi(s_t;\theta)) \tag{8.23}$$

where the advantage part is expressed as $\sum_{i=1}^{n}\gamma^{i-1} r_{t+1} + v(s_{t+n}) - v(s_t)$ with n-step return, and the weighting factor is denoted by β. Advantage actor–critic (AAC) is a synchronized version of A3C, delivering comparable or enhanced performance while utilizing GPUs more efficiently [10]. Actor–Critic with Experience Replay (ACER) [19] is an off-policy actor–critic method suited for both discrete as well as continuous action spaces, acting as A3C's off-policy equivalent. Off-policy training can cause instability due to discrepancies between the target and behavior policies. To address this, ACER employs three strategies: (i) uses Q-value estimation from Retrace [20] for low variance and efficient data use; (ii) safely incorporates data from behavior policies; and (iii) applies importance sampling to modify the policy gradient's weight.

8.3.2.4 Deep Deterministic Policy Gradient (DDPG)

Prior policies that have been analyzed as probability density functions or probability distributions are always considered stochastic policies because, given any state, every action has a probability of being sampled. On the other hand, the policy is deterministic for various tasks. When the agent comes across the identical situation under deterministic policy, the selected course of action is the same and is written as $a = \mu(s)$. Silver et al. [21] deduced the Deep Deterministic Policy Gradient (DDPG) algorithm with the following expression

$$\nabla_\theta J(\theta) = E\left[\nabla_a Q_\mu(s,a)\nabla_\theta \mu(s)\Big|_{a=\mu_\theta(s)}\right] \tag{8.24}$$

The development of deterministic actor–critic algorithms introduced both on-policy and off-policy methods by integrating them into a broader policy gradient outline. The DDPG algorithm extends deterministic policies into continuous action spaces using an actor–critic structure. DDPG incorporates two key techniques from DQN:

- **Experience Replay:** Since the value function depends only on the state, off-policy methods are feasible. DDPG uses a replay buffer to store samples and randomly picks a portion to optimize the value function, increasing data efficiency.
- **Target Network:** Like DQN, DDPG avoids overestimating Q-values by using target networks. Unlike DQN, it employs "soft replacement," ensuring gradual updates toward the main network for more stable training. The synchronized part is written as

$$\omega' \leftarrow \tau\omega + (1-\tau)\omega', \tau \in (0,1) \tag{8.25}$$

$$\theta' \leftarrow \tau\theta + (1-\tau)\theta' \tag{8.26}$$

  where ω and ω' are the primary critic network and target critic network parameters, respectively, along with the update coefficient being τ. The primary actor network and target actor network parameters are denoted through θ and θ'.

The DDPG algorithm often overestimates the Q-value in the critic network, which can lead to accumulated errors, causing the agent to become trapped in local optima or experience catastrophic forgetting. To mitigate this issue, Twin Delayed Deep Deterministic Policy Gradient (TD3PG) was introduced [22]. TD3PG aims to reduce overestimation bias by using two critic networks (the "twin"), applying action noise regularization, and incorporating delayed updates to the actor (referred to as "delay") [23].

8.3.3 Multi-agent DRL

A multi-agent system can be modeled as a Markov game, where agents simultaneously select actions based on the current state, with each agent's behavior affecting the environment's transitions. In complex scenarios, agents may only observe partial information. Before deep learning, RL was used for fully cooperative, competitive, and mixed tasks. Multi-agent deep reinforcement learning (MADRL) extends these capabilities by handling high-dimensional inputs and continuous action spaces. Current MADRL algorithms can be categorized into several groups:

- **Enhanced Experience Replay Mechanism:** Traditional experience replay mechanisms assume a stationary environment, which is not the case in MADRL settings. To address this, some methods add information to experience tuples to better reflect the non-stationary environment. An example is Lenient-DQN, which assigns leniency values to experiences in the replay buffer, decaying over time to prioritize recent experiences that more accurately represent the environment dynamics.

- **Centralized Training with Decentralized Execution (CTDE):** To address the non-stationary environment issue, agents are trained centrally with knowledge of other agents' policies but execute their policies independently. A notable algorithm, Multiple-Agent DDPG (MADDPG), uses a centralized critic with global observations and decentralized actors using local information. Attention mechanisms can further enhance the performance of these frameworks.
- **Recurrent Neural Networks (RNNs):** RNNs enhance memory capabilities and are used in MADRL to address partially observable environments and long-term credit assignment problems. By extending RNNs into the multi-agent domain, agents can better handle partially observable Markov games.
- **Parameter Sharing:** In this approach, agents share network parameters during training, although they make different decisions based on unique input information, improving efficiency while allowing for diverse actions.

8.4 Reinforcement Learning in Power Electronics Applications

Applications of RL algorithms in power electronics have been rapidly expanding, addressing challenges in areas such as control optimization of converters, fault detection, energy management, and grid stability. RL is particularly valuable in managing the complexities of power electronics systems, where traditional control methods struggle with nonlinearities and dynamic operating conditions. In power converters, RL optimizes switching strategies to enhance efficiency, reduce switching losses, and adapt to varying load and input conditions in real time. Additionally, RL enhances energy management in systems like microgrids. In grid-connected power electronics, RL helps maintain voltage stability and improve power flow, especially in systems integrating renewable energy sources. As renewable energy penetration grows, RL plays a critical role in managing the variability and unpredictability of sources like solar and wind, while optimizing the performance of energy storage and inverter systems to maintain grid reliability. The division based on different application part in electronic systems is discussed in detail in this section.

8.4.1 Control Optimization of Power Converters

Power converters are integral components in electrical systems, responsible for transforming and regulating electrical energy across various forms – DC to DC, AC to DC, or vice versa. RL offers a novel approach to optimizing the control of these converters by learning to maximize efficiency, improve performance, and adapt to changing conditions. In traditional control methods like Proportional–Integral–Derivative (PID) or Model Predictive Control (MPC), the control strategies are typically predefined based on mathematical models of the system. These methods work well in predictable environments but may struggle when the system becomes more complex or nonlinear, or when the operating conditions vary dynamically. RL, however, allows the system to learn optimal control policies by interacting with the environment. Instead of relying on predefined models, RL agents can adapt by exploring various control actions and learning from the consequences of these actions. Over time, the RL agent can develop an optimal control strategy that maximizes performance metrics like efficiency, stability, or reduced switching losses. An example of RL-based control of power electronics is presented by authors in [24]. The proposed control uses measured data to repeatedly solve for the best control solution, eliminating the requirement to derive an accurate system/plant model. Originating from the linear quadratic regulator (LQR), the proposed optimization approach comprises the Algebraic Riccati equation's (ARE) iterative solution. A detailed review of RL techniques utilized for control of power converters is shown in [25]. The authors have sorted the power converters that are controlled using RL into Renewable Energy Generation (REG) converters, DC/DC converters, DC/AC converters, and modular converters.

The best example of REG converters utilizing RL is wind power generation (WPG) converters. The authors in [26] were the first to combine the RL algorithm with maximum power point tracking (MPPT) control for a WPG Q-learning (QL) control, which improved the capacity to capture maximum power. The schematic algorithm is shown in Figure 8.4. Notably, the authors introduced a time-varying temperature coefficient and learning rate to dynamically change the RL algorithm's exploration propensity and accelerated the algorithm's convergence

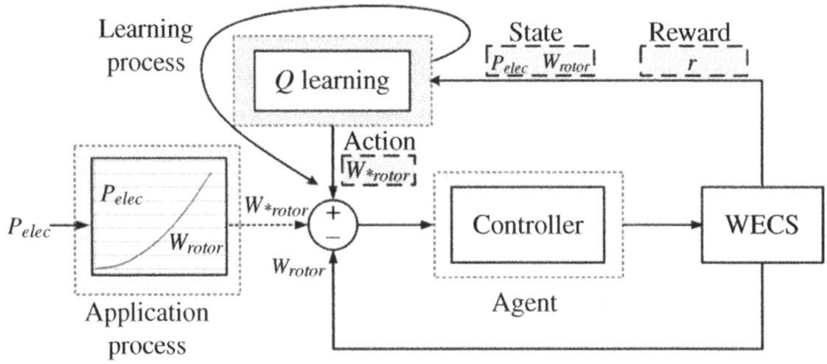

Figure 8.4 Control strategy for WPG converter.

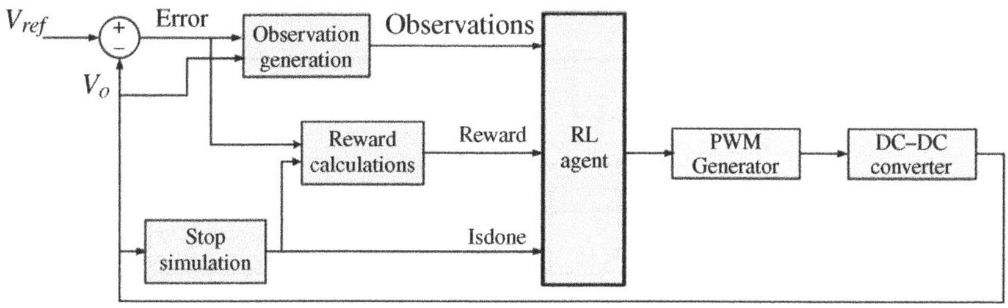

Figure 8.5 Control strategy for a DC–DC converter with independent RL control.

speed by utilizing the Boltzmann exploration approach. The power generation and rotational speed of the wind turbine were separated into discrete values to generate the state space, while the action space was built as tiny variations in the turbine's rotational speed. This is because the QL method is only applicable to discrete spaces. The reward function was only based on whether the power generation was positive or negative. The results show that, in terms of wind power usage efficiency, the MPPT control strategy based on the QL algorithm performs better than the conventional perturb and observe (P&O) approach.

Rather than using complex converters, several researchers use DC/DC converters as typical instances to validate the suggested RL control algorithms because of their simplicity. DC/DC converters are the focus of the majority of RL control schemes that are currently being suggested. A simple example of buck converter utilizing independent RL control is proposed in [27]. Based on watching the state space, this control algorithm directly creates the duty cycle of the converter. Figure 8.5 displays an independent RL control technique that regulates its output voltage together with a typical control block design.

Another type of converter where RL is extensively used for control is a DC–AC converter. Authors in [28] utilized a Deep *Q*-learning (DQN)-based independent RL control method for a three-phase neutral-point-clamped (NPC) multilevel converter used in the power grid. According to this research, the addition of a pre-processing module for the state space, as shown in Figure 8.6, facilitates algorithm convergence. The proposed RL control approach shows high resilience and dynamic response capabilities in handling variations in ambient parameters, parameter uncertainties, and sensor measurement difficulties, according to comparative experiments conducted with MPC of converter. The study additionally emphasizes how difficult it is to create the reward function and how much training time this control approach demands. An example of an independent RL control of an inverter is proposed in [29]. The authors compared independent DDPG algorithm and PI control with RL tuning. DDPG combines ideas from Deterministic Policy Gradient (DPG) and DQN. It learns a deterministic policy while also using a *Q*-value function critic to estimate the value of the optimal

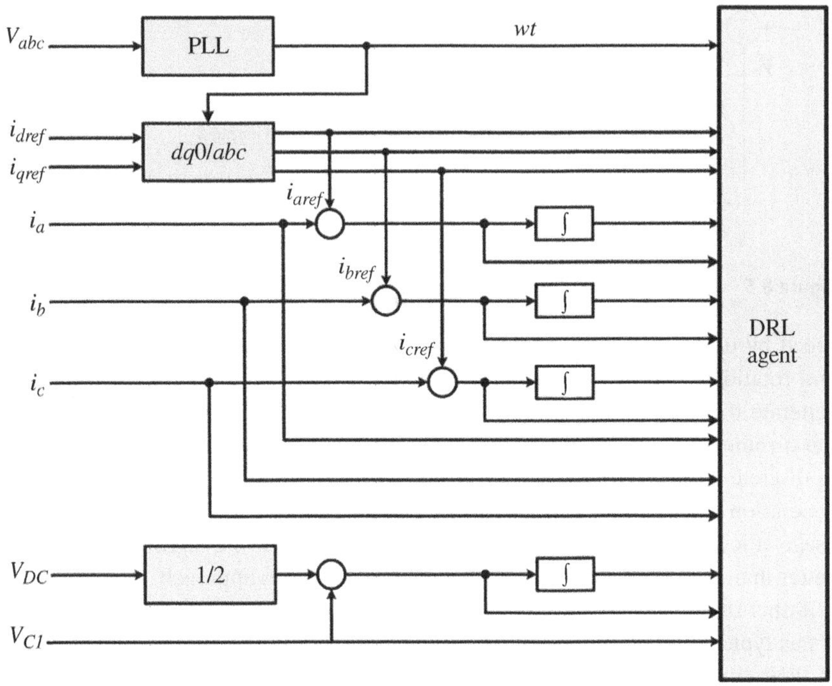

Figure 8.6 Pre-processing module for state space in NPC converter control.

policy. The authors demonstrated the benefits of these two approaches by simulation results with varying converter parameters. It was shown that the independent DDPG control performs better than PI + DDPG tuning in terms of enhancing dynamic performance.

When compared to individual converters, the RL control method for modular converters is typically more complex. The state space, action space, and reward function will become much more complicated and dimensional if all of the modular converter's modules are viewed as a single agent. However, if every module in the modular converter is viewed as a separate agent, it could result in intricate coupling connections between the control objectives of various agents, calling for further study of multi-objective game issues. Currently, a number of modular converters using RL control techniques are documented in the literature. These include, among others, interleaved Buck converters, cascaded H-bridge (CHB), modular multilevel converters (MMC), and input-series–output-parallel Dual Active Bridge converter (ISOP-DAB).

Generally, a multi-agent RL algorithm is utilized to study RL control techniques for the modular converters. In [30], an ISOP-DAB is taken as a modular converter

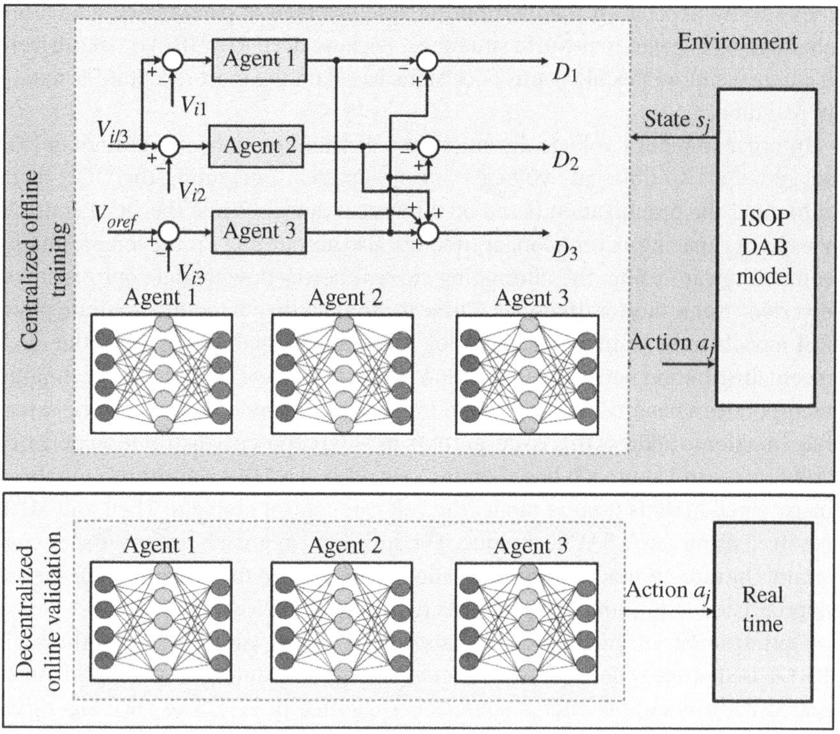

Figure 8.7 Control framework for ISOP DAB model.

with each DAB module viewed as an intelligent agent. There is a hierarchical structure in place, with the slave agents handling input voltage balancing between modules and the master agent controlling the output voltage. Using a centralized training with distributed execution (CTDE) control architecture, the discrete action space is handled by the MADDPG algorithm. Figure 8.7 shows the control framework of [30].

8.4.2 Energy Management

Applications of RL algorithms for energy management have been growing in recent years, including the optimization of distribution network and Integrated Energy System (IES) management and optimization of microgrid. The operation of the distribution network is greatly challenged by the voltage fluctuation and power quality problems brought on by the growing penetration of Distributed Energy Resources (DERs) and EVs in distribution networks. The conventional techniques are unable to handle the highly uncertain environment. Furthermore, they mostly depend on the distribution system's precise parameters, which are challenging to get in real-world situations. Because deep RL (DRL) is data-driven, it can make more flexible control decisions based on the most recent information in real time.

In order to reduce voltage deviation in a distribution network authors in [31] suggested a two-timescale voltage control approach integrating the DQN algorithm and the optimization based on a physical model. While the DQN controls the shunt capacitors over a longer timeline also accounting for the long-term discounted reward value, the alternating current power flow model controls smart inverters over a shorter timescale. These techniques do not require accurate physical models to be trained; instead, they can make judgments based on the most recent distribution network information in real time. Nevertheless, this technique incorporates a penalty into the reward function and therefore doesn't work in real life. In order to address this issue, authors in [32] proposed a distribution network volt–var control approach based on the safe off-policy DRL algorithm. Initially, a constrained MDP is used to model the volt–var control problem. Then, the MDP is solved using a safe SAC technique. The suggested approach better satisfies constraints by directly modeling the operation constraints in the MDP, making it more appropriate for optimization problems requiring high levels of security.

Optimization of microgrid has also been done using RL algorithms. A DRL-based strategy for the cost-effective energy scheduling of microgrids integrated with renewable energy sources is suggested in [33]. The DRL algorithm can be utilized in an uncertain environment because it doesn't require the system model or knowledge about uncertainty. Authors in [34] suggested using RL to manage the energy of multi-microgrids. First, supervised training is used to teach DNN how various microgrids behave. Subsequently, a near-optimal pricing plan is developed using the MC RL algorithm, which aims to simultaneously decrease the peak-to-average ratio and maximize income. This approach works well for issues involving large search spaces and concealed data.

RL plays a significant role in IES management. The integrated system of energy generation, supply, and marketing during the planning, building, and operating phases is referred to as an IES. As shown in Figure 8.8, it is primarily made up of an energy supply network that includes power, gas, cooling, and heating networks; an energy exchange unit, an energy storage link, a terminal integrated energy supply unit (microgrid); and a large number of customers. Energy management of IES is difficult because it necessitates cooperatively optimizing supply and demand, which is random, and coordinating the regulation and control of many energy supply units.

In order to reduce operating costs and carbon emissions, authors in [35] developed an optimization model based on the RL approach and created an intelligent energy management system with combined heat and power unit (CHP). The

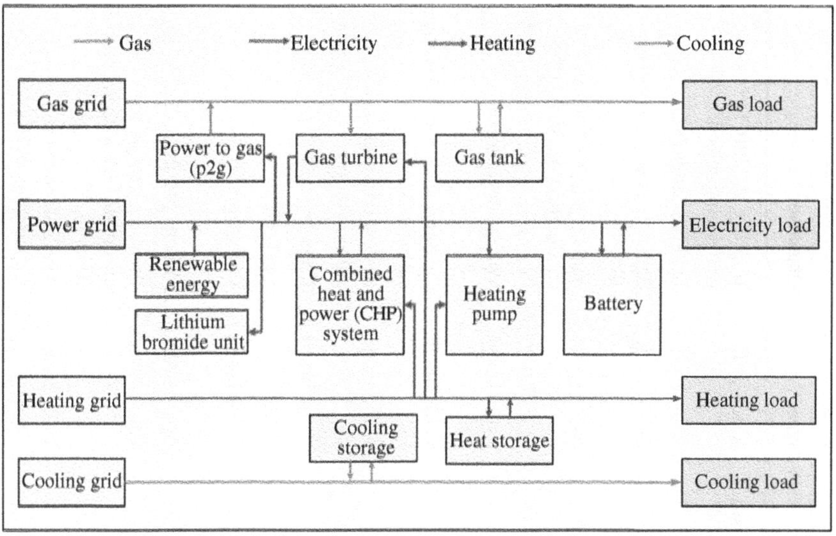

Figure 8.8 Framework of IES.

research proposed to use a dynamic energy conversion and management approach based on the DDPG algorithm to smooth the net load curve while taking system economics into account. In order to lower peak load and encourage demand response from consumers, [36] suggested an RL-based energy management plan for networks that supply natural gas and electricity.

8.4.3 Demand-Side Management

RL plays a key role in demand-side management (DSM) by optimizing how and when electrical loads consume energy. In power electronics, DSM is particularly important in applications like EV charging, residential and industrial energy management, and integrating energy storage systems (ESSs). RL algorithms can dynamically adjust the power consumption of flexible loads in response to changing electricity prices, grid conditions, and renewable energy availability. For example, RL is used to optimize smart charging of EVs, where it learns to allocate charging power based on factors like grid demand, electricity pricing, and the state of charge of multiple vehicles [37]. This ensures that EVs are charged efficiently without overloading the grid, and charging costs are minimized. A batch RL approach is proposed in [38] to lower EV charging expenses. The suggested charging strategy can result in cost savings for the EV owner ranging from 10% to 50%, according to the results. A unique DQN-based energy management technique for hybrid electric buses is examined by authors in [39]. The

Table 8.1 Various applications of reinforcement learning in power electronics.

Applications	Reference	Field	Algorithm	Type	Objective
Control optimization	[42]	Isolated DC–DC converter	Twin delayed deep deterministic policy gradient (TD3)	Actor–critic	Improve stability by optimizing selection of hyperparameters
	[43]	Three phase inverter	Deep deterministic policy gradient (DDPG)	Actor–critic	Speed or torque control of motor
	[44]	Modular converter	Multiagent twin-delayed deep deterministic policy gradient (MA-TD3)	Actor–critic	Voltage control and optimal modulation of ISOP-DAB
Energy management	[31]	Distribution Network	DQN	Value based	Voltage deviation
	[34]	Microgrid	Monte Carlo	Value based	Demand-side peak-to-average ratio (PAR)
	[45]	IES	DDPG	Actor–critic	System cost and peak load shifting target
Demand-side management	[37]	Demand response	DQN	Value based	Charging cost
	[46]	Demand response	Learning automation	Others	Fuel economy
	[47]	Demand response	Constrained policy optimization (CPO)	Value based	Charging cost
Grid stability	[48]	Operational control	DDPG	Actor–critic	Voltage profiles
	[49]	Operational control	*Q* learning	Value based	Frequency target for automatic generation control (AGC)
	[50]	Reliable grid control	DQN	Value based	Available transfer Capabilities (ATCs)

training results show that in terms of fuel economy and time consumption, the DQN-based technique outperforms Q-learning and dynamic programming.

8.4.4 Grid Stability

Maintaining grid stability is a critical challenge, especially with the growing integration of RESs, which introduce variability and unpredictability into the system. RL offers innovative solutions to enhance grid stability by optimizing the operation of power electronics devices that interface with the grid, such as inverters, converters, and ESSs. Maintaining stable voltage and frequency levels is crucial to ensure reliable operations. RL can be applied to optimize the control of grid-tied inverters that regulate both voltage and frequency. For example, in scenarios where REG fluctuates (e.g., solar or wind power), RL can learn to dynamically adjust the output of inverters or ESSs to help maintain stable voltage levels and correct frequency deviations [40]. This is especially beneficial in systems with a high share of intermittent energy sources, where conventional control methods might not respond quickly enough to stabilize the grid. An example of controlling an ESS using RL is proposed in [41]. In the research, an RL-based optimal technique is utilized for controlling the ESS in an AC–DC microgrid. One NN is used as an identifier to estimate the system's status. The other NN is then trained to discover the best control strategy for the ESS, which can lessen the disruptions brought on by its charging and discharging.

Some recent research work based on the applications of RL in power electronics is tabulated in Table 8.1. It shows the different fields in power electronics where RL is utilized, the algorithm used, and the objective achieved in each case.

8.5 Case Study – RL-based Control of Buck Converter

8.5.1 Problem Statement and Significance

DC–DC converters are essential components in power electronics, responsible for converting DC voltage levels to meet the requirements of various electronic systems. Traditional control methods, such as PID controllers, often struggle with maintaining optimal performance under varying loads and input conditions. These methods can be rigid and may require frequent manual tuning to adapt to changes in system parameters.

RL offers a dynamic and adaptive approach to control DC–DC converters, addressing the limitations of traditional methods. RL-based controllers can learn optimal control policies through interaction with the system, adapting to changes in real-time. This capability makes RL particularly suitable for

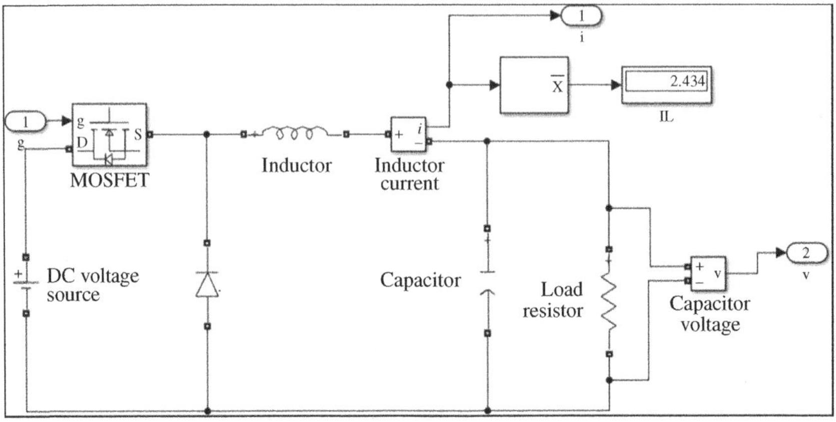

Figure 8.9 Schematic diagram of Buck DC–DC converter in Simulink.

applications requiring high efficiency and stability under varying conditions, such as renewable energy systems and EVs.

8.5.2 Methodology

- **System Model:** The system here consists of a simple Buck Converter, which is a step-down DC–DC converter. The converter is modeled in MATLAB Simulink. The input to the converter is 100 V DC, and it is desired to step down the voltage to 25 V DC, and therefore using the expression below, the duty cycle is found to be 25% (0.25).

$$V_o = D.V_{in} \tag{8.27}$$

The converter is modeled to include key parameters such as input voltage, output voltage, inductor current, and load resistance as shown in Figure 8.9. Further parameters for the converter are shown in Table 8.2.

8.5.2.1 Reinforcement Learning Framework

The Buck DC–DC Converter closed loop control is implemented using RL technique. The steps toward implementing this closed loop control include the following:

- **Define the Observation and Action Spaces**: The state (observation) space and action space for the RL agent is defined in the initial step. The state space typically includes the measured output voltage and the error signal. The error signal here is the difference between the reference voltage (25 V) and the measured output voltage. The action space consists of generating the duty cycle

Table 8.2 Buck converter parameters.

S.No.	Parameter	Value
1.	Input voltage	100 V
2.	Inductor	1 mH
3.	Capacitor	450 μF
4.	Load (resistive load)	10 Ω
5.	Switching frequency	25 kHz

for the pulse width modulated (PWM) signal. All of these are defined in the Simulink model too before feeding them to the required RL agent as observed in Figure 8.10.

- **Define the Reward Function:** The next step is to define the reward function in the Simulink model. The reward function should penalize deviations from the desired output voltage and encourage stability. Hence, the reward function is expressed as

$$R = 5(|err| < 0.1) - 1(|err| \geq 0.1) \tag{8.28}$$

- **Select the RL Agent:** For the precise duty cycle control of the buck converter in this chapter, DDPG-based agent was selected since DDPG offers stable convergence and model-free learning, which suits the complexity of Buck converters without requiring an accurate system model. Furthermore, it helps to avoid issues like false convergence and overfitting. Once the RL agent block is imported in Simulink, the required inputs are fed to it. It is important to define the parameters for the RL agent. DDPG agents evaluate the policy's value by applying a parametrized Q-value function approximator. A Q-value function critic accepts input as the present observation and action, and provides the output as a scalar, which is the estimated cumulative long-term reward (discounted). DDPG agents evaluate the policy's value by applying a parametrized Q-value function approximator. One scalar (the estimated discounted cumulative long-term reward for which one obtains the action from the state corresponding to the current observation, and following the policy afterward) is the output of a Q-value function critic, which accepts the current observation and an action as inputs. A neural network with two input layers – one for the action channel, as defined by "actInfo," and one for the observation channel, as defined by "obsInfo," and one output layer, which yields the scalar value. This scalar value is used to simulate the parametrized Q-value function inside the critic. Further code for defining the critic is written below.

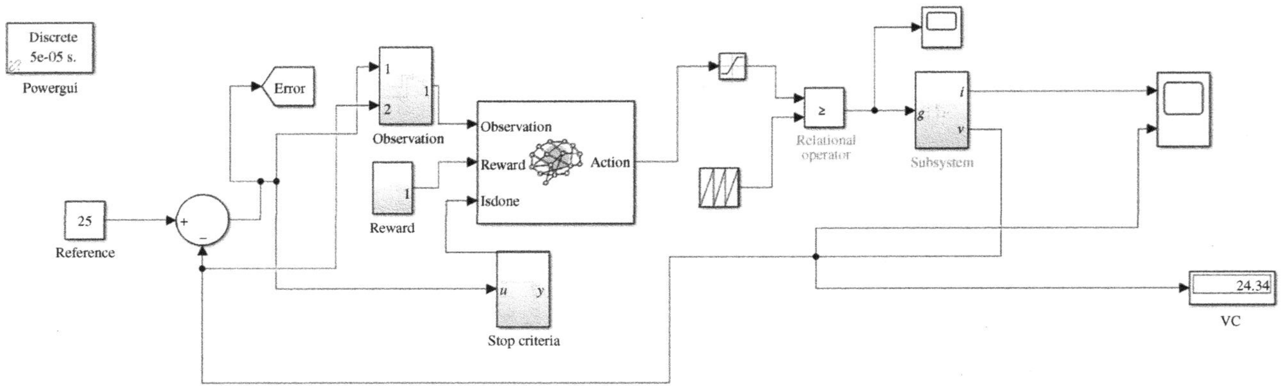

Figure 8.10 Complete schematic in Simulink with RL agent.

```
statePath = [
     featureInputLayer(numObservations,'Normalization',
'none','Name','State')
     fullyConnectedLayer(30,'Name','CriticStateFC1')
     reluLayer('Name','CriticRelu1')
     fullyConnectedLayer(15,'Name','CriticStateFC2')];
actionPath = [
     featureInputLayer(numActions,'Normalization',
'none','Name','Action')
     fullyConnectedLayer(15,'Name','CriticActionFC1')];
commonPath = [
     additionLayer(2,'Name','add')
     reluLayer('Name','CriticCommonRelu')
     fullyConnectedLayer(1,'Name','CriticOutput')];

criticNetwork = layerGraph();
criticNetwork = addLayers(criticNetwork,statePath);
criticNetwork = addLayers(criticNetwork,actionPath);
criticNetwork = addLayers(criticNetwork,commonPath);
criticNetwork = connectLayers(criticNetwork,
'CriticStateFC2','add/in1');
criticNetwork = connectLayers(criticNetwork,
'CriticActionFC1','add/in2');
criticOpts = rlRepresentationOptions('LearnRate',1e-03,
'GradientThreshold',1);
critic = rlQValueRepresentation(criticNetwork,obsInfo,
actInfo,'Observation',{'State'},'Action',{'Action'},
criticOpts);
```

After defining the critic, actor network is also defined with the required hyper-parameters as below

```
%Actor
actorNetwork = [
     featureInputLayer(numObservations,'Normalization',
'none','Name','State')
     fullyConnectedLayer(3, 'Name','actorFC')
     tanhLayer('Name','actorTanh')
     fullyConnectedLayer(numActions,'Name','Action')
     ];

actorOptions = rlRepresentationOptions('LearnRate',
1e-04,'GradientThreshold',1);
```

```
actor = rlDeterministicActorRepresentation
(actorNetwork,obsInfo,actInfo,'Observation',{'State'},
'Action',{'Action'},actorOptions);
```

Once the critic and actor are defined, it is important to configure the DDPG agent, which is named "Newagent" for this case study, and the MATLAB code for setting up the DDPG agent with the hyperparameters including the number of episodes and stop training criteria [51] is given below

```
agentOpts = rlDDPGAgentOptions(...
    'SampleTime',Ts,...
    'TargetSmoothFactor',1e-4,...
    'DiscountFactor',0.99, ...
    'MiniBatchSize',48, ...
    'ExperienceBufferLength',1e6);
agentOpts.NoiseOptions.Variance = 0.3;
agentOpts.NoiseOptions.VarianceDecayRate = 1e-5;
Newagent = rlDDPGAgent(actor,critic,agentOpts);
maxepisodes = 200;
maxsteps = ceil(Tf/Ts);
trainingOpts = rlTrainingOptions(...
    'MaxEpisodes',maxepisodes,...
    'MaxStepsPerEpisode',maxsteps,...
    'Verbose',false,...
    'Plots','training-progress',...
    'StopTrainingCriteria','EpisodeReward',...
    'StopTrainingValue',10);
```

The final step is to train the agent and validate the trained agent; the following code can be observed here

```
doTraining = true;
if doTraining
    % Train the agent.
    trainingStats = train(Newagent,env,trainingOpts);
else
    %Load the pretrained agent for the example.
    load('Buck_conv_DDPG.mat','agent')
end
%Validate Trained agent
simOpts = rlSimulationOptions( ...
    MaxSteps=ceil(Tf/Ts), ...
```

```
    StopOnError="on");
experiences = sim(env,Newagent,simOpts);
```

- **Results:** Multiple training episodes are conducted, where the agent learns to optimize the control policy through trial and error. The training curve obtained from the Simulink model is illustrated in Figure 8.11 for up to 110 episodes. The control performance of the Buck DC–DC converter was evaluated using a DDPG RL agent. The training curve is converging toward an optimal reward, where the error is minimized. The trained DDPG agent successfully regulated the output voltage to the desired reference value. As displayed in Figure 8.12, the output voltage closely follows the reference voltage with minimal steady-state error and fast transient response. The voltage remains stable throughout demonstrating the agent's ability to dynamically adjust the duty cycle and maintain regulation. The inductor current and the gating pulses for the converter are also provided in Figures 8.13 and 8.14, respectively. Overall, the DDPG agent showed a robust control performance by maintaining stable output voltage and inductor current. The agent demonstrated superior adaptability and efficiency while minimizing overshoot and undershoot in both voltage and current waveforms.

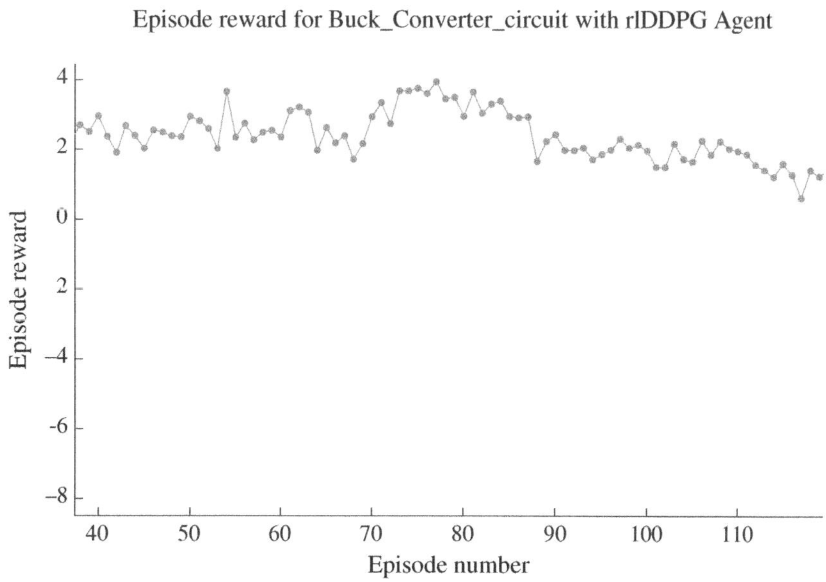

Figure 8.11 Training curve for DDPG-RL agent.

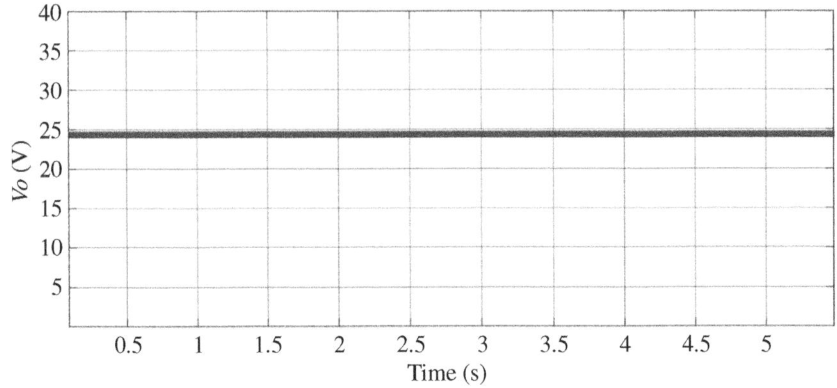

Figure 8.12 Output voltage diagram for buck DC–DC converter.

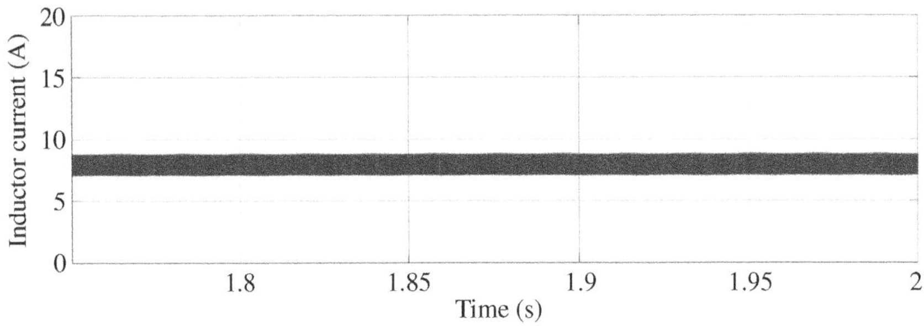

Figure 8.13 Output inductor current for buck DC–DC converter.

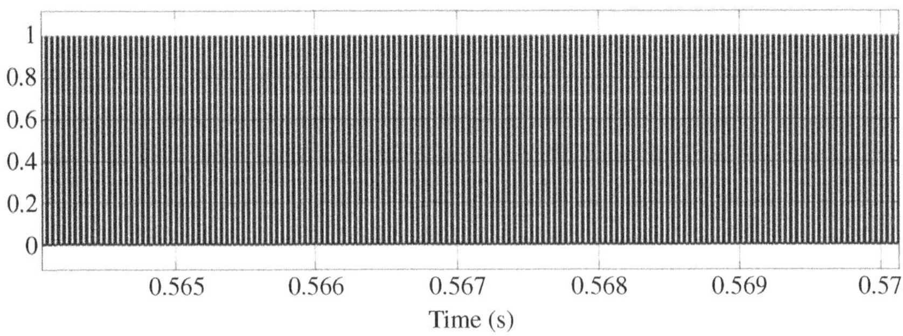

Figure 8.14 The gating pulses for the switching device in buck DC–DC converters.

8.6 Future Research Directions

Integrating RL with other AI techniques like machine learning and deep learning is a growing trend aimed at improving RL's capabilities. Combining RL with supervised or unsupervised learning helps leverage historical data for pre-training, reducing the data required for RL training. Deep learning models can approximate complex functions and environments, making RL more efficient in handling high-dimensional state spaces. Contemporary power electronics systems are increasingly complex, making single-agent RL inadequate for large-scale systems. MADRL helps address this by using centralized training and decentralized execution, though challenges remain in scaling to very large systems, necessitating further research into advanced MADRL algorithms with population scaling mechanisms. Future of RL lies in implementing more MADRL structure for complex and advanced issues with regards to power electronics applications.

With the increasing complexity and uncertainty in modern power electronics systems, alongside the extensive deployment of advanced sensors, RL emerges as a viable alternative for control applications. In tasks with sparse rewards or multiple objectives, conventional learning algorithms may struggle due to vast exploration spaces and challenging reward settings. A promising solution is meta-reinforcement learning (meta-RL), which enhances learning speed and generalization by applying knowledge from multiple tasks to new ones. Offline RL allows training without environment interaction, using previously collected data, while transfer learning (TL) improves efficiency by leveraging knowledge from previous tasks. These approaches drive DRL research toward autonomous systems with enhanced understanding.

8.7 Conclusion

RL has emerged as a powerful tool for advancing the field of power electronics, providing notable enhancements over conventional control techniques. Inverter control, DC–DC converter control, MPPT, motor drive control, optimization, anomaly detection, and cybersecurity are just a few of the power electronics applications that RL can handle using adaptive and optimal control schemes. This technology offers reliable solutions that can adjust to uncertainties and changes in real time, in addition to improving performance and efficiency. Long training times, high computational complexity, and significant data requirements are some of the obstacles that RL in power electronics must overcome despite its exciting potential. Current research is focused on overcoming these limitations by developing more efficient algorithms, integrating RL with other AI techniques, and exploring hybrid control systems that combine RL with traditional methods. The goal of these initiatives is to make RL more scalable and useful for a range of power electronics applications.

With emerging applications in smart grids, EVs, and other cutting-edge technology, RL in power electronics appears to have a bright future. Power electronic systems should become more reliable, effective, and adaptable as a result of ongoing research and development. RL can make a substantial contribution to the development of power electronics by resolving existing issues and utilizing new developments. This will advance energy management, the integration of renewable energy sources, and system optimization in general. In conclusion, RL represents a transformative approach in the field of power electronics, capable of achieving superior control and optimization. As research progresses, the integration of RL into power electronic systems will continue to unlock new possibilities and drive the development of smarter, more efficient, and resilient power solutions.

Exercises

8.1 What advantages does reinforcement learning (RL) have over conventional control techniques in the field of power electronics?

8.2 What distinguishes reinforcement learning from supervised learning in its approach to training agents?

8.3 How is the MDP important in the reinforcement learning framework?

8.4 In reinforcement learning, what are the primary components and how does each component contribute to agent decision-making?

8.5 How does deep learning enhance reinforcement learning capabilities in handling complex, high-dimensional problems?

8.6 What is the primary difference between on-policy and off-policy training methods in reinforcement learning?

8.7 How do dynamic programming methods like policy iteration and value iteration solve reinforcement learning problems?

8.8 How does the Monte Carlo (MC) method converge, and how does it differ from other RL techniques?

8.9 Explain how Temporal Difference (TD) learning uses the bootstrapping approach and describe its key components.

8.10 How do policy gradient methods differ from value-based methods like dynamic programming, TD, and MC in reinforcement learning?

8.11 How does the combination of RL with MPPT control improve wind power generation converters?

8.12 What role do time-varying temperature coefficients and the Boltzmann exploration approach play in enhancing converter performance?

8.13 Why are DC–DC converters commonly used as test cases for RL algorithms, and how does independent RL control contribute to optimizing the output voltage of these converters?

8.14 What challenges arise when applying RL to modular converters, and how do multi-agent RL algorithms, like the MADDPG approach, address these complexities?

8.15 How does RL contribute to energy management in integrated energy systems (IES), and what role does it play in optimizing supply and demand across different energy sources?

References

1 A. Haque and A. Malik, *Machine Learning in Renewable Energy Systems for Smart Cities*. CRC Press, 2024, p. 28. [Online]. Available: https://www.taylorfrancis.com/chapters/edit/10.1201/9781032669809-13/machine-learning-renewable-energy-systems-smart-cities-ahteshamul-haque-azra-malik.

2 M. Schenke and O. Wallscheid, "A deep Q-learning direct torque controller for permanent magnet synchronous motors," *IEEE Open J. Ind. Electron. Soc.*, vol. 2, pp. 388–400, June 2021. doi: 10.1109/OJIES.2021.3075521.

3 V. S. B. Kurukuru, A. Haque, M. A. Khan, S. Sahoo, A. Malik, and F. Blaabjerg, "A review on artificial intelligence applications for grid-connected solar photovoltaic systems," *Energies*, vol. 14, no. 15, p. 4690, Aug. 2021. doi: 10.3390/en14154690.

4 M. Schenke, W. Kirchgassner, and O. Wallscheid, "Controller design for electrical drives by deep reinforcement learning: A proof of concept," *IEEE Trans. Ind. Inform.*, vol. 16, no. 7, pp. 4650–4658, 2020. doi: 10.1109/TII.2019.2948387.

5 C. Xiang *et al.*, "A novel deep deterministic policy gradient assisted learning based control algorithm for three-phase DC/AC inverter with an RL load," *IEEE J. Emerg. Sel. Top. Power Electron.*, vol. 11, no. 6, pp. 5529–5539, 2022. doi: 10.1109/JESTPE.2022.3174638.

6 C. Li, Y. A. Chen, C. Jin, R. Sharma, and J. Kleissl, "Online PV smart inverter coordination using deep deterministic policy gradient," *Electr. Power Syst. Res.*, vol. 209, p. 107988, Aug. 2022. doi: 10.1016/j.epsr.2022.107988.

7 Y. Luo, "Application of reinforcement learning algorithm model in gas path fault intelligent diagnosis of gas turbine," *Comput. Intell. Neurosci.*, vol. 2021, pp. 1–10, 2021. doi: 10.1155/2021/3897077.

8 G. Seo, S. Yoon, J. Song, E. Srivastava, and E. Hwang, "Label-free fault detection scheme for inverters of PV systems: Deep reinforcement learning-based dynamic threshold," *Appl. Sci.*, vol. 13, no. 4, p. 2470, 2023. doi: 10.3390/app13042470.

9 K. Beyer, R. Beckmann, S. Geißendörfer, K. von Maydell, and C. Agert, "Adaptive online-learning volt-var control for smart inverters using deep reinforcement learning," *Energies*, vol. 14, no. 7, p. 1991, 2021. doi: 10.3390/en14071991.

10 X. Wang *et al.*, "Deep reinforcement learning: A survey," *IEEE Trans. Neural Network Learn. Syst.*, vol. 35, no. 4, pp. 5064–5078, Apr. 2024. doi: 10.1109/TNNLS.2022.3207346.

11 D. Cao *et al.*, "Reinforcement learning and its applications in modern power and energy systems: A review," *J. Mod. Power Syst. Clean Energy*, vol. 8, no. 6, pp. 1029–1042, 2020. doi: 10.35833/MPCE.2020.000552.

12 T. T. Nguyen, N. D. Nguyen, and S. Nahavandi, "Deep reinforcement learning for multiagent systems: A review of challenges, solutions, and applications," *IEEE Trans. Cybern.*, vol. 50, no. 9, pp. 3826–3839, Sep. 2020. doi: 10.1109/TCYB.2020.2977374.

13 H. van Hasselt, A. Guez, and D. Silver, "Deep reinforcement learning with double Q-learning," *Proc. AAAI Conf. Artif. Intell*, vol. 30, pp. 1–7, 2016. doi: 10.1109/ICRA40945.2020.9196684.

14 Z. Wang, T. Schaul, M. Hessel, H. Van Hasselt, M. Lanctot, and N. De Frcitas, "Dueling network architectures for deep reinforcement learning," *33rd International Conference on Machine Learning (ICML 2016)*, vol. 4, no. 9, pp. 2939–2947, 2016.

15 B. Iii, D. Avenue, W. Dc, W. Patterson, and A. F. B. Oh, "Advantage updating," *Security*, vol. 94, pp. 1–47, 1993.

16 J. Schulman, S. Levine, P. Moritz, M. Jordan, and P. Abbeel, "Trust region policy optimization," *32nd International Conference on Machine Learning (ICML 2015)*, vol. 3, pp. 1889–1897, 2015.

17 J. Schulman, F. Wolski, P. Dhariwal, A. Radford, and O. Klimov, "Proximal policy optimization algorithms," pp. 1–12, 2017, [Online]. Available: http://arxiv.org/abs/1707.06347.

18 V. Mnih *et al.*, "Asynchronous methods for deep reinforcement learning," *33rd International Conference on Machine Learning (ICML 2016)*, vol. 4, pp. 2850–2869, 2016.

19 Z. Wang *et al.*, "Sample efficient actor-critic with experience replay," *5th International Conference on Learning Representation ICLR 2017 – Conference on Track Proceedings*, no. 2016, 2017.

20 R. Munos, T. Stepleton, A. Harutyunyan, and M. G. Bellemare, "Safe and efficient off-policy reinforcement learning," *Adv. Neural Inf. Proces. Syst.*, vol. 29, pp. 1054–1062, 2016.

21 D. Silver, G. Lever, N. Heess, T. Degris, D. Wierstra, and M. Riedmiller, "Deterministic policy gradient algorithms," *31st International Conference on Machince Learning (ICML 2014)*, vol. 1, pp. 605–619, 2014.

22 O. E. Egbomwan, S. Liu, and H. Chaoui, "Twin delayed deep deterministic policy gradient (TD3) based virtual inertia control for inverter-interfacing DGs in microgrids," *IEEE Syst. J.*, vol. 17, no. 2, pp. 2122–2132, 2023. doi: 10.1109/JSYST.2022.3222262.

23 A. Malik, A. Haque, V. S. B. Kurukuru, and S. Mekhilef, "Fault-resilient control of parallel PV inverters using multi-agent twin-delayed deep deterministic policy gradient approach," *Int. J. Circuit Theor. Appl.*, vol. 52, no. 7, pp. 3230–3254, Jul. 2024. doi: 10.1002/cta.3911.

24 D. Alfred, D. Czarkowski, and J. Teng, "Reinforcement learning-based control of a power electronic converter," *Mathematics*, vol. 12, no. 5, p. 671, Feb. 2024. doi: 10.3390/math12050671.

25 P. Chen *et al.*, "A review on the applications of reinforcement learning control for power electronic converters," *IEEE Trans. Ind. Appl.*, vol. 60, pp. 1–24, 2024. doi: 10.1109/TIA.2024.3435170.

26 C. Wei, Z. Zhang, W. Qiao, and L. Qu, "Reinforcement-learning-based intelligent maximum power point tracking control for wind energy conversion systems," *IEEE Trans. Ind. Electron.*, vol. 62, no. 10, pp. 6360–6370, Oct. 2015. doi: 10.1109/TIE.2015.2420792.

27 P. S. V. Kishore, N. Jayaram, and J. Rajesh, "Performance enhancement of buck converter using reinforcement learning control," in *2022 IEEE Delhi Section Conference (DELCON)*, Feb. 2022, pp. 1–5. doi: 10.1109/DELCON54057.2022.9752938.

28 P. Qashqai, M. Babaie, R. Zgheib, and K. Al-Haddad, "A model-free switching and control method for three-level neutral point clamped converter using deep reinforcement learning," *IEEE Access.*, vol. 11, pp. 105394–105409, 2023. doi: 10.1109/ACCESS.2023.3318264.

29 J. Bo, R. Wei, and H. Cai, "Controller parameterization for grid-connected power converters through reinforcement learning," in *PESS + PELSS 2022; Power and Energy Student Summit*, Kassel, Germany, 2022, pp. 1–6.

30 Y. Zeng *et al.*, "Deep reinforcement learning based input voltage sharing method for input-series output-parallel dual active bridge converter in DC microgrids," in *2021 IEEE Energy Conversion Congress and Exposition (ECCE)*, Oct. 2021, pp. 3348–3352. doi: 10.1109/ECCE47101.2021.9595137.

31 Q. Yang, G. Wang, A. Sadeghi, G. B. Giannakis, and J. Sun, "Two-timescale voltage control in distribution grids using deep reinforcement learning," *IEEE Trans. Smart Grid*, vol. 11, no. 3, pp. 2313–2323, May 2020. doi: 10.1109/TSG.2019.2951769.

32 W. Wang, N. Yu, Y. Gao, and J. Shi, "Safe off-policy deep reinforcement learning algorithm for volt-VAR control in power distribution systems," *IEEE Trans. Smart Grid*, vol. 11, no. 4, pp. 3008–3018, Jul. 2020. doi: 10.1109/TSG.2019.2962625.

33 E. Kuznetsova, Y.-F. Li, C. Ruiz, E. Zio, G. Ault, and K. Bell, "Reinforcement learning for microgrid energy management," *Energy*, vol. 59, pp. 133–146, Sep. 2013. doi: 10.1016/j.energy.2013.05.060.

34 Y. Du and F. Li, "Intelligent multi-microgrid energy management based on deep neural network and model-free reinforcement learning," *IEEE Trans. Smart Grid*, vol. 11, no. 2, pp. 1066–1076, Mar. 2020. doi: 10.1109/TSG.2019.2930299.

35 Q. Sun, D. Wang, D. Ma, and B. Huang, "Multi-objective energy management for we-energy in Energy Internet using reinforcement learning," in *2017 IEEE Symposium Series on Computational Intelligence (SSCI)*, Nov. 2017, pp. 1–6. doi: 10.1109/SSCI.2017.8285243.

36 A. Sheikhi, M. Rayati, and A. M. Ranjbar, "Demand side management for a residential customer in multi-energy systems," *Sustain. Cities Soc.*, vol. 22, pp. 63–77, Apr. 2016. doi: 10.1016/j.scs.2016.01.010.

37 Z. Wan, H. Li, H. He, and D. Prokhorov, "A data-driven approach for real-time residential EV charging management," in *2018 IEEE Power & Energy Society General Meeting (PESGM)*, Aug. 2018, pp. 1–5. doi: 10.1109/PESGM.2018.8585945.

38 A. Chis, J. Lunden, and V. Koivunen, "Reinforcement learning-based plug-in electric vehicle charging with forecasted price," *IEEE Trans. Veh. Technol.*, vol. 66, no. 5, pp. 3674–3684, 2016. doi: 10.1109/TVT.2016.2603536.

39 T. Liu, Y. Zou, D. Liu, and F. Sun, "Reinforcement learning of adaptive energy management with transition probability for a hybrid electric tracked vehicle," *IEEE Trans. Ind. Electron.*, vol. 62, no. 12, pp. 7837–7846, Dec. 2015. doi: 10.1109/TIE.2015.2475419.

40 G. Zhang *et al.*, "A data-driven approach for designing STATCOM additional damping controller for wind farms," *Int. J. Electr. Power Energy Syst.*, vol. 117, p. 105620, May 2020. doi: 10.1016/j.ijepes.2019.105620.

41 J. Duan, Z. Yi, D. Shi, C. Lin, X. Lu, and Z. Wang, "Reinforcement-learning-based optimal control of hybrid energy storage systems in hybrid AC–DC microgrids," *IEEE Trans. Ind. Inform.*, vol. 15, no. 9, pp. 5355–5364, Sep. 2019. doi: 10.1109/TII.2019.2896618.

42 Y. Tang *et al.*, "Deep reinforcement learning aided variable-frequency triple-phase-shift control for dual-active-bridge converter," *IEEE Trans. Ind. Electron.*, vol. 70, no. 10, pp. 10506–10515, Oct. 2023. doi: 10.1109/TIE.2022.3220893.

43 A. Traue, G. Book, W. Kirchgassner, and O. Wallscheid, "Toward a reinforcement learning environment toolbox for intelligent electric motor control," *IEEE Trans. Neural Network Learn. Syst.*, vol. 33, no. 3, pp. 919–928, Mar. 2022. doi: 10.1109/TNNLS.2020.3029573.

44 Y. Zeng *et al.*, "Autonomous input voltage sharing control and triple phase shift modulation method for ISOP-DAB converter in DC microgrid: A multiagent deep reinforcement learning-based method," *IEEE Trans. Power Electron.*, vol. 38, no. 3, pp. 2985–3000, Mar. 2023. doi: 10.1109/TPEL.2022.3218900.

45 B. Zhang *et al.*, "Dynamic energy conversion and management strategy for an integrated electricity and natural gas system with renewable energy: Deep reinforcement learning approach," *Energy Convers. Manag.*, vol. 220, p. 113063, Sep. 2020. doi: 10.1016/j.enconman.2020.113063.

46 H. Kazmi, F. Mehmood, S. Lodeweyckx, and J. Driesen, "Gigawatt-hour scale savings on a budget of zero: Deep reinforcement learning based optimal control of hot water systems," *Energy*, vol. 144, pp. 159–168, Feb. 2018. doi: 10.1016/j.energy.2017.12.019.

47 H. Li, Z. Wan, and H. He, "Constrained EV charging scheduling based on safe deep reinforcement learning," *IEEE Trans. Smart Grid*, vol. 11, no. 3, pp. 2427–2439, May 2020. doi: 10.1109/TSG.2019.2955437.

48 J. Duan *et al.*, "Deep-reinforcement-learning-based autonomous voltage control for power grid operations," *IEEE Trans. Power Syst.*, vol. 35, no. 1, pp. 814–817, Jan. 2020. doi: 10.1109/TPWRS.2019.2941134.

49 T. Yu, B. Zhou, K. W. Chan, L. Chen, and B. Yang, "Stochastic optimal relaxed automatic generation control in non-Markov environment based on multi-step Q learning," *IEEE Trans. Power Syst.*, vol. 26, no. 3, pp. 1272–1282, Aug. 2011. doi: 10.1109/TPWRS.2010.2102372.

50 T. Lan *et al.*, "AI-based autonomous line flow control via topology adjustment for maximizing time-series ATCs," in *2020 IEEE Power*

& Energy Society General Meeting (PESGM), Aug. 2020, pp. 1–5. doi: 10.1109/PESGM41954.2020.9281518.

51 "Train TD3 Agent for PMSM Control," 2023. [Online]. Available: https://in.mathworks.com/help/reinforcement-learning/ug/train-td3-agent-for-pmsm-control.html.

9

Implementation of Machine Learning for Power Electronics Application Using MATLAB

Manauwar Hussain, Ahteshamul Haque, and Md Zafar Khan

Advance Power Electronics and Research Laboratory, Jamia Millia Islamia (A Central University), New Delhi, India

9.1 Introduction

Power electronics is a domain of electrical engineering that emphasizes on the conversion, control, and conditioning of electric power using electronic circuits and components like diodes, MOSFETs, IGBTs, and thyristors, among others. These systems effectively and dependably handle the conversion of electrical energy between different forms (such as DC to AC, AC to DC, and voltage-level adjustments). The block diagram for power electronics working is depicted in Figure 9.1.

The significance of power electronics is highlighted in the following points [1].

- **Efficient Energy Conversion**: For effective energy conversion in a variety of applications, power electronics is essential. It minimizes energy losses and enhances system performance by enabling highly efficient conversion of electrical energy between different forms.
- **Regulation and Control**: Reliable regulation and control of electrical power become feasible by power electronics. In applications where accurate and constant power regulation is required for optimal performance, such as power supplies, motor drives, and renewable energy systems, this is crucial.
- **Renewable Energy Integration**: Power electronics is essential for incorporating renewable energy sources into the power grid. It enables the effective conversion of power from sources like solar panels and wind turbines, allowing their use in the mainstream power supply.

Artificial Intelligence for Power Electronics, First Edition.
Edited by Ahteshamul Haque, Saad Mekhilef, and Azra Malik.

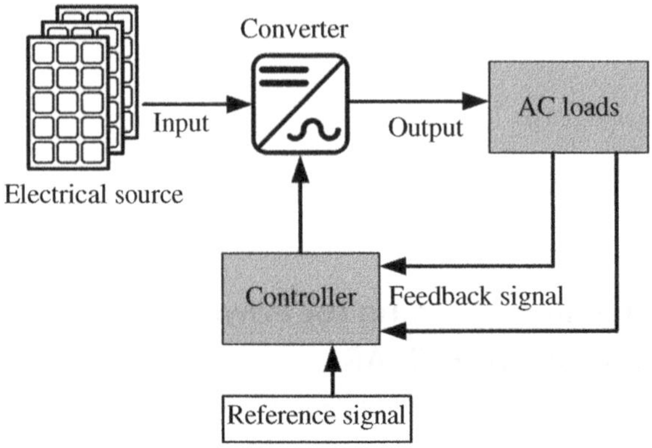

Figure 9.1 Block diagram of power electronics.

- **Enhanced Power Quality**: By reducing problems like harmonics, voltage sags, and surges, power electronic devices contribute to an improved power quality. This results in reliable and stable power supply, which is necessary for industrial processes and sensitive electronic equipment.
- **Compact and Lightweight Systems**: Systems can be made lighter and more compact by utilizing power electrical devices. This is especially crucial for applications like portable electronics, automobiles, and aircraft where size and weight are critical.
- **Automation and Control**: Power electronics devices are essential components in modern automation and control systems. It enables precise control of equipment and processes, boosting productivity and efficiency in industrial applications.
- **Electric Vehicles**: The key technological advancement in the creation and functioning of electric vehicles (EVs) is power electronics devices. It enhances EV technology via its usage in power management, battery charging systems, and electric motor control.
- **Energy Savings**: Power electronics makes significant improvements to energy savings through enhancing the electrical systems' efficiency. This minimizes the consumption of energy and related expenses, which benefits the economy and our surroundings.
- **Smart Grid:** The advancement of smart grids depends mainly on power electronics, which strive to maximize the generation, transmission, distribution, as well as utilization of electrical energy. It enables cutting-edge functions including distributed energy resource integration, demand response, and real-time monitoring [2].

9.2 Machine Learning

Frequently, when people think of machine learning (ML), they imagine robots. While some consider them as beneficial, others consider them to be dangerous. But ML isn't limited to the big screen. It is genuine and has already been around for a long time. Reading text from images was one of its early uses; this process is done via optical character recognition (OCR) technology. The first machine-learning technique, the spam filter in the 1990s, was broadly implemented. Spam filters are straightforward but efficient. Learn how to identify and delete unwanted emails. There are numerous more ML tools available now. These accessories significantly improve a lot of everyday products. Voice search and personalized recommendations are two examples [3].

Where does ML get started and end? For a machine to learn means? Does downloading a copy of Wikipedia make your computer smarter or learn something?

Before we start exploring ML, we'll first get to know the basics. We'll examine the key areas and concepts, such as model-based versus instance-based learning, online versus batch learning, and supervised versus unsupervised learning. After that, we'll go over the procedures associated with a normal ML project, discuss typical problems, and outline how to assess and enhance a ML system.

9.2.1 Being Familiar with Machine Learning

Programming computers to learn from data is the science and art of ML.

> In general, an engineering-oriented definition states that if a computer program performs better on task T as measured by P after exposure to experience E, then it has learned from that experience. —**Tom Mitchell**, 1997.

A ML example to be your spam filter. It develops the ability to distinguish between spam and normal emails (sometimes known as "ham") by using examples of spam that users have identified. These examples teach the system how to identify and mark fresh spam emails. The term "training instance" refers to each unique example or sample, and the collection of examples utilized for learning is termed the training set [4].

Here, the objective is to detect spam, and the experience serves as a training set. The accuracy, a measure of how well the system classifies emails, is the performance indicator. For categorization tasks, accuracy is a common parameter.

Simply downloading a copy of Wikipedia doesn't mean your computer has learned anything useful for tasks. It has more data, but it doesn't become better at specific tasks without learning from it. So, downloading Wikipedia isn't ML.

9.2.2 Importance of Machine Learning

Think about how you would create a spam filter using traditional programming methods [5].

- You would first consider the standard form of spam. You may have observed that some words or phrases (such "4U," "credit card," "free," and "amazing") keep appearing in the subject line. The email's body, other sections, and the sender's identity may also reveal a few more trends to you.
- You would design a detecting algorithm for every pattern you observed, and if numerous similarities were found in emails, your program would mark them as spam.
- After testing your program, you would go through steps 1 and 2 again until it was ready to be launched [6].

Because the problem is complex, your program may end up being a lengthy and difficult-to-maintain set of intricate rules. On the other hand, by identifying remarkably common word patterns in spam as opposed to ordinary emails, a spam filter that using ML automatically discovers which words and phrases are good indicators of spam. As a result, the software is small, easier to maintain, and probably more accurate (Figure 9.2).

What happens if spammers discover that "4U" is being blacklisted from all of their emails? Probably they should begin writing "For U" instead. To flag "For U" emails, A filter for email spam that employs conventional programming techniques might require modifications.

The ML approach can be understood from Figure 9.3. We will constantly need to create new rules if spammers manage to eliminate your spam filter [4].

On the other hand, a spam filter that employs ML techniques automatically recognizes that the term "For U" has begun to show up in spam that users have

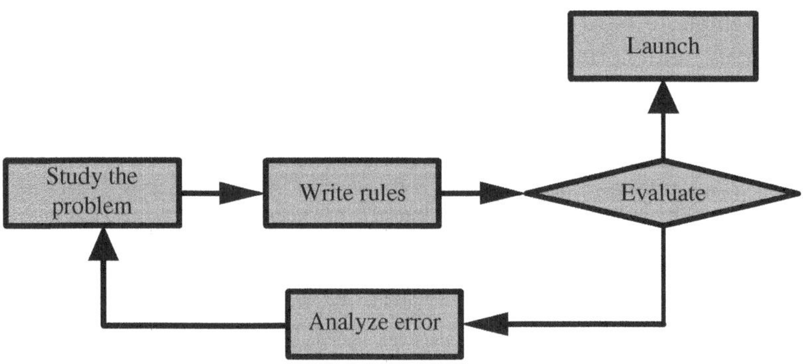

Figure 9.2 The traditional approach.

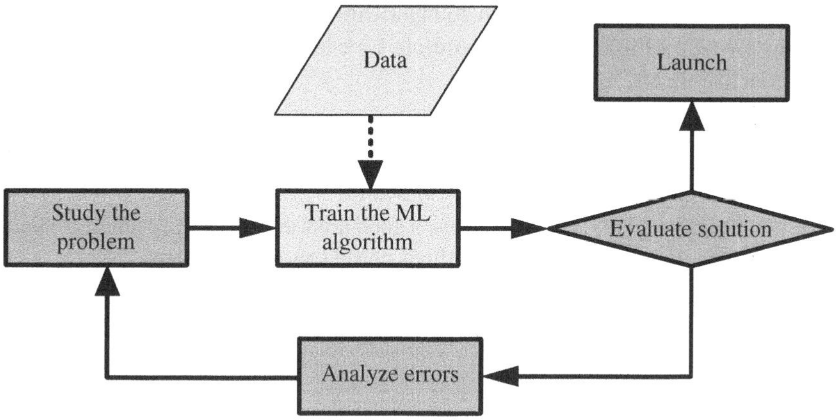

Figure 9.3 The machine learning approach.

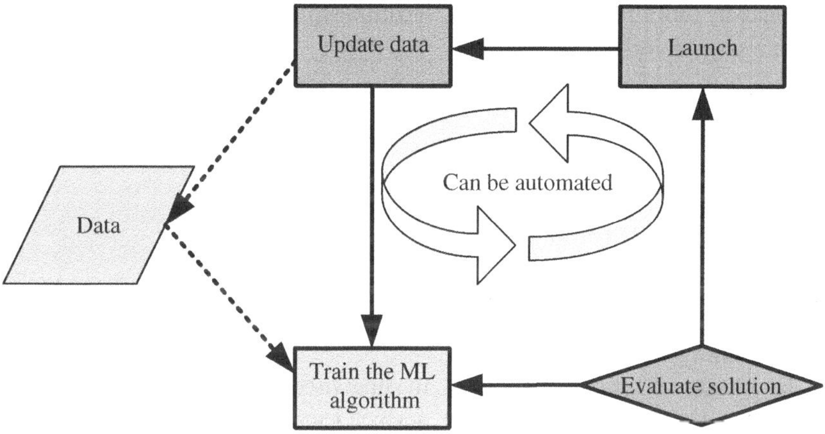

Figure 9.4 Automatically adapting to changes.

reported more often than usual. This filter begins flagging messages without your interaction (Figure 9.4).

When it comes to situations that are either too challenging for traditional techniques or lack a reliable algorithm, ML excels. Think about speech recognition, for instance. Assume for this instance that you wish to start writing a program that can distinguish between the phrases "one" and "two." You could exploit the fact that the word "two" starts with a high-pitched sound ("T") to hardcode an algorithm that distinguishes between ones and twos based on the intensity of high-pitched noises. But clearly, this approach is not scalable to thousands of phrases spoken in thousands of languages by millions of diverse people in loud situations. Right now, the best method is to develop an algorithm that is capable of learning on its own by giving multiple example recordings for each word.

In conclusion, ML is a great tool for:

- Issues whose current remedies need a lengthy list of rules of fine-tuning: algorithm can frequently be made simpler and performed better with a single ML algorithm than with a conventional algorithm.
- For complex problems that don't have a good answer when approached conventionally the most advanced ML methods could solve the problem.
- ML algorithms adjust to new data or information in dynamic contexts.
- Acquiring understanding of complicated problems and large amounts of information

9.3 Types of Machine Learning

Let's classify the different kinds of ML according to whether or not human supervision was used during training:

- Supervised learning
- Unsupervised learning
- Semi-supervised learning

9.3.1 Supervised Learning

In supervised learning, the training data given to the algorithm consists of input–output pairs, with the output representing the target solution or label that the algorithm aims to predict.

Classification is a fundamental task in supervised learning. A classic example is a spam filter, illustrated in Figure 9.5, which learns to classify new emails into

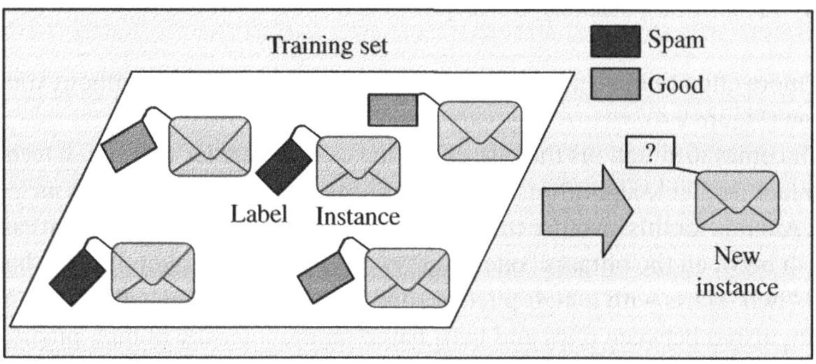

Figure 9.5 Labeled training set for spam (supervised learning).

Figure 9.6 An unlabeled training set for supervised learning.

categories such as spam or ham based on a large dataset of labeled emails. It's important to note that some regression techniques can also be applied to classification problems and vice versa. For instance, logistic regression, despite its name, is often used for classification tasks. It provides a probability indicating the likelihood of an instance belonging to a particular class, such as a 20% chance that an email is spam. Key techniques for supervised learning include *k*-Nearest Neighbors, Linear Regression, Logistic Regression, Support Vector Machines (SVMs), Decision Trees, Random Forests, and Neural Networks.

9.3.2 Unsupervised Learning

In unsupervised learning, the training data lacks labels. The system endeavors to identify patterns and structures within the data without any explicit guidance or predefined labels, essentially learning independently. An unlabeled training set can be observed from Figure 9.6.

Key Characteristics include, no Labels, pattern discovery, and exploratory nature: The algorithm accepts unlabeled data. It helps in finding patterns, clusters, or structures in the data. It is frequently used to examine data and discover hidden patterns.

These are a few of the most significant algorithms for unsupervised learning.

- **Clustering**: *K*-Means, Hierarchical Cluster Analysis (HCA), and DBSCAN
- **Anomaly and Novelty Detection**: One-class SVM and Isolation Forest
- **Visualization and Dimensionality Reduction**: Principal Component Analysis (PCA), Kernel PCA, t-Distributed Stochastic Neighbor Embedding (t-SNE), Locally Linear Embedding (LLE)
- **Association Rule Learning**: Eclat and Apriori

9.3.3 Semi-supervised Learning

Labeling data can be both time-consuming and expensive, leading to scenarios where there are numerous unlabeled instances and only a few with labels. To

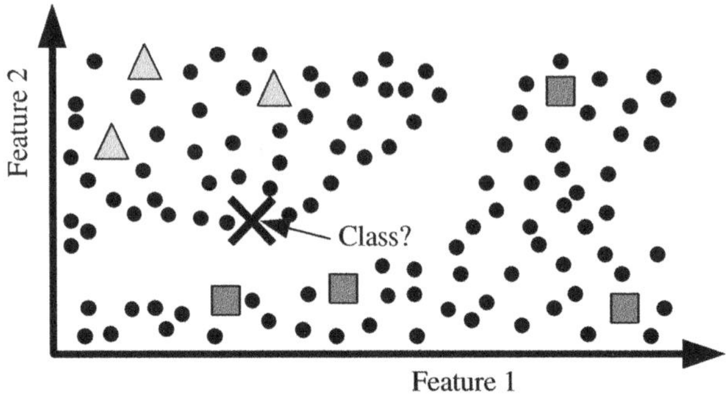

Figure 9.7 Supervised learning with two classes.

address this, some algorithms can work with partially labeled data through a method called semi-supervised learning (see Figure 9.7). This technique utilizes the wealth of unlabeled data in conjunction with the limited labeled data to enhance learning accuracy and efficiency.

9.3.4 Reinforcement Learning

Reinforcement Learning involves an agent that observes its environment, selects, and performs actions, and receives rewards or penalties in return (Figure 9.8). The agent learns the best strategy, or policy, to maximize rewards over time. A policy defines the action the agent should take in a given situation to achieve the highest reward [7].

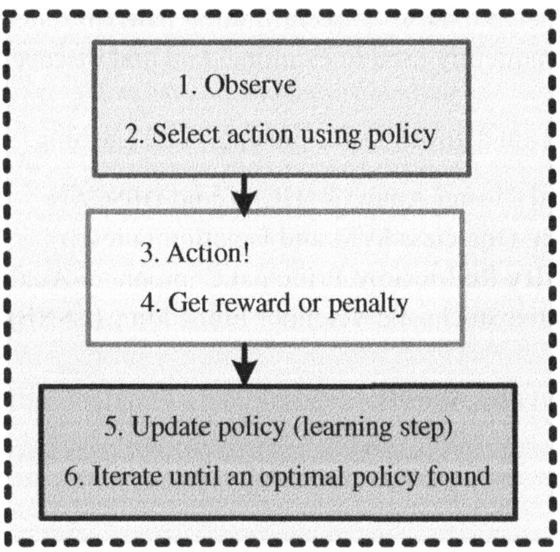

Figure 9.8 Reinforcement learning.

9.4 ML in Power Electronics

ML is the process of using data-driven algorithms to boost the functionality and performance of electrical systems. In this section, we'll look at some of the approaches and uses of ML in the field of power electronics (Figure 9.9). The following are common applications of ML in power electronics:

- Predictive Maintenance
- Anomaly Detection
- Controller Design
- Deep Learning
- Adaptive Control
- Optimization

- **Predictive Maintenance**: Power electronic components, such as inverters and converters, can use ML to monitor their health and condition, enabling the anticipation of potential failures. This predictive maintenance approach reduces downtime and helps avoid costly issues. Neural networks, Random Forests, and SVMs are commonly used supervised learning methods in this context, effectively predicting when these components are likely to fail.
- **Anomaly Detection:** Also, irregularities and deviations in power electronic systems, such as overloads, short circuits, and harmonics, can be detected using ML. This way, system security and Reliability could be maintained. Unsupervised learning techniques such as PCA and clustering are utilized in anomaly identification.
- **Controller Design:** Power electronic equipment, including power converters and inverters, can benefit from ML's advanced control techniques. These systems can adapt to changing operating conditions while maximizing stability and energy efficiency. Reinforcement learning techniques, such as policy

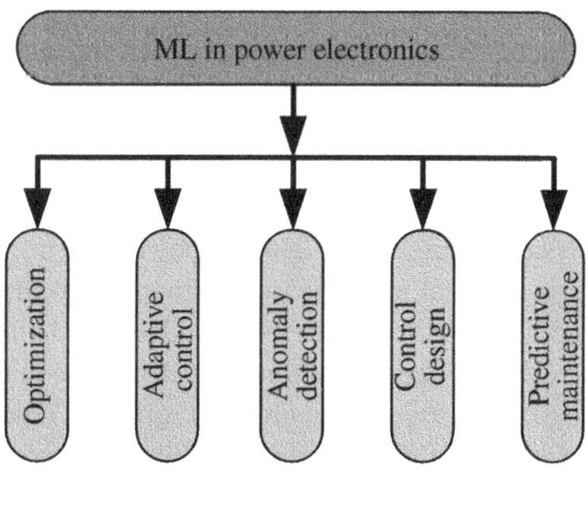

Figure 9.9 Machine learning application in power electronics.

gradient and Q-learning, are optimized for designing controllers that achieve these objectives.

- **Deep Learning:** Deep learning, a type of ML that utilizes multi-layered artificial neural networks (ANNs), processes and learns from large datasets to perform complex tasks. In power electronics, deep learning can be applied to signal processing for behavioral modeling, time series analysis for prediction, and image analysis for fault identification. Techniques such as convolutional neural networks (CNNs) and recurrent neural networks (RNNs) are commonly used in these applications.
- **Adaptive Control:** Adaptive control refers to a control mechanism that dynamically adjusts its parameters in response to changes in the environment or system. ML can facilitate adaptive control in power electronic systems, such as microgrids, electric cars, and renewable energy sources, by using data to learn and update control parameters. Neural networks and fuzzy logic are two methods commonly used in implementing adaptive control.
- **Optimization:** Optimization is the process of finding the best solution to a specific problem or goal. ML can aid in determining the optimal values for parameters such as duty cycle, modulation index, and switching frequency, enhancing the performance of power electronic devices like motors, inverters, and converters. Evolutionary algorithms, such as genetic algorithms, particle swarm optimization, and ant colony optimization, are commonly used methods in this optimization process.

9.5 Current Trends and Research in the Integration of ML with Power Electronics

9.5.1 Control Techniques in Power Electronics

To maximize power electronic converters' performance, stability, and efficiency, both linear and non-linear control techniques are required. They are used in several types of converters, such as AC–DC rectifiers, DC–AC inverters, DC–DC converters, and AC–AC converters.

9.5.1.1 Linear Control Techniques

Linear controllers are designed and utilized based on the linear model of the system being controlled. In power electronic systems, linear control methods are essential for reducing fluctuations in the output voltage due to disturbances or variations in load. Four widely used linear control strategies are deadbeat control, proportional–integral–derivative (PID) control, pole placement control, and proportional–integral (PI) control. These methods help maintain the desired system performance and stability. PI control is commonly employed in power electronic systems due to its simplicity and effectiveness. This approach integrates a proportional term (P) that reacts to the immediate error between the target and actual output voltages with an integral term (I) that sums up this error over time. Together, these components work to minimize steady-state errors, enhancing the overall performance of the system. While the integral term makes sure the system produces the expected result by settling any persistent disparities, the proportional term assures a rapid response by modifying the control input proportionate to the fault. Because of this combination, PI control is very beneficial in power electronic applications to maintain accurate and stable output. PI controllers perform effectively in conditions with modest disturbances and slow dynamics. In the continuous domain, the PI controller's transfer function can be written as

$$G(s) = K_p + (K_i \div s) \tag{9.1}$$

where the integral and proportional benefits are denoted, respectively, by K_i and K_p.

Three-phase, two-level voltage source converters (VSCs) are frequently used as the controlled plant in power electronic systems. The PI controller's control diagram, which regulates the current in the VSC, is depicted in Figure 9.10. Here, Z_f represents the filter impedance, which could be an LCL filter or a basic L filter, Z_g stands for the impedance of the weak grid; V_{abc} is the converter voltage; V_{dc} is the DC voltage; and C is the DC capacitance. The grid voltage and grid current of the point of common coupling are U_{uabc} and I_{uabc}, respectively.

The PI control method is enhanced by adding a derivative (D) term to form the PID control. The derivative component is proportional to the rate of change of the error and helps the controller better understand and respond to the system's dynamics. A PID controller adjusts the control signal based on the combined

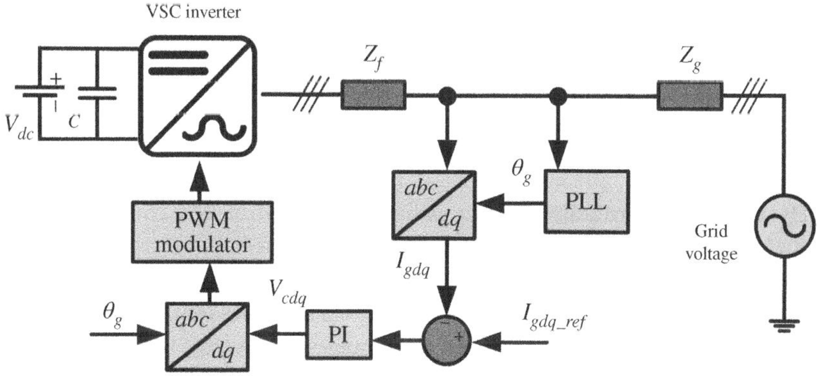

Figure 9.10 Control diagram of PI controller applied in a 3-phase VSC.

effect of the proportional, integral, and derivative terms. The derivative term specifically improves transient response by enabling the controller to respond more quickly to changes in the system. PID control is often employed in power electronic systems that require accurate voltage regulation and enhanced dynamic performance.

Pole placement is a feedback control technique that enables the dominant closed-loop poles of a system to be positioned at specific locations in the s-plane. This is accomplished by employing state feedback and selecting a suitable gain matrix, provided that all system state variables are measurable and controllable. The locations of these poles determine the roots of the system's characteristic equation, which in turn influences the overall behavior of the system. This approach enables designers to fine-tune the dynamic and steady-state performance by adjusting the damping ratio and natural frequency through the placement of the dominant closed-loop poles.

A discrete-time control method called deadbeat control aims to reduce the controlled current's tracking error down to zero in a short period of time—usually one or two sampling intervals. Based on the discrepancy between the desired and actual current values, it chooses the control action along with system parameters. This method simplifies the design of filtering elements by offering a set switching frequency and a fast transient response. Deadbeat control still is impacted by noise in measurements and changes in parameters. Robust control methods and precise system modeling are necessary for ensuring stability and performance.

9.5.1.2 Non-linear Control Techniques

In challenging situations such as when handling distorted or uneven grid voltage, linear control techniques commonly fail. Especially when taking disturbance uncertainty and changes in system parameters into account, nonlinear controllers are more resilient and flexible. In power electronic systems, nonlinear control methods are essential. By ensuring stability and optimizing performance through precise current regulation across different components. This section explores three widely used nonlinear control methods: Model Predictive Control (MPC), Hysteresis Control, and Artificial Intelligence (AI).

- **Hysteresis Control,** also known as bang-bang control, is a commonly used current control method in power electronics. This technique works by keeping the current within a specified hysteresis band. It does so by comparing the actual current to a reference current. If the current goes beyond the upper or lower limits of this band, the controller changes the state of the power device to return the current to within the set boundaries. This method is straightforward and offers a quick response. However, its variable switching frequency

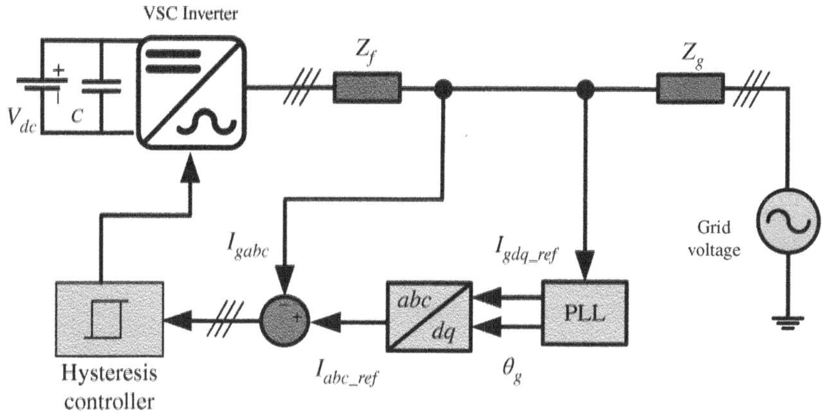

Figure 9.11 Hysteresis control scheme applied on 3-phase VSC.

can impact performance, potentially complicating filter design or introducing harmonic distortion. Diagrams like Figure 9.11 are often used to depict the hysteresis controller's control logic for each phase.

- **Model Predictive Control (MPC)** is an advanced control technique that predicts the future behavior of a power electronic system using its mathematical model. Based on these predictions, the controller optimizes control actions to minimize a cost function, which could involve tracking errors or reducing switching losses. While MPC offers excellent performance and can adapt to changes in system parameters, it can be more complex to implement than other control methods and typically requires substantial computational resources.
- **Artificial Intelligence (AI)** approaches that have become increasingly popular for controlling Power Electronic Systems are Fuzzy Logic Control and Neural Network Control. These techniques provide robust yet adaptable control, successfully tackling the complexity and unpredictability present in power electronics.

  Fuzzy Logic Control uses rule-based systems to adjust for uncertain inputs, whereas Neural Network Control uses learning algorithms to adapt to changing environments. Both methods enhance the system's functionality in a range of complex and dynamic conditions. Fuzzy Logic Control offers a systematic method for managing systems that are imprecise and uncertain, which are often difficult to model mathematically. This approach enables approximate reasoning by utilizing degrees of truth, which is particularly valuable for handling complex power electronic systems. By using membership functions and linguistic rules, Fuzzy Logic Control can effectively manage nonlinear and dynamic power electronic systems, transforming fuzzy sets into specific control actions.

Neural Network Control, inspired by the neural architecture of the human brain, utilizes ANNs to manage complex and nonlinear systems, particularly in power electronics where precise mathematical models may be hard to define. ANNs are advantageous in these situations because they can learn and adapt based on experience. By analyzing input–output data pairs, ANNs can effectively determine system behavior and generate appropriate outputs for specific inputs, offering dependable solutions for dynamic challenges in power electronics system control [8].

9.6 Machine Learning in Power Electronics Using MATLAB

9.6.1 Overview of MATLAB

MATLAB (Matrix Laboratory) is a popular high-level programming language and interactive environment for numerical computing, visualization, and programming, which was created by MathWorks. It offers engineers, scientists, and researchers in a variety of fields a flexible tool by linking a potent computational engine, excellent visualization, and a user-friendly scripting language.

Table 9.1 below shows some of the key features, capabilities, and applications of MATLAB.

9.6.2 MATLAB Toolboxes for ML and Power Electronics

MATLAB offers a variety of ML toolboxes that facilitate the creation, elaboration, and implementation of ML models [10]. Some essential blocks can be seen in Table 9.2. Among the essential MATLAB toolboxes for ML are the following:

- Statistics and Machine Learning Toolbox
- Deep Learning Toolbox
- Reinforcement Learning Toolbox
- Computer Vision Toolbox
- Text Analytics Toolbox
- Automated Driving Toolbox
- Sensor Fusion and Tracking Toolbox

These toolboxes simplify the process of creating and implementing ML applications by offering pre-configured functions, algorithms, and interfaces that smoothly coordinate with the computational environment of MATLAB.

Table 9.1 Key Features, Capabilities, and Applications of MATLAB.

Key Features and Capabilities	Description	Applications
Numerical computation	Offers a broad range of mathematical functions for numerical integration, statistics, Fourier analysis, filtering, and optimization in addition to linear algebra. Created to solve problems quickly using arrays and matrices	Solving complex numerical problems, data analysis, and financial modeling
Algorithm development and programming	Provides rich libraries and toolboxes appropriate to certain applications, in addition to an intuitive and easy-to-learn scripting language that supports procedural, object-oriented, and functional programming paradigms	Addressing complex numerical problems, performing data analysis, and conducting financial modeling
Data analysis and visualization	Excels in analyzing and visualizing data, offering resources for importing, displaying, and analyzing big data sets. Facilitates the creation of graphs, animations, and charts in two and three dimensions for data representation	Visual data exploration, large data analytics, and statistical analysis
Simulink	Provides an environment for multi-domain simulation and model-based design of dynamic systems using block diagrams. Widely used in control system design, signal processing, and communication systems [9]	Enables the model-based design of dynamic systems utilizing block diagrams and multi-domain simulation in an environment. Widely employed in the design of communication, signal processing, and control systems
Toolboxes	Provides specific toolboxes in domains like as control systems, power electronics, signal processing, computer vision, machine learning, and deep learning. Each toolbox has capabilities and applications customized for tasks	Machine learning, deep learning, computer vision, signal processing, control systems
Application development and deployment	MATLAB provides tools such as App Designer to facilitate the development of software components and independent applications. It facilitates GUI development, software deployment, and web application development by enabling the design of GUIs and their deployment	Software deployment, GUI development, and web applications
Integration and connectivity	Integrates well with a variety of environments and programming languages, including C, C++, Java, .NET, and Python. Allows for connecting to databases, web services, and hardware platforms	Data integration, cross-platform development, hardware interfacing

Table 9.2 Description and Symbol of Various Blocks.

<table>
<tr><th>Block</th><th>Description</th><th>Symbol</th></tr>
<tr><td>Predict</td><td>A trained deep learning neural network is used to predict responses (since R2020b)</td><td>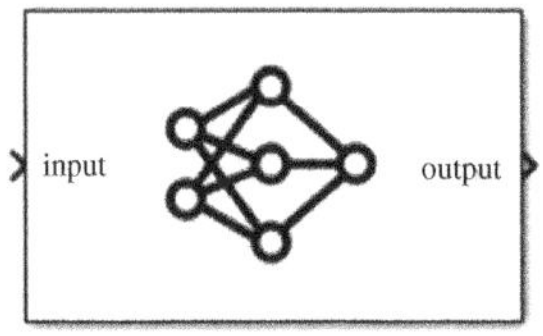</td></tr>
<tr><td>Image classifier</td><td>Application of a trained deep learning neural network, classify data (since R2020b)</td><td>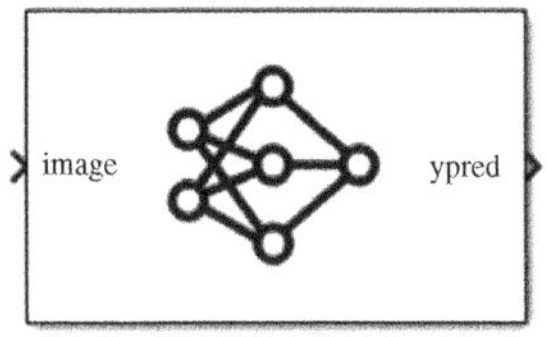</td></tr>
<tr><td>Stateful classify</td><td>Implementing a trained recurrent neural network for deep learning, classify data (since R2021a)</td><td>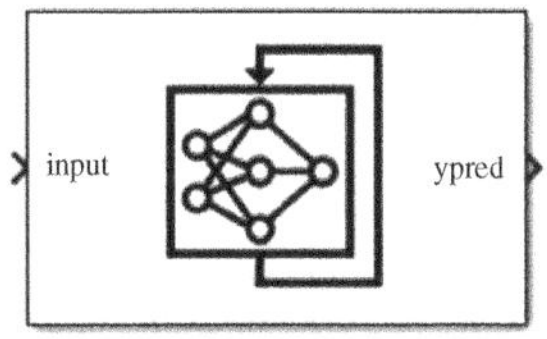</td></tr>
<tr><td>Stateful predict</td><td>Utilizing a trained recurrent neural network, forecast responses (since R2021a)</td><td>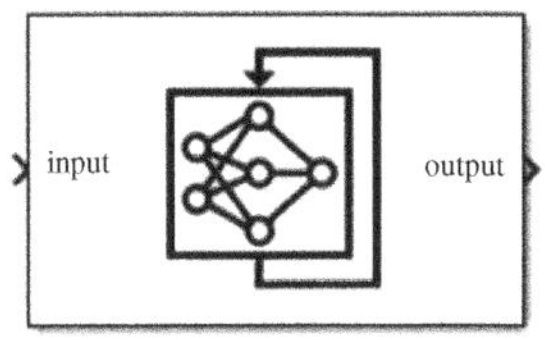</td></tr>
<tr><td>TensorFlow model predict</td><td>Using a pretrained Python TensorFlow model, predict responses (since R2024a)</td><td>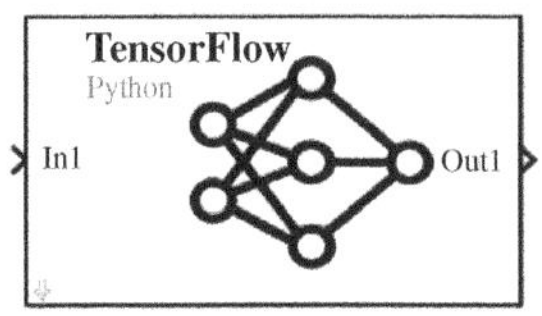</td></tr>
<tr><td>PyTorch model predict</td><td>Predict responses by using a Python PyTorch model that has been pretrained (since R2024a)</td><td>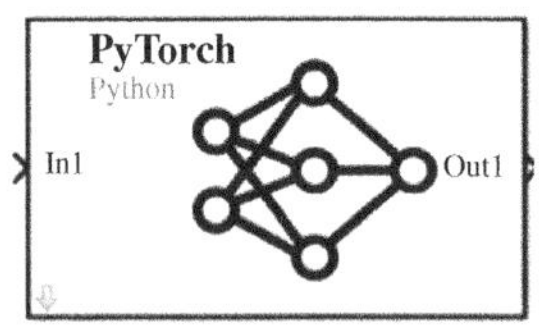</td></tr>
<tr><td>ONNX model predict</td><td>Use the pretrained Python ONNX model (since R2024a) to predict information.</td><td>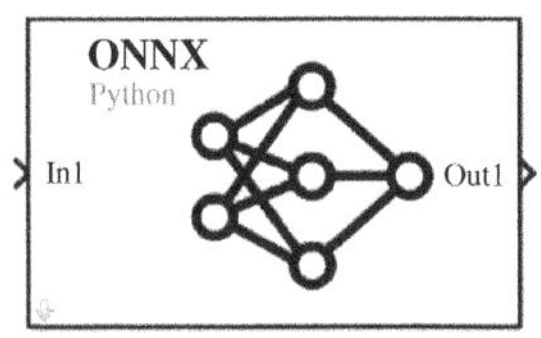</td></tr>
</table>

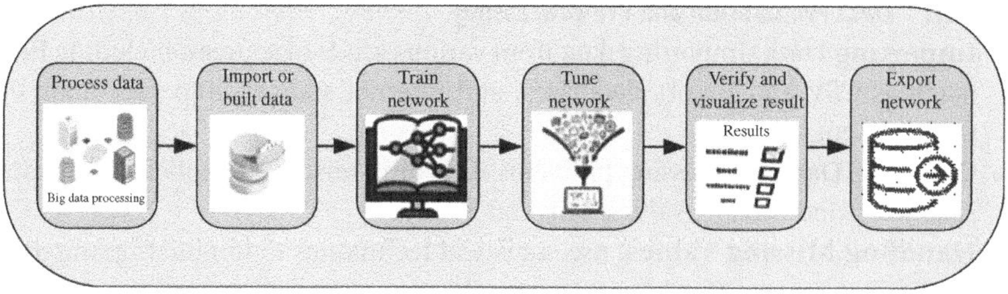

Figure 9.12 Deep learning.

This chapter mainly focuses on familiarizing one of the ML toolboxes, i.e. Deep Learning Toolbox. The working of a deep learning model can be understood from Figure 9.12.

Designing, building, and simulating deep neural networks is made easier with the help of Simulink® blocks, apps, and Deep Learning Toolbox™. The toolbox offers a framework for building and utilizing a variety of networks, including transformers and CNNs. Verify network attributes, visualize, and comprehend predictions, and compress networks via quantization, projection, or pruning.

The Deep Network Designer software allows you to import pre-trained models, export networks to Simulink, and construct, edit, and evaluate networks interactively. Working with several deep learning frameworks is made possible by the toolbox. Models for inference, transfer learning, simulation, and deployment can be imported using PyTorch®, TensorFlow, and ONNX. Models can also be exported to ONNX and TensorFlow.

Tools are available for every stage of the deep learning workflow in Deep Learning Toolbox™.

- Utilizing interactive applications and command-line functions, prepare data for deep network training.
- Use MATLAB® or third-party systems like TensorFlow™ 2, TensorFlow-Keras, PyTorch®, and ONNXTM to import pretrained networks.
- Design networks interactively using the Deep Network Designer app or with command-line tools.
- Select training settings and use custom training loops or built-in training cycles to train networks.
- Increase network performance by adjusting hyperparameters or by using the Experiment Manager app to run several trials.
- During and after training, observe and validate network behavior.
- Networks can be exported to other platforms like ONNX and TensorFlow 2.

9.6.2.1 Data Preparation and Pre-processing

- **Importing Data:** Importing data from various kinds of sources, including files (such as CSV and Excel), databases, and internet sources, into your analysis environment.
- **Cleaning Data:** Addressing problems including mistakes, duplication, inconsistencies, and missing values. This can include:
- **Handling Missing Values**: use statistical techniques to impute missing data or eliminate entries that have missing values.
- **Removing Duplicates**: identifying and removing redundant data to prevent duplication.
- **Correcting Errors**: correcting errors, deviations, and abnormal events in data entry.
- **Normalization/Standardization**: To enhance model performance, numerical quantities might be shifted to a common scale or distribution.

9.6.2.2 Feature Selection and Extraction

Selecting the most pertinent attributes (variables) for your model will enhance its functionality and simplify its design. Methods consist of: Filter Methods, Wrapper Methods, and Embedded Methods. Feature extraction involves improving the capabilities of current data to generate new features that better reveal underlying

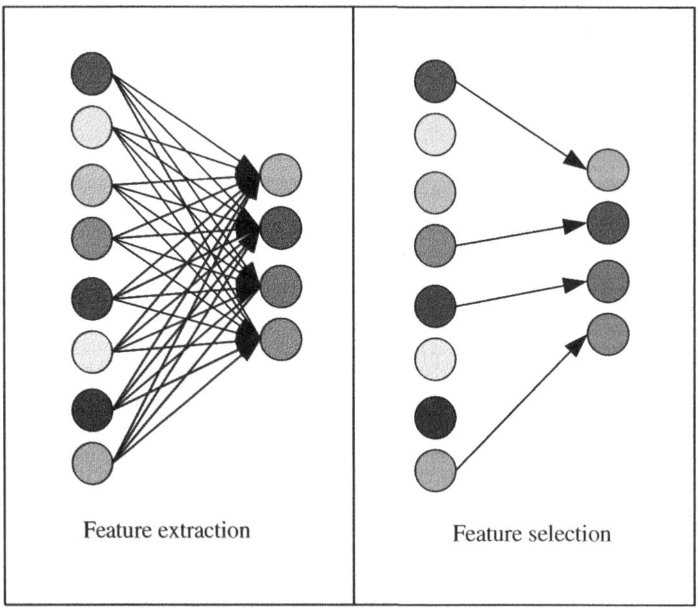

Figure 9.13 Neural network data extraction and selection.

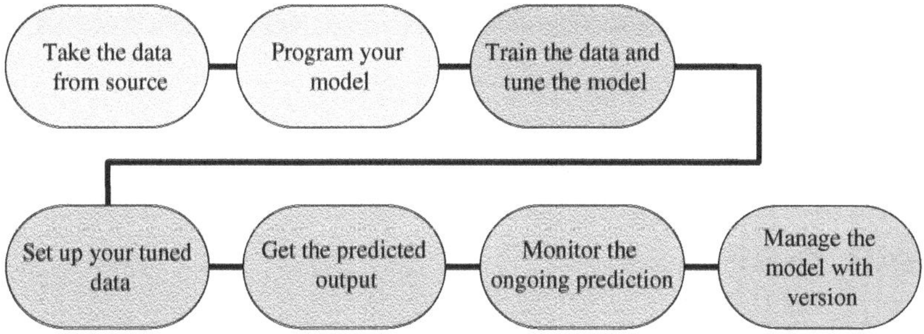

Figure 9.14 Steps of machine learning.

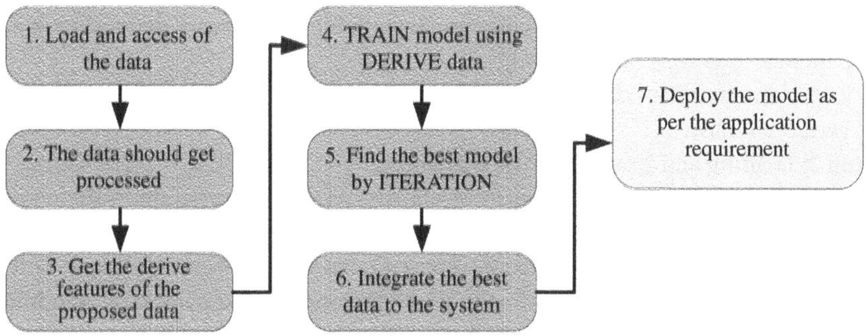

Figure 9.15 Stepwise machine learning approach.

trends. This can involve Dimensionality Reduction, Transformations, and Decomposition. The process can be understood from Figure 9.13.

9.6.2.3 Developing ML Models in MATLAB

Figure 9.14 provides a high-level summary of the phases in an ML workflow. The first two boxes in the framework indicate the places where AI Platform provides managed services and APIs:

The systematic ML workflow outlined below in Figure 9.15 can assist you in addressing ML challenges, and you can accomplish the entire process using MATLAB.

9.7 Case Study

9.7.1 Case Study 1: Predicting Efficiency of a DC–DC Converter

To predict a DC–DC converter's efficiency which is based on input voltage, output voltage, and load current, we'll implement a simple linear regression model. Data preparation, model training, validation, and testing are all steps in the workflow.

```
Command Window
>> % Create the dataset
data = table([12; 12; 12; 12; 12; 12; 24; 24; 24; 24; 24; 24; 48; 48; 48; 48; 48; 48], ...
             [5; 5; 5; 3.3; 3.3; 3.3; 12; 12; 12; 5; 5; 5; 24; 24; 24; 12; 12; 12], ...
             [0.5; 1; 1.5; 0.5; 1; 1.5; 0.5; 1; 1.5; 0.5; 1; 1.5; 0.5; 1; 1.5; 0.5; 1; 1.5], ...
             [0.87; 0.85; 0.83; 0.90; 0.88; 0.86; 0.89; 0.87; 0.85; 0.91; 0.89; 0.87; 0.92; 0.90; 0.88; 0.94; 0.92; 0.90], ...
             'VariableNames', {'InputVoltage', 'OutputVoltage', 'LoadCurrent', 'Efficiency'});

% Save the dataset to a CSV file
writetable(data, 'converter_data.csv');
```

Figure 9.16 Create the data set.

Step 1: Create and Save the Dataset in MATLAB:

We develop a program in the command window that is shown below in Figure 9.16 and run it to create and save the dataset.

The above program automatically creates a table at the workspace after it is completed, as seen in Figure 9.17.

The demonstrated data is now found in a CSV file identified as converter_data.csv.

Step 2: Loading and Examining the Dataset

To load and inspect the dataset in MATLAB: We write a program in the command window that is shown below in Figure 9.18.

The above program automatically creates a table with column title Input Voltage, Output Voltage, Load Current, and Efficiency at the command window after it is finished (run), as seen in below Figure 9.19.

Workspace

Name	Value	Size	Class
data	*18x4 table*	18x4	table

Figure 9.17 Workspace created for dataset.

```
Command Window
writetable(data, 'converter_data.csv');
>> % Load the data
data = readtable('converter_data.csv');

% Display the first few rows of the data
head(data);
```

Figure 9.18 Load data set in MATLAB.

```
Command Window
head(data);
    InputVoltage    OutputVoltage    LoadCurrent    Efficiency
    ____________    _____________    ___________    __________

         12                5             0.5           0.87
         12                5               1           0.85
         12                5             1.5           0.83
         12              3.3             0.5            0.9
         12              3.3               1           0.88
         12              3.3             1.5           0.86
         24               12             0.5           0.89
         24               12               1           0.87

>> [0.87; 0.85; 0.83; 0.90; 0.88; 0.86; 0.89; 0.87; 0.85; 0.91; 0.89; 0.87; 0.92; 0.90; 0.88; 0.94; 0.92; 0.90], ...
```

Figure 9.19 Columns of data created.

```
Command Window
>> [0.87; 0.85; 0.83; 0.90; 0.88; 0.86; 0.89; 0.87; 0.85; 0.91; 0.89; 0.87; 0.92; 0.90;
% Extract features and target variable
features = data(:, {'InputVoltage', 'OutputVoltage', 'LoadCurrent'});
efficiency = data.Efficiency;
```

Figure 9.20 Extraction feature.

Step 3: Data Preparation

To extract relevant features and the target variable, we develop a program in the command window that is shown below in Figure 9.20.

After the completion of the above program, we successfully extracted relevant features and the target variable in the Command window and workspace, respectively, which is shown below in Figure 9.21.

Step 4: Model Training

A key phase in the ML process is model training, which involves teaching an algorithm to make predictions or perform a specific task. Here we write a program to Split the data into training and testing sets as shown in Figure 9.22.

After the completion of the above program, we successfully Split the data into training and testing sets in the workspace, which is shown below in Figure 9.23.

Step 5: Train a linear regression model:

The objectives of training a linear regression model are to gain insights, understand correlations, develop accurate predictions, and offer an easily understood framework for data analysis. It is a fundamental technique in ML and statistics that provides the foundation of more complex models. The code is displayed in Figure 9.24.

After linear regression model training, the command window shows the following outcomes (Figure 9.25).

```
Command Window
ans =

    0.8700
    0.8500
    0.8300
    0.9000
    0.8800
    0.8600
    0.8900
    0.8700
    0.8500
    0.9100
    0.8900
    0.8700
    0.9200
    0.9000
    0.8800
    0.9400
    0.9200
    0.9000
```

Workspace

Name	Value	Size	Class
ans	*18x1 double*	18x1	double
data	*18x4 table*	18x4	table
efficiency	*18x1 double*	18x1	double
features	*18x3 table*	18x3	table

Figure 9.21 Extracting feature from workspace.

```
Command Window
>> % Split data into training and testing sets
cv = cvpartition(size(data, 1), 'HoldOut', 0.2);
trainingData = features(training(cv), :);
trainingLabels = efficiency(training(cv));
testData = features(test(cv), :);
testLabels = efficiency(test(cv));
```

Figure 9.22 Splitting the data.

Step 6: Model Testing

Testing the model is an important step to ensure a ML model is reliable, robust, as well as able to provide correct predictions on new data; in this step, we test and validate the model using the test dataset as provided in Figure 9.26:

Final Result: After completion of the Model Testing Process the below outcomes in Figure 9.27 will appear; here X-axis shows Actual Efficiency, the Y-axis shows Predicted Efficiency and Root Mean Squared Error (RMSE) = 0.02.

Workspace

Name	Value	Size	Class
data	*18x4 table*	18x4	table
features	*18x3 table*	18x3	table
testData	*3x3 table*	3x3	table
trainingData	*15x3 table*	15x3	table

Figure 9.23 Training and test data in workspace.

Command Window

```
>>
>>
>> % Train a linear regression model
model = fitlm(trainingData, trainingLabels);

% Display the model summary
disp(model);
```

Figure 9.24 Code for training the model.

Command Window

```
Linear regression model:
    y ~ 1 + InputVoltage + OutputVoltage + LoadCurrent

Estimated Coefficients:
                      Estimate        SE         tStat       pValue
                     __________    __________    ______    __________

    (Intercept)        0.88321     0.0084274     104.8     7.4609e-18
    InputVoltage     0.0021413    0.00025327    8.4546     3.8475e-06
    OutputVoltage   -0.0021465    0.00059377    -3.615      0.0040619
    LoadCurrent      -0.037469     0.0058928   -6.3584     5.3816e-05


Number of observations: 15, Error degrees of freedom: 11
Root Mean Squared Error: 0.00899
R-squared: 0.933,  Adjusted R-Squared: 0.915
F-statistic vs. constant model: 51.4, p-value = 9.27e-07
>>
```

Figure 9.25 Regression model results.

```
Command Window
>> % Predict efficiency on the test set
predictedEfficiency = predict(model, testData);

% Calculate performance metrics
rmse = sqrt(mean((predictedEfficiency - testLabels).^2));
fprintf('Root Mean Squared Error: %.2f\n', rmse);

% Plot actual vs predicted efficiency
figure;
plot(testLabels, predictedEfficiency, 'o');
xlabel('Actual Efficiency');
ylabel('Predicted Efficiency');
title('Actual vs Predicted Efficiency');
```

Figure 9.26 Model testing code.

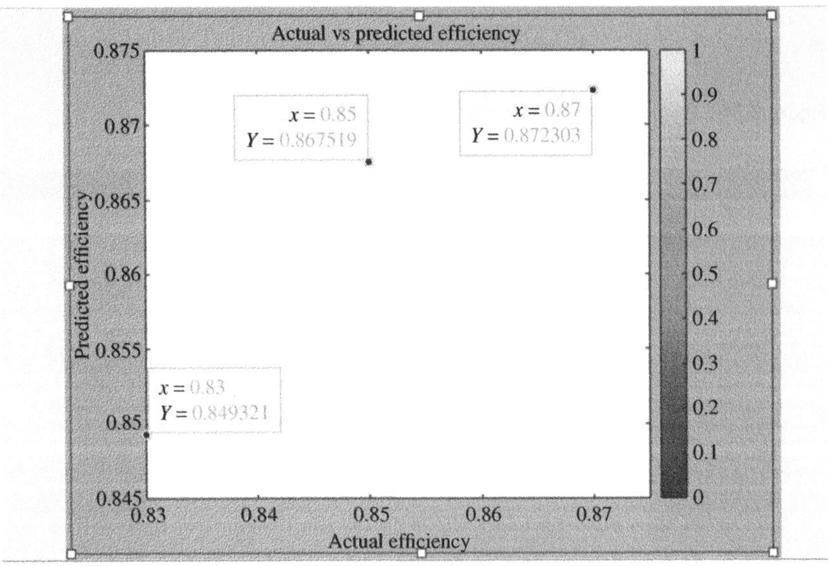

Figure 9.27 Plot of actual versus predicted efficiency.

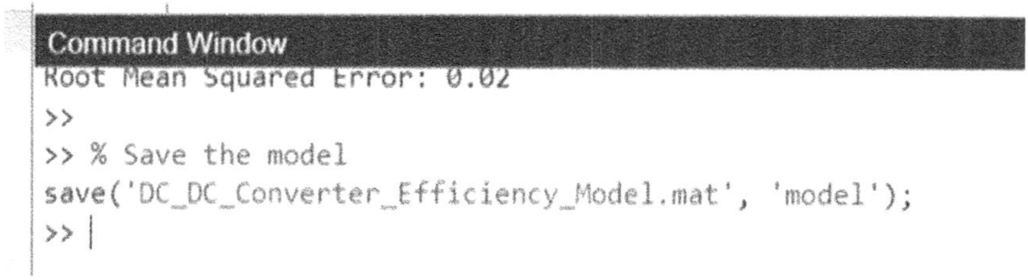

Figure 9.28 Model saving.

Step 7: Model Deployment

If we want to Save the trained model for future use then perform the below command in the command window (Figure 9.28).

Summary of Results:

- **Model Summary**: Provides statistical data and the coefficients for the linear regression model.
- **Root Mean Squared Error (RMSE)**: A performance measure illustrating the typical discrepancy between actual and expected efficiency levels.
- **Scatter Plot**: A graphic comparison of the expected and actual efficiency levels.

These approaches will enable us to train a MATLAB machine-learning model to anticipate and evaluate the performance of a DC–DC converter.

9.7.2 Case Study 2: Predicting the Temperature of a Power Semiconductor Device Using a Neural Network

To predict a Temperature of a Power Semiconductor Device which is based on input voltage, output voltage, and load current, we'll implement by using Neural Network. Generate data, model training, define prediction function, and testing are all steps in the workflow.

Step 1: Generate and Save the Dataset in MATLAB:

We develop a program in the command window that is shown below in Figure 9.29 and run it to create and save the dataset.

The above program automatically creates a table at the workspace after it is completed, as seen in Figure 9.30.

The demonstrated data is now found in a CSV file identified as converter_data.csv.

```
Command Window
fx >> % Filename: generate_data.m

% Number of samples
numSamples = 1000;

% Randomly generate data
rng(0); % For reproducibility
input_voltage = 10 + 5*rand(numSamples, 1); % Range: 10 to 15 V
input_current = 2 + 1*rand(numSamples, 1); % Range: 2 to 3 A
ambient_temperature = 20 + 10*rand(numSamples, 1); % Range: 20 to 30 °C
power_dissipation = 5 + 3*rand(numSamples, 1); % Range: 5 to 8 W

% Generate device temperature with some noise
device_temperature = 25 + 0.5*input_voltage + 1.5*input_current + ...
                     0.3*ambient_temperature + 0.2*power_dissipation + ...
                     2*randn(numSamples, 1); % Adding random noise

% Create a table
data = table(input_voltage, input_current, ambient_temperature, ...
             power_dissipation, device_temperature);

% Save the dataset as CSV
writetable(data, 'semiconductor_temperature_data.csv');
```

Figure 9.29 Data generation.

Workspace

Name ▲	Value
ambient_tem...	1000x1 double
data	1000x5 table
device_tempe...	1000x1 double
input_current	1000x1 double
input_voltage	1000x1 double
numSamples	1000
power_dissip...	1000x1 double

Figure 9.30 Table creation at the workspace.

Step 2: Load, Pre-process Data, and Train Model

To load, pre-process data, and train model in MATLAB: We write a program in the command window that is shown below in Figure 9.31.

The above program automatically trains the neural network a table with column title Input Voltage, Output Voltage, Load Current, and Efficiency at the command window after it is finished (run), as seen in below Figure 9.32.

```
Command Window
% Load the dataset
data = readtable('semiconductor_temperature_data.csv');
% Extract features and labels
X = data{:, {'input_voltage', 'input_current', 'ambient_temperature', 'power_dissipation'}};
y = data.device_temperature;
% Split data into training and testing sets
cv = cvpartition(height(data), 'HoldOut', 0.3);
XTrain = X(training(cv), :);
yTrain = y(training(cv));
XTest = X(test(cv), :);
yTest = y(test(cv));
% Save training and testing data to .mat files for reuse
save('temperature_train_data.mat', 'XTrain', 'yTrain');
save('temperature_test_data.mat', 'XTest', 'yTest');
% Define the neural network architecture
layers = [
    featureInputLayer(4)  % Number of features
    fullyConnectedLayer(10)
    reluLayer
    fullyConnectedLayer(1) % Output layer for regression
    regressionLayer];
% Specify training options
options = trainingOptions('adam', ...
    'MaxEpochs', 100, ...
    'MiniBatchSize', 32, ...
    'Plots', 'training-progress', ...
    'Verbose', false);
% Train the neural network
net = trainNetwork(XTrain, yTrain, layers, options);
% Save the trained model
save('temperature_model.mat', 'net');
% Predict using the test set
predictions = predict(net, XTest);
% Calculate and display the performance metrics
rmse = sqrt(mean((predictions - yTest).^2));
fprintf('Root Mean Squared Error (RMSE): %.2f\n', rmse);
```

Figure 9.31 MATLAB code for pre-processing as well as training.

Step 3: Define the Prediction Function

To extract relevant features and the target variable, we develop a program in the editor window that is shown below in Figure 9.33.

After the completion of the above program, we successfully extracted relevant features and the target variable in the Command window and workspace, respectively, which is shown below in Figures 9.34 and 9.35.

Step 4: Test the Prediction Function

A key phase in the machine learning (ML) process is model training, which involves teaching an algorithm to make predictions or perform a specific task. Here we write a program to Split the data into training and testing sets. After the completion of the above program, we successfully Split the data into training and testing sets in the workspace, which is shown below in Figure 9.36.

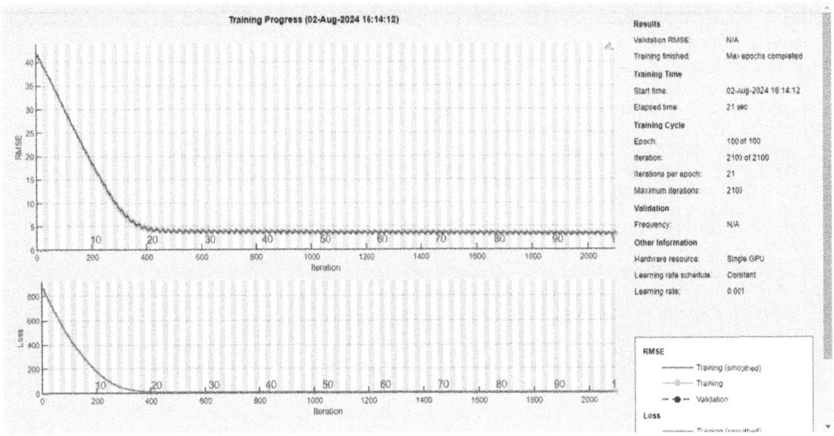

Figure 9.32 Obtained training curve.

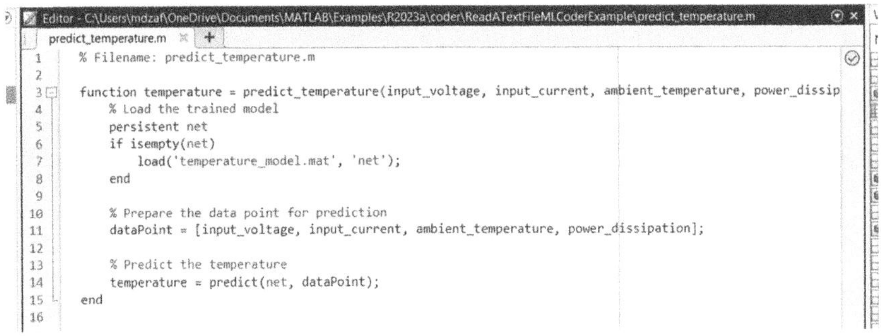


```
% Filename: predict_temperature.m

function temperature = predict_temperature(input_voltage, input_current, ambient_temperature, power_dissip
    % Load the trained model
    persistent net
    if isempty(net)
        load('temperature_model.mat', 'net');
    end

    % Prepare the data point for prediction
    dataPoint = [input_voltage, input_current, ambient_temperature, power_dissipation];

    % Predict the temperature
    temperature = predict(net, dataPoint);
end
```

Figure 9.33 Code for feature extraction.

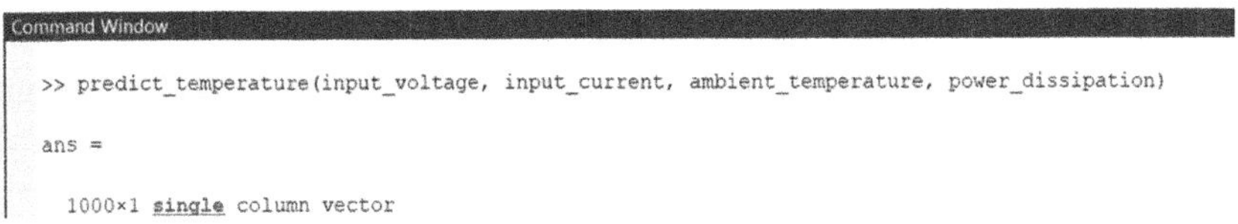


```
>> predict_temperature(input_voltage, input_current, ambient_temperature, power_dissipation)

ans =

  1000×1 single column vector
```

Figure 9.34 Command window display.

Case 1 provides the predicted temperature for the given input parameters as shown in Figure 9.37.

Case 2 results for different set of parameters are shown in Figure 9.38.

Case 3 has different input parameters than the previous two cases, and the prediction outcome is displayed in Figure 9.39.

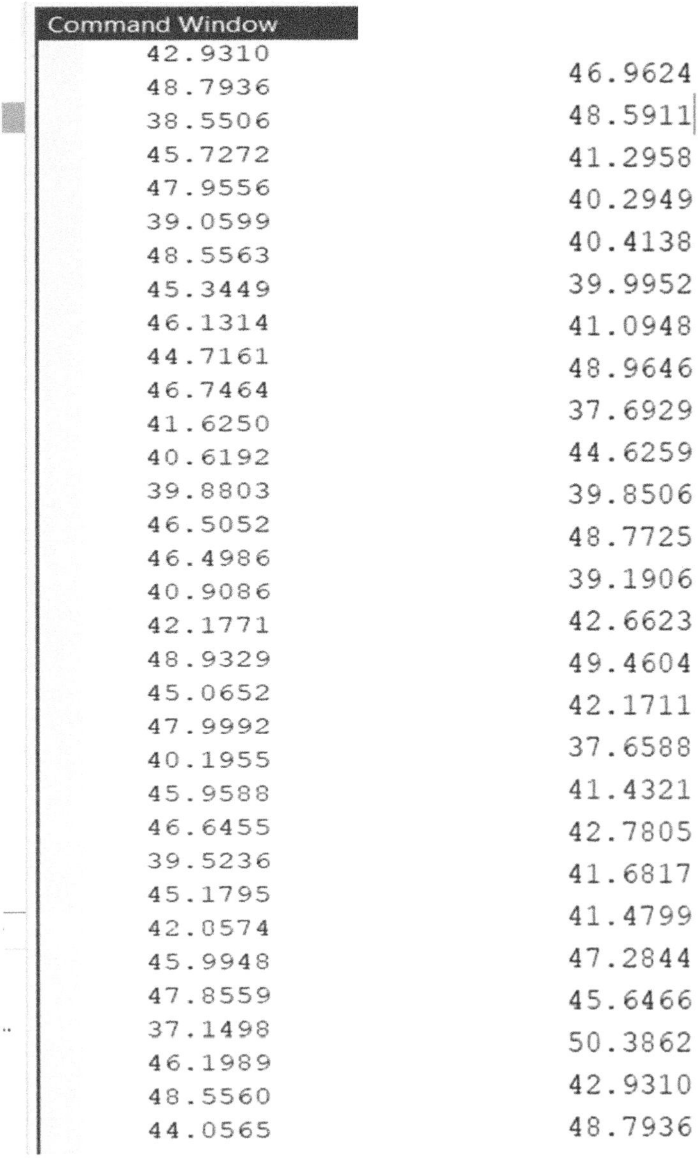

Figure 9.35 Results after running the code.

```
fx >> % Filename: test_prediction.m

% Example usage
input_voltage = 12;
input_current = 5;
ambient_temperature = 25;
power_dissipation = 10;

temperature = predict_temperature(input_voltage, input_current, ambient_temperature, power_dissipation)
fprintf('Predicted Device Temperature: %.2f°C\n', temperature);
```

Figure 9.36 Testing the ML model.

```
>> % Filename: test_prediction.m

% Example usage
input_voltage = 12;
input_current = 5;
ambient_temperature = 25;
power_dissipation = 10;

temperature = predict_temperature(input_voltage, input_current, ambient_temperature, power_dissipation)
fprintf('Predicted Device Temperature: %.2f°C\n', temperature);
Predicted Device Temperature: 47.62°C
>>
```

Figure 9.37 Prediction of temperature for case 1.

```
>> % Filename: test_prediction.m

% Example usage
input_voltage = 20;
input_current = 7;
ambient_temperature = 35;
power_dissipation = 15;

temperature = predict_temperature(input_voltage, input_current, ambient_temperature, power_dissipation);
fprintf('Predicted Device Temperature: %.2f°C\n', temperature);
Predicted Device Temperature: 67.85°C
fx >>
```

Figure 9.38 Prediction of temperature for case 2.

```
>> % Example usage
input_voltage = 25;
input_current = 6;
ambient_temperature = 22;
power_dissipation = 9;

temperature = predict_temperature(input_voltage, input_current, ambient_temperature, power_dissipation);
fprintf('Predicted Device Temperature: %.2f°C\n', temperature);
Predicted Device Temperature: 47.36°C
fx >>
```

Figure 9.39 Prediction of temperature for case 3.

9.8 Conclusion

The application of ML in power electronics, as demonstrated through the case studies on predicting DC–DC converter efficiency using regression and forecasting semiconductor device temperature with neural networks, underscores the transformative potential of these techniques for enhancing converter performance and reliability. The reliability and lifespan of power electronics systems depend on these solutions, which simplify predictive maintenance, optimize power converter efficiency, and enhance thermal control by utilizing MATLAB's comprehensive ML toolbox.

In the future, ML is going to have a bigger part in power electronics. Real-time optimization, adaptive control, and fault detection may be made possible by sophisticated ML models, making converters more resilient to variable loads and environmental factors. Furthermore, using cutting-edge methods like reinforcement learning may enable adaptive systems to automatically adjust control settings for optimal performance. Combining ML (ML) with physics-informed models and digital twin frameworks will create new opportunities for dynamic system modeling and predictive diagnostics as computing power increases, bringing the industry one step closer to autonomous, self-optimizing power electronics.

Exercises

9.1 What is the importance of power electronics in improving power quality? Mention two power quality issues that power electronics can help mitigate.

9.2 In what way does power electronics impact the development of smart grids? Provide two examples of functions it enables in smart grids.

9.3 Discuss two major reasons why machine learning is beneficial in complex problem-solving.

9.4 Describe an example of machine learning in everyday life, such as a spam filter. How does it "learn" to differentiate between spam and regular emails?

9.5 What is the difference between supervised and unsupervised learning? Provide examples where each type would be used effectively.

9.6 What is a labeled training set, and why is it essential in supervised learning? Provide an example.

9.7 What is clustering, and which algorithms are typically used for clustering in unsupervised learning?

9.8 Provide an example scenario where reinforcement learning would be more applicable than supervised or unsupervised learning.

9.9 Discuss how machine learning can improve the efficiency and reliability of power converters in renewable energy systems.

9.10 What role does deep learning play in power electronics? Provide examples of tasks where deep learning techniques, such as CNNs or RNNs, are effective.

9.11 Explain the use of reinforcement learning in controller design for power electronics. Why is it beneficial for these controllers to adapt to changing conditions?

9.12 List and briefly describe four linear control strategies used in power electronics.

9.13 Describe the key features of deadbeat control and its application in power electronics.

9.14 How does machine learning enhance predictive maintenance in power electronic components?

9.15 What are the key features of MATLAB that make it suitable for machine learning applications?

9.16 What steps are involved in the data preparation process when developing machine learning models in MATLAB?

9.17 Discuss how data quality can affect the outcomes of machine learning models in predicting efficiency.

9.18 In the context of predicting the efficiency of a DC–DC converter, what factors must be considered during data preparation?

9.19 What techniques can be used to validate the performance of the linear regression model in the case study?

9.20 Describe how you would collect and preprocess data for training the neural network. What steps would you take to ensure data quality?

9.21 How would you implement the trained neural network for real-time temperature prediction in a power semiconductor device? What considerations must be made for deployment?

9.22 Compare and contrast the neural network approach with traditional modeling techniques for predicting the temperature of power semiconductor devices. What are the advantages and disadvantages of each?

References

1 P. S. Bimbhra, "Power electronics by PS BIMBHRA (PDFDrive.com).pdf."

2 M. H. Rashid, M. F. Rahman, D. Patterson, A. Cheok, and R. Betz, "Motor drives," in *Power Electronics Handbook*, Fifth Edition. Oxford: Butterworth-Heinemann, 2023. 10.1016/B978-0-323-99216-9.00042-1.

3 Y. LeCun, Y. Bengio, and G. Hinton, "Deep learning," *Nature*, vol. 521, no. 7553, pp. 436–444, May 2015. 10.1038/nature14539.

4 C. M. Bishop, *Pattern Recognition and Machine Learning*. New York: Springer, 2006. 10.1007/978-0-387-45528-0.

5 L. Morán and J. Dixon, "Active filters," in *Power Electronics Handbook*. Oxford: Elsevier, 2011, pp. 1193–1228. 10.1016/B978-0-12-382036-5.00041-0.

6 P. Domingos, "A few useful things to know about machine learning," *Commun. ACM*, vol. 55, no. 10, pp. 78–87, Oct. 2012. 10.1145/2347736.2347755.

7 R. S. Sutton and A. G. Barto, "Reinforcement learning: An introduction," *IEEE Trans. Neural Netw.*, vol. 9, no. 5, pp. 1054–1054, Sep. 1998. 10.1109/TNN.1998.712192.

8 A. Krizhevsky, I. Sutskever, and G. E. Hinton, "ImageNet classification with deep convolutional neural networks," *Commun. ACM*, vol. 60, no. 6, pp. 84–90, May 2017. 10.1145/3065386.

9 P. Yadav and L. M. Saini, "Powerline FSK communication using signal wire technique," in *2017 Conference on Emerging Devices and Smart Systems (ICEDSS)*, IEEE, Mar. 2017, pp. 210–213. 10.1109/ICEDSS.2017.8073682.

10 G. James, D. Witten, T. Hastie, and R. Tibshirani, *An Introduction to Statistical Learning*, Springer Texts in Statistics, vol. 103. New York, NY: Springer, 2013. 10.1007/978-1-4614-7138-7.

10

Implementation of Machine Learning for Power Electronics Application Using PYTHON

Mohammad Amir, Izhar Ahmad Saifi, and Ahteshamul Haque

Advance Power Electronics and Research Laboratory, Jamia Millia Islamia (A Central University), New Delhi, India

10.1 Introduction

The PYTHON language has numerous significant advantages for engineers working in power electronics (PE) industries and other R&D centers, where sophisticated algorithms with complex datasets are frequently utilized [1]. These allow them to design machine learning (ML) models and integrate them into real-world systems more efficiently [2]. This chapter discusses the ML classification, such as supervised learning, unsupervised learning, reinforcement learning, and neural networks (NNs) as shown in Figure 10.1. Further, useful ML algorithms for PE applications are studied. Each of these strategies targets a distinct difficulty or opportunity in PE applications, such as control optimization, fault diagnostics, predictive maintenance, and increased efficiency.

10.1.1 Supervised Learning

It is a type of ML in which models are trained with characterized data to predict an output variable based on input attributes as depicted in Figure 10.2. This supervised approach is commonly used in PE for applications that need classification, regression, or faults detection.

- **Classification of Fault Detection**: The ability to identify issues early in systems like power converters/inverters is crucial for avoiding instability and maintaining operational safety. Classification methods, such as Support Vector Machines (SVMs) and Decision Trees, are widely employed to detect various types of faults [3–5]. For example, SVMs may identify operating states based on voltage, current, and temperature data, allowing potential errors to be identified before they cause grid system breakdowns.

Artificial Intelligence for Power Electronics, First Edition.
Edited by Ahteshamul Haque, Saad Mekhilef, and Azra Malik.

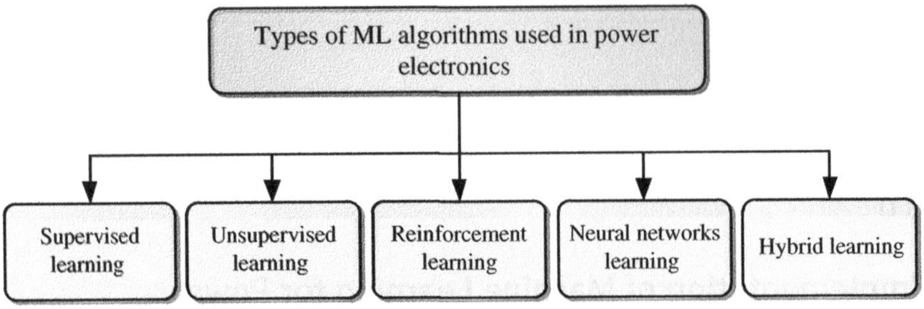

Figure 10.1 Types of ML algorithms deployed in power electronics systems.

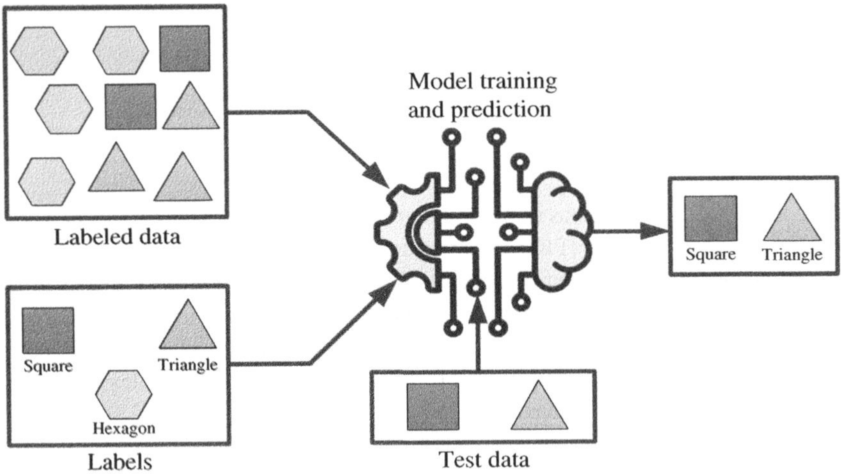

Figure 10.2 Supervised learning approach.

- **Regression for Predictive Maintenance**: Predictive maintenance is necessary for PE systems to maintain reliability and performance over time. Regression methods, such as Linear Regression or Random Forest Regression, can be used to forecast the remaining usable life of components based on past performance data [6, 7]. These regression-based models examine characteristics like temperature changes, voltage/current levels, and switching patterns to give predictive maintenance insights, decreasing unexpected outages, and increasing component longevity.

10.1.2 Unsupervised Learning

Unlike supervised learning, it does not require labeled data. Instead, it analyses patterns and structures within unlabeled data, making it well-suited for

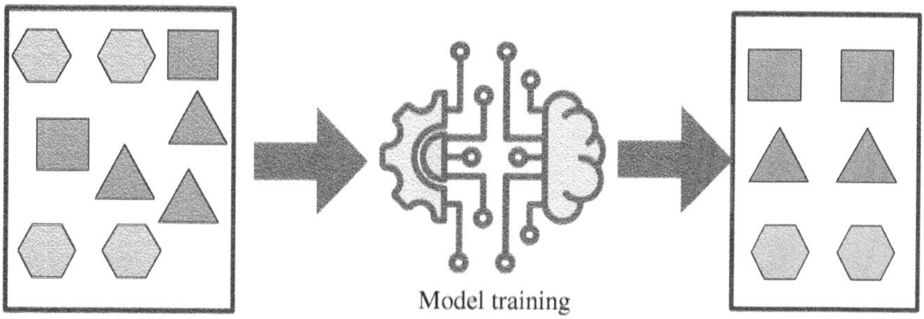

Figure 10.3 Unsupervised learning approach.

applications like clustering and anomaly detection in PE. Figure 10.3 shows the unsupervised learning approach.

- **Clustering for Fault Detection**: Usually, clustering methods, like *K*-Means and Hierarchical Clustering, can detect irregularities in system performance. These algorithms can discover deviations that may signal problems or inefficiency by clustering data into typical operational states [8, 9]. In a power converter, for instance, clustering can be utilized to identify anomalous temperature or voltage levels that indicate possible overheating or energy losses.

- **Principal Component Analysis for Data Dimension Reduction**: PE systems create a lot of operating data, which can be difficult to analyze. Principal component analysis (PCA) is a dimensionality reduction approach that facilitates data analysis by breaking down complicated datasets into a smaller number of primary components that preserve the majority of the original data's variability [10]. This streamlines data processing and allows for faster, more efficient decision-making in real-time PEs applications.

10.1.3 Reinforcement Learning (RL)

The Reinforcement Learning (RL) is a ML paradigm where models learn optimal actions through interaction with an environment. In PE, RL can be used for applications requiring real-time control and adaptive decision-making, as it is capable of learning policies that optimize system behavior over time.

- **Adaptive Control of Power Converters**: Traditional power converter control methods are based on set parameters that may not adjust to changing situations. RL algorithms, notably *Q*-learning and Deep Q Networks (DQNs), can improve control techniques by learning from operational data [11]. For example, an RL model in a DC–DC converter might dynamically alter voltage levels according

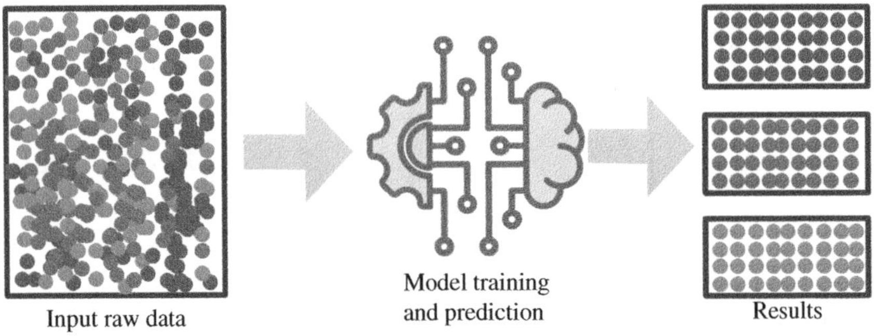

Figure 10.4 Reinforcement learning approach.

on load circumstances, increasing energy efficiency and prolonging the life of power electronic components [12, 13].

- **Optimization of Power Grid Systems**: RL is especially useful for controlling power distribution in smart grids, where circumstances change often. By modifying factors such as power flow and energy storage levels, RL models may improve grid stability and efficiency. RL framework presents in Figure 10.4. This method has advantages in renewable energy systems, where RL can assure optimal energy distribution even when supply from sources like as solar or wind varies [14].

10.1.4 Deep Learning for Fault Diagnosis

Convolutional Neural Networks (CNNs) and Recurrent Neural Networks (RNNs) have proven effective in detecting and diagnosing faults in PE [15]. CNNs, commonly used in image processing, are useful for identifying patterns in time-series data. For example, a CNN model can analyze waveforms from power converters to identify signal anomalies that indicate potential faults. RNNs, with their ability to remember sequential information, are well-suited for monitoring time-dependent behavior in systems like PEs inverters or motor drives.

- **Neural Networks**: NNs, especially deep learning models, are highly effective for handling complex, non-linear problems in PE. These models, inspired by the structure of the human brain, consist of multiple interconnected layers that can learn complex relationships within large datasets [16]. NNs are especially beneficial for advanced applications, such as predictive maintenance, fault diagnosis, and system optimization.
- **Energy Efficiency Optimization with Neural Networks**: PEs systems, such as inverters and converters, often require high efficiency and minimal power loss [17]. Deep neural networks (DNNs) can analyze data from sensors

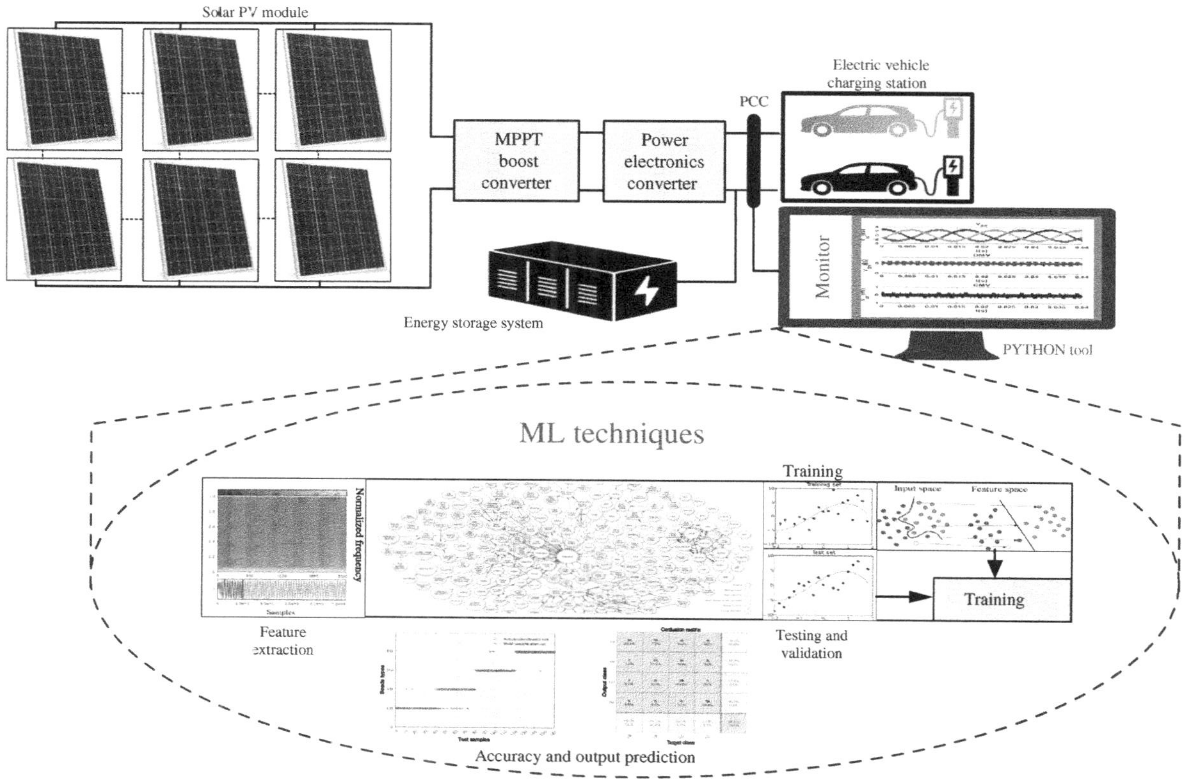

Figure 10.5 Applications of ML using PYTHON for independent system operator.

to identify patterns in energy usage and recommend operational adjustments. By predicting optimal switching frequencies or voltage levels, DNNs can help reduce power loss and improve energy efficiency [18].

10.1.5 Hybrid Approach

In many PEs applications, combining multiple ML algorithms can yield better results than using a single method. Hybrid approaches, such as integrating supervised learning for classification with reinforcement learning for adaptive control, can provide enhanced performance, especially in complex systems with diverse operational conditions. A hybrid system could combine anomaly detection using unsupervised learning with supervised classification to determine fault types. Once a fault is identified, a reinforcement learning model could then implement adaptive control measures to mitigate the fault's impact on the system [19]. Hybrid approaches are particularly useful in complex systems like electric vehicles (EVs) and renewable-energy-based setups, where system adaptability and resilience are crucial [20].

In recent years, the PYTHON language has witnessed a transformative shift with the interface of ML techniques for control through independent system operator (ISO), shown in Figure 10.5. PYTHON tool, committed to sustainability and fostering intelligent systems, has the potential to harness these advancements to maximize the efficiency of PEs applications [21].

This chapter delves into the diverse techniques of ML using PYTHON, covering aspects such as fault classification (low voltage [LV]/high voltage [HV]) strategies for optimal sizing of grid and power quality. The focus is on tailoring these applications to the enhanced monitoring and controlling through PYTHON and highlighting the applications of PE such as Fault detection in grid.

10.2 ML Algorithms Used in Power Electronics Utilizing PYTHON Platform

ML is a branch of artificial intelligence that emphasizes the development of systems that can learn from data and make decisions with little human intervention [22]. ML algorithms utilize pre-historical data to identify patterns, forecast outcomes, and enhance decision-making methods as time progresses. Various PEs industries such as transportation and engineering have taken notice of these techniques due to their capacity to enhance performance, lower costs, and boost automation. In the domain of PE, ML presents significant opportunities to improve the design, control, and functionality of the electronic systems utilized in power conversion and smart energy management.

- **ML in PV-Inverter Control for Sustainable Energy**: Optimizing inverter control approach is pivotal for photovoltaic (PV) systems, where sunlight is abundant but environmental conditions vary in nature. Different ML algorithms, powered by real-time data, can adaptively analyze factors such as temperature and humidity specific to different climate conditions. In a scenario, ML-driven inverter control resulted in a 15% increase in energy yield compared to traditional control methods, showcasing the potential for higher efficiency in the different climatic conditions and contributing to the sustainable energy solutions.

10.2.1 Key Domains of ML in Power Electronics Using PYTHON

Python has become the preferred programming language for ML applications across industries due to its ease of use, readability, and extensive ecosystem of libraries and frameworks [23]. Furthermore, Python's connection with hardware systems, such as embedded systems and Internet of Things (IoT) devices, allows for the easy deployment of ML models in real-world PE applications.

- **Control Systems**: By optimizing control parameters and offering adaptive control techniques, ML algorithms may greatly enhance the performance of control systems in PEs system. Conventional control systems may not take system fluctuations or outside disturbances into account since they are based on pre-set models and control principles. ML-based controllers can improve the stability and efficiency of power electrical systems by learning from real-time data and adapting to changing circumstances [24].
- **Fault Detection and Condition Monitoring**: One of the most major applications of ML in PEs field is predictive maintenance and fault detection. ML techniques such as anomaly detection and classification of these abnormality can be used to identify early signs of faults in power converters, inverters, and other power electronic devices. By analyzing the system behavior and operational data, ML models can predict potential failures, allowing for timely maintenance and efficient condition monitoring of the PE converters. Python makes it easy to develop and implement these models using libraries like "*Scikit-learn*" for classification and anomaly detection, which can identify abnormal patterns in power electronic components like converters or transformers [21, 25]. This leads to reduced downtime and increased reliability of large-scale power systems.
- **Optimized Efficient**: PEs systems are supposed to be very efficient, but load, temperature, and other external variables can all have an impact on performance. ML algorithms may be used to dynamically optimize the functioning of PE systems by constantly learning from data and altering system parameters

in real time [26]. For example, ML models can optimize inverter switching frequencies or voltage levels, resulting in higher energy efficiency and lower power losses. Libraries like "*Tensor Flow*" can be used to design reinforcement learning models that optimize the operation of inverters, motor drives, or energy storage systems in real time [27, 28].

- **Energy Management in Smart Grids**: With the growing use of hybrid renewable energy sources, efficiently integrating these intermittent distributed energy sources into the grid has become an obstacle. ML is essential for improving energy management systems because it predicts energy consumption, forecasts renewable energy output, and allows for real-time decision-making [29]. This results in better use of renewable energy, greater grid stability, and lower operating expenses [30].
- **Thermal Management**: The operation of the PEs systems creates a lot of heat, which can lower performance and shorten component lifespan. ML approaches may be used to track thermal behavior, forecast overheating, and optimize cooling solutions. By detecting thermal anomalies and modifying system parameters, ML helps to maintain optimal operating conditions and protects sensitive electronic components.

10.3 PYTHON Library and Model Development

Usually, Python's straightforward syntax and readability make it easy to use, even for engineers with less computer science experience. Its simplicity enables developers to focus on issue solutions rather to get mired down in complicated coding frameworks [31]. This ease-of-use speeds up the development process and allows PE engineers to swiftly deploy ML solutions without having to understand complex programming languages.

Python's extensive ecosystem of ML libraries has made it the language of choice for engineers and researchers in PEs. These libraries provide ready-to-use tools for data analysis, ML model development, optimization, and deployment, significantly reducing the time and expertise required to build sophisticated ML applications. The workflow of the ML algorithm utilizing Python is shown in Figure 10.6. By leveraging Python's ML libraries, engineers can address a range of challenges in PE, from predictive maintenance and fault detection to efficiency

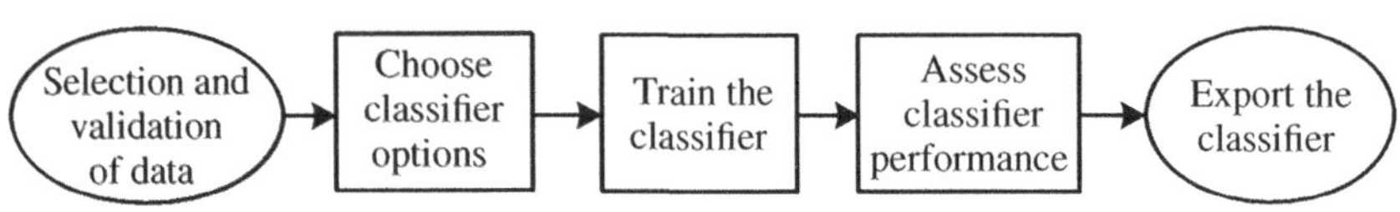

Figure 10.6 Flow chart of the ML classifier using PYTHON.

optimization and real-time control. Below is an overview of some essential Python libraries used in ML applications within PEs.

10.3.1 Extensive ML Libraries and PYTHON Framework

One of Python's most significant features is its extensive ecosystem of libraries created expressly for ML and data research. Key libraries include:

- **Scikit-Learn**: Scikit-Learn is a widely utilized ML library that provides effective tools for complex data analysis, particularly for data classification, regression, as well as clustering. It contains a broad range of ML algorithms, including classification, regression, clustering, and dimensionality reduction, making it an ideal choice for general-purpose ML applications in PE.
- **Tensor Flow and Keras**: These syntaxes are powerful tools for creating and deploying deep learning models. Tensor Flow, in particular, is performance-optimized and can be implemented across a variety of platforms, including embedded systems and cloud infrastructure. "*Tensor Flow*" is a popular open-source library for deep learning, offering tools for building and training NNs. It's highly efficient for large-scale ML tasks and supports both CPU and GPU processing, enabling fast training and deployment. Tensor Flow is highly flexible and can be used for real-time applications in PE, such as dynamic system control and optimization [32]. "*Keras*" is an application programming interface (API) built on top of Tensor Flow that make simpler the process of designing deep learning models. Additionally, its user-friendly interface and modular design make it ideal for rapid prototyping. Engineers, working with deep learning for complex PE applications like fault classification in converters or efficiency optimization in motor drives, find Keras useful for quickly building, training, and deploying deep learning models.
- **NumPy and SciPy for Numerical Computation**: NumPy is vital for scientific computing in Python and offers assistance for large data, multi-layer arrays, and complex matrices, along with a vast gathering of high-level dynamic mathematical functions to activate on related arrays. PE applications often involve processing sensor data, waveform signals, and other forms of operational data that require efficient numerical computation. NumPy offers the efficiency needed for such operations, including Fourier transforms and linear algebra computations, essential for signal processing and analysis in PE. "*SciPy*" as SciPy builds on the NumPy's capabilities with supplementary tools for majorly technical computing. These includes modules for complex optimization, system integration, data interpolation, practical eigenvalue glitches, and more. In PE applications, SciPy can be used to fine-tune control systems, optimize circuit parameters, and simulate electrical systems under various conditions.

- **Pandas for Data Manipulation**: Pandas is a commanding library for complex data manipulation and high-level analysis, designed to work with structured data like time-series data, which is common in PE applications. Pandas makes it easy to clean, filter, aggregate, and transform large datasets. Engineers use Pandas to pre-process and analyze data collected from PE systems, such as current and voltage measurements, sensor readings, and operational logs. This library is essential for preparing data before feeding it into ML models, allowing for accurate and effective model training and evaluation.
- **PyTorch for Advanced NN Applications**: PyTorch is more popular deep learning library that proposes greater flexibility for custom ML architectures, often preferred for research and development. Known for its dynamic computation graph, PyTorch is useful for complex and adaptive applications that require real-time updates and interactive experimentation. PyTorch's flexible architecture allows the implementation of advanced architectures like auto encoders for anomaly detection, which is useful for identifying and diagnosing faults in PEs systems.

10.3.2 Cross-Platform and Hardware Compatibility

Python is platform independent, which means it can operate on any operating system (Windows, macOS, or Linux) without change. This adaptability is essential for engineers who operate in contexts that demand various systems for design, simulation, and testing [33]. Furthermore, Python works well with hardware systems, such as those used in PE. Python, for example, can communicate with embedded systems, microcontrollers, and IoT devices via libraries like PySerial or bespoke APIs, making it an adaptable tool for incorporating ML into hardware applications [34].

The combination of ML and PEs, aided by Python, is generating advances that allow smarter, more efficient, and more reliable power systems. ML improves decision-making, automates system control, and anticipates faults, all of which help to revolutionize current PE [35].

10.4 Stepwise Developing a Power Electronics Classification Model in PYTHON

For developing a classification mode collection of the data from the sensors at the PE then writing the model in the Python libraries. The rest of the steps have been shown in Figure 10.7.

To begin the simulation, click on "*New Session*," after that choose the desired datasets needed to execute and then directed from the workspace. Then we have

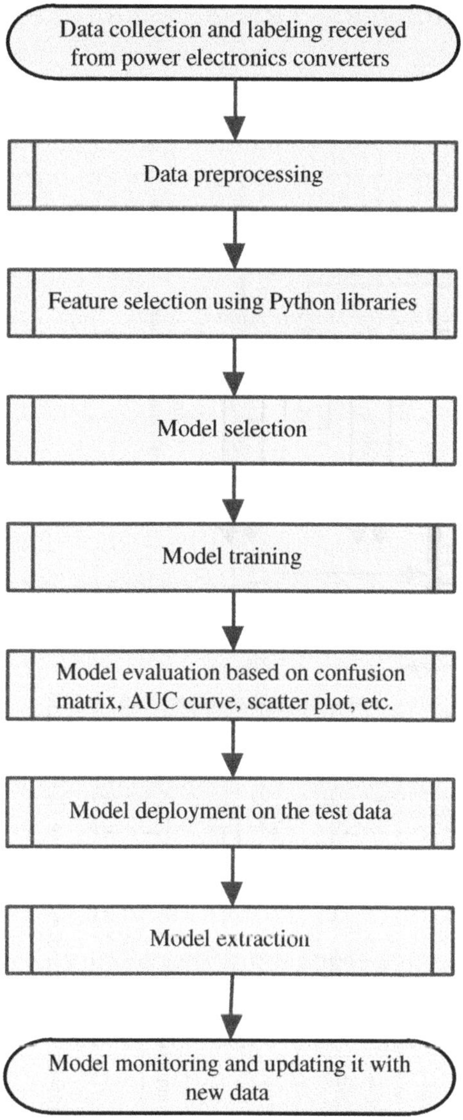

Figure 10.7 Flow chart showing the development of classification model in Python.

to click on the classification learner tool for training after selecting the model type (different artificial neutral network [ANN]/SVM/K-nearest neighbor [KNN]). Additionally, utilized PCA, if required. Export the executed data to a workspace once training concludes, or export it as new-save file. Within the model's window, it appears an assortment of model types available for different selection. After

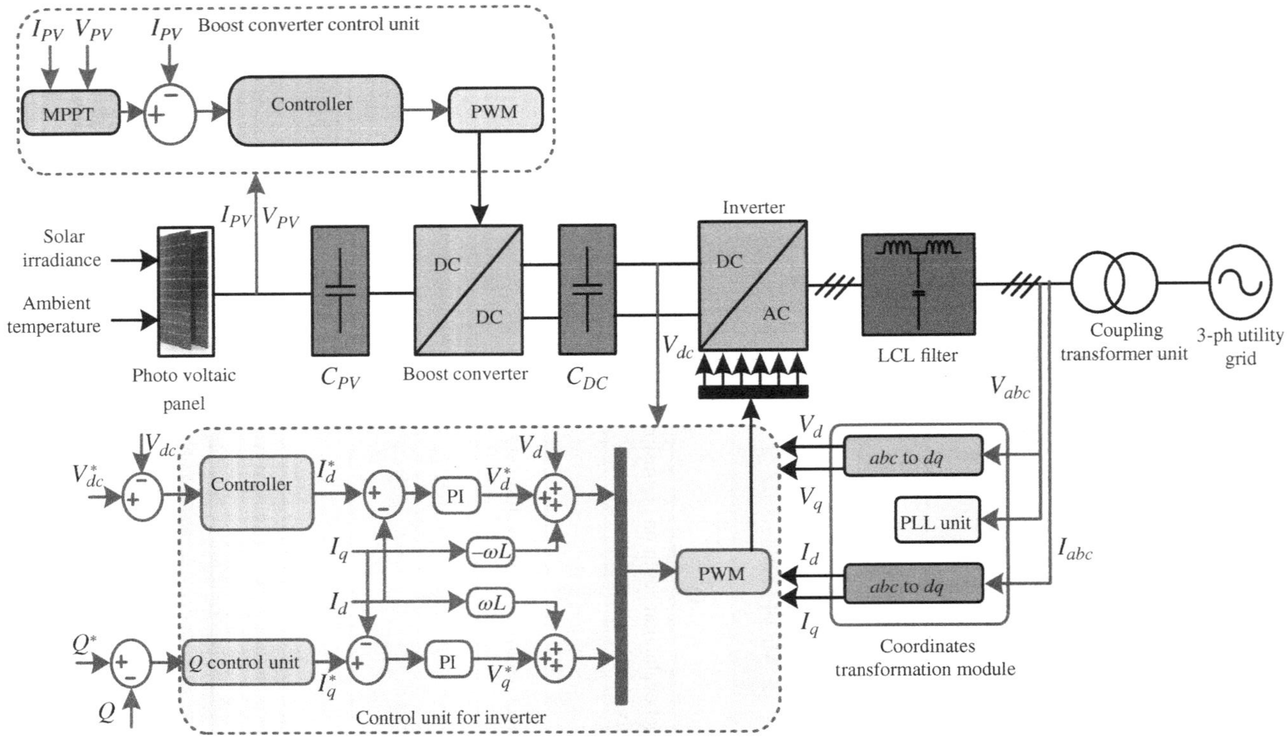

Figure 10.8 Control architecture of proposed 3-phase GCPV system.

training, the model with the maximum validation accuracy score will appear in a scope-tool.

10.4.1 PYTHON Used to Control 30-ph Grid Connected System

Typically, solar PV irradiance levels and ambient temperature are utilized to control the PEs converter to maximize power output from solar panels if the maximum power point tracking (MPPT) is needed for the higher power output from the PV systems [36]. To achieve the desired MPPT operation under standard grid operating conditions, the PEs converter is managed using a dynamic controller and the desired MPPT-based algorithm, as shown in Figure 10.8. This concept is further clarified with an example of a three-phase grid-connected photovoltaic (GCPV) system. A line-to-ground fault, along with voltage and frequency ride-through, is introduced at the point of common coupling (PCC), after which relevant fault data points are extracted. Figure 10.8 presents the block diagram of the three-phase GCPV system, illustrating the control block configuration. It highlights the integration of proportional-based controllers, a current controller, an orthogonally based signal generator, and a phase-locked loop (PLL) subsystem to ensure precise and effective controller in a three-phase PEs setup. This control approach is essential for maintaining system stability and enhancing GCPV performance.

Figure 10.9 shows the Python-based measurement system for a three-phase grid-connected setup. It illustrates a two-stage, three-phase grid connected system with PEs converters controls, where the DC-to-DC converter serves as the first stage. The controller for the three-phase voltage source converter (VSC)-based inverter includes a power balancing system associated with a DC-link, a voltage-feedback-based controller, a transformation system with their own PLL and direct quadrature (DQ) control unit for coordinate transformations, and proportional integral (PI)-based controllers. The control strategies used in the three-phase GCPV system are outlined as follows. Through dynamic gate trigger signals with all six SCRs operate as controlled switches, activating as required operation [37]. This form of PEs is broadly implemented in industrial-based PEs applications, mainly when working with total output power levels of up to

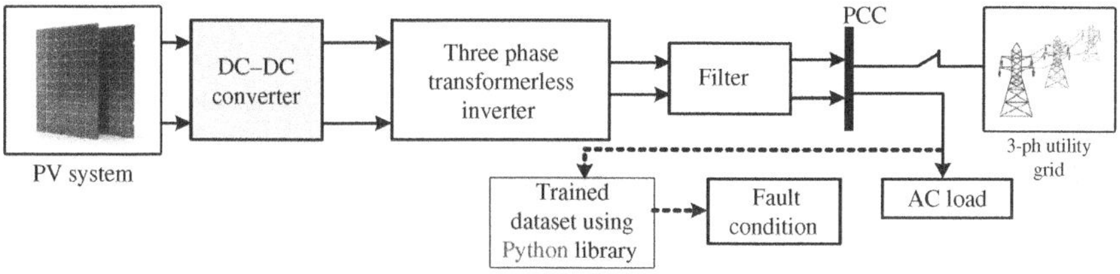

Figure 10.9 Power flow diagram of the 3-phase GCPV system.

120 kW and needing two quadrant operations. The triggering of the thyristors occurs at regular intervals of $\pi/3$ rad, which is equivalent to an interval of 30°. This approach ensures grid synchronization as well as the effective control of the dynamic switching method. A voltage/current feature of the three-phase power electronics converter (PEC) is the frequency of the ripple in its output voltage, which is detected to be 3 times the fundamental frequency [38]. Although this ripple voltage exists, it is effectively controlled by the use of switches and well-timed triggering signals. Due to its efficiency and adaptability, this type of PEC is widely used in industrial-based applications requiring dependable and precise power regulation. Its capability to handle higher power levels and perform 2-quadrant operations makes it ideal for various industrial PEs systems [39].

- **Control of PEs VSC**: A fundamental controller, such as a PI/proportional integral derivative (PID) controller, is utilized as the outer loop in the system control unit. The major function is to establish an adjusted current reference by directly detecting the DC-link voltage and comparing it to a predetermined reference value [40]. The output from this controller serves as the direct current reference for the inner loop. Within this inner loop, a current controller manages the direct current reference to deliver the requisite current. PI controllers are utilized in conjunction with an orthogonal signal generator for accurate regulation in a three-phase GCPV system. This control technique facilitates precise management of the injected current, especially when the system functions within the synchronously rotating reference frame, commonly known as DQ-control [41]. A PLL technology is employed to align the injected grid current with the grid voltage. The PLL ensures accurate alignment and synchronization between the grid current and voltage, which is crucial for the system's reliable operation [42].
- **Voltage Dips**: A uncertain short-circuit-based failure at a specific location in the grid system may source a voltage dip in which a lower-level voltage than the reference voltage that appears at the distribution or far end terminals [43]. Furthermore, voltage dips can occur due to motor starting, overloads, or sudden transitions of heavy loads. Although these dips last only a brief period, they can damage sensitive equipment, especially power electronic devices.
- **Low-Voltage Ride Through (LVRT) Techniques**: The drawback of VSCs is their vulnerability to voltage disturbances, such as voltage dips. When the controller attempts to maintain a constant power level at the DC-link, there is a sudden grid voltage dropping, which often leads to an increase in current for the VSC, which may trigger the VSC to circuit breaker (CB) trip in order to protect the other systems. Additionally, most of the dips are triggered by uncertain faults through the power lines. These imbalances can create unwanted low-order frequency power oscillations, resulting in current harmonics and poor control of the DC link voltage [41]. Faults may also cause the PEs converters to sudden trip,

which protects the PE device from damage caused by higher overvoltage. Voltage sag can occur on the grid during anomalous conditions, with magnitudes ranging from 0.52 to 0.91 p.u. Most of the LVRT techniques, including those mentioned below, have been proposed to keep the VSC operating adequately under these conditions [44].

- Preventing the operating nature of the MPPT-based controller under faulty scenarios.
- Discharging the accumulated dynamic energy in an energy storage system, including buffer storage, supercapacitors, and battery energy storage systems, among others.
- Advanced control tactics utilizing symmetrical components of the system (management of positive and negative sequences).

Additionally, several methods have been implemented as in [45, 46] for detecting positive and negative sequences, though these approaches encounter stability issues because of the delays in P/N sequence extraction, sometimes which can compromise the overall efficiency of the control unit. These intricacies might cause uncertainty challenges and complicate the control method [47, 48].

10.5 Development of ML Classification Model Using PYTHON for PEs Converters

When assessing nonlinear-based loads, the reliability of PEs becomes a critical factor. To tackle these crucial challenges, effective monitoring is essential for extending the operational lifespan and maintaining optimal system performance. In the context of maintenance, this involves a series of strategic procedures performed on industrial products. The goal is to restore them from a degraded state to one where they can efficiently perform their intended functions, ensuring continuous and reliable operation of the grid. When creating models for binary or a multiclass based classification problem, the first step is to select a suitable ML algorithm for testing as well as validation, followed by writing the Python code using TensorFlow. Numerous algorithms are available. Furthermore, it can generate Python code to facilitate programmatic classification or export a model to the workspace for application with new data. In the Classification Learner, the training process follows a dual approach:

- **Validation of a Model:** Design a model employing a testing and validation strategy. By default, the program utilizes its cross-validation to prevent over fitting. Alternatively, holdout validation is also available. The classifier app displays the proposed ML validated model.

- **Performance Evolution:** Train the ML model using the entire dataset without validating. The ML model is trained in conjunction with the validated model by the Python. Also, the ML model trained/tested on the basis of normalized dataset. When selecting a desired classifier app for export to the system workspace, the Classifier Learner tool exports the full ML model.

In this study, the ML techniques used Weighted KNN and Fine Gaussian SVM. The model has been trained and predicted utilizing the test data. The model has achieved a significant improved accuracy to detect the abnormality in PE converters. Following section details about the performance evaluation of the outcomes obtained from the ML classification model.

10.5.1 Case Study: Fault Classification Using ML (SVM and KNN) Techniques

We implement faults analysis as Low-Voltage Ride-Through (LVRT) and High-Voltage Ride-Through (HVRT) with respective thresholds and evaluate the performance of classifiers (Fine Gaussian SVM and Weighted KNN) as present in the Python code. Also, we implement LVRT and HVRT thresholds (0.7, 0.8, 0.9 for LVRT and 1.1, 1.2, 1.3 for HVRT) using sample voltage data as shown in Table 10.1. Firstly, we generate the overall ML performance plots (scatter plots for visualization). Further, we evaluate desired classifiers and visualize results via confusion matrix and Region of Convergence (ROC) curves.

Step 1: LVRT and HVRT Implementation

In this study, sample data [47] simulates the LVRT and HVRT performance evaluations and visualize. Table 10.1 illustrates the obtained confusion matrix of the normalized datasets points for the ML-based (SVM/KNN) classifier.

Step 2: Model Training and Performance Evaluation

The data is used for training [47], then tests the proposed Weighted KNN and Fine Gaussian SVM models. Afterward, we plot the confusion matrices and ROC curves can be achieved.

Confusion Matrix: This is a tool used to see how well the classifier is working for each category. After training, we need to check it out in the Plots section of the Classification Learner tab by selecting Confusion Matrix with Validation Data (like cross-validation). The matrix helps spot where the classifier is missing the mark. Rows represent the actual classes, and columns show the predicted ones, making it easy to see how well the model's doing. In cross-validation or holdout, the matrix is based on predictions made on the set-aside validation data. Diagonal cells show where the predicted and actual classes match up – blue shading on these cells usually means the model's getting things right. Each cell shows how many data points fall into each category by default.

Table 10.1 PYTHON-panda code for define data sets (LVRT and HVRT).

```
import numpy as np
import pandas as pd
from sklearn.metrics import confusion_matrix, roc_curve, auc
from sklearn.neighbors import KNeighborsClassifier
from sklearn.svm import SVC
from sklearn.model_selection import train_test_split
import seaborn as sns

# Generate synthetic voltage data
np.random.seed(42)
n_samples = 1000
voltage_levels = np.random.uniform(0.5, 1.5, n_samples) #
Random voltages
labels = np.where((voltage_levels <= 0.9) & (voltage_levels
>= 0.7), "LVRT",
         np.where((voltage_levels >= 1.1) & (voltage_levels
<= 1.3), "HVRT", "Normal"))

# Scatter plot of LVRT and HVRT regions
plt.figure(figsize=(10, 6))
plt.scatter(range(n_samples), voltage_levels,
c=np.where(labels=="LVRT", 'blue', np.where(labels=="HVRT",
'red', 'green')), alpha=0.6)
plt.axhline(0.7, color='blue', linestyle='--', label='LVRT
(0.7)')
plt.axhline(0.9, color='blue', linestyle='--', label='LVRT
(0.9)')
plt.axhline(1.1, color='red', linestyle='--', label='HVRT
(1.1)')
plt.axhline(1.3, color='red', linestyle='--', label='HVRT
(1.3)')
plt.xlabel("Sample Index")
plt.ylabel("Voltage Level")
plt.title("Voltage Levels with LVRT and HVRT Thresholds")
plt.legend()
plt.show()
```

Table 10.2 Fault classification using python.

```
# Encoding the labels for classifier use
label_map = {"LVRT": 0, "HVRT": 1, "Normal": 2}
labels_encoded = np.vectorize(label_map.get)(labels)
data = pd.DataFrame({'voltage_levels': voltage_levels,
'labels': labels_encoded})

# Train-test split
X_train, X_test, y_train, y_test =
train_test_split(data[['voltage_levels']], data['labels'],
test_size=0.3, random_state=42)

# Initialize classifiers
knn = KNeighborsClassifier(n_neighbors=3, weights="distance")
svm = SVC(kernel="rbf", probability=True)

# Train the models
knn.fit(X_train, y_train)
svm.fit(X_train, y_train)

# Predictions
knn_preds = knn.predict(X_test)
svm_preds = svm.predict(X_test)

# Confusion Matrix Visualization
fig, axes = plt.subplots(1, 2, figsize=(12, 6))
sns.heatmap(confusion_matrix(y_test, knn_preds), annot=True,
fmt="d", ax=axes[0], cmap="Blues", cbar=False)
axes[0].set_title("Confusion Matrix for Weighted KNN")
sns.heatmap(confusion_matrix(y_test, svm_preds), annot=True,
fmt="d", ax=axes[1], cmap="Reds", cbar=False)
axes[1].set_title("Confusion Matrix for Fine Gaussian SVM")
plt.show()
```

(Continued)

Table 10.2 (Continued)

```
# Calculate TPR, FNR, PPV, FDR values from confusion matrix
def calculate_metrics(cm):
  TN, FP, FN, TP = cm.ravel()
  TPR = TP / (TP + FN)
  FNR = FN / (FN + TP)
  PPV = TP / (TP + FP)
  FDR = FP / (FP + TP)
  return TPR, FNR, PPV, FDR

# Calculate and print metrics for KNN and SVM
print("Performance Metrics for KNN:")
TPR_knn, FNR_knn, PPV_knn, FDR_knn =
calculate_metrics(confusion_matrix(y_test, knn_preds))
print(f"TPR: {TPR_knn}, FNR: {FNR_knn}, PPV: {PPV_knn}, FDR:
{FDR_knn}")

print("\nPerformance Metrics for SVM:")
TPR_svm, FNR_svm, PPV_svm, FDR_svm =
calculate_metrics(confusion_matrix(y_test, svm_preds))
print(f"TPR: {TPR_svm}, FNR: {FNR_svm}, PPV: {PPV_svm}, FDR:
{FDR_svm}")
```

Scatter Plot: This plot visually depicts the relationship between two variables within a dataset, positioning each data point in a two-dimensional space. The *X*-axis typically represents the independent variable (or feature), while the *Y*-axis corresponds to the dependent variable. Scatter plots are essential for uncovering trends, clusters, or possible correlations between variables. By plotting these data points, patterns and potential associations between variables become evident, facilitating an intuitive understanding of the data's characteristics and supporting data-driven interpretations.

True-Positive Rate and False-Negative Rate: To check how well the model is doing for each category, you can look at True-Positive Rate (TPR) and False-Negative Rate (FNR) in the Plot section. TPR tells you the percentage of times the classifier got it right for each actual class, while FNR shows the rate of misses relative to correct classifications.

Table 10.2 presents the Python code for the Performance Matrix under LVRT, HVRT, and normal scenarios.

Table 10.2 presents the confusion matrices for the Fine Gaussian SVM and Weighted KNN classifiers, along with their corresponding TPR and FNR values. These summaries offer valuable insights into the effectiveness of the classifiers in distinguishing between classes, aiding in the evaluation and optimization of the classification process. By analyzing TPR, FNR, and class-specific summaries, informed decisions can be made, allowing for fine-tuning of the classifiers to achieve optimal results across various classes in the dataset.

Step 3: ROC Curve for Classifier Comparison

This process includes plotting the ROC curves to evaluate the performance of both classifiers across classes using data from [47] using PYTHON code. After

Table 10.3 PYTHON code for ROC curve.

```
# Get probability scores for ROC
knn_probas = knn.predict_proba(X_test)
svm_probas = svm.predict_proba(X_test)

# Plot ROC curve
fig, ax = plt.subplots(1, 1, figsize=(10, 6))
for i, label in enumerate(label_map.keys()):
  fpr_knn, tpr_knn, _ = roc_curve(y_test == i, knn_probas[:, i])
  fpr_svm, tpr_svm, _ = roc_curve(y_test == i, svm_probas[:, i])
  auc_knn = auc(fpr_knn, tpr_knn)
  auc_svm = auc(fpr_svm, tpr_svm)

  ax.plot(fpr_knn, tpr_knn, label=f"KNN - {label}
(AUC={auc_knn:.2f})", linestyle="--")
  ax.plot(fpr_svm, tpr_svm, label=f"SVM - {label}
(AUC={auc_svm:.2f})")

ax.plot([0, 1], [0, 1], "k--", label="Random Classifier")
ax.set_xlabel("False Positive Rate")
ax.set_ylabel("True Positive Rate")
ax.set_title("ROC Curve Comparison for Fine Gaussian SVM and
Weighted KNN")
ax.legend()
plt.show()
```

training a model, you can display the ROC curve by selecting ROC Curve and Validation Data in the Plots section of the Classifier Learner tab. The ROC curve shows the TPR and false positive rate (FPR) for the currently used trained classifier. You can choose from various classes to analyze. The efficiency of the selected classifier parameters is represented by a response on the plot, which indicates the TPR and FPR values for that classifier. For example, in this case, the FPR for the with PCA enabled SVM is 0.02, meaning that 2% of the datasets are incorrectly classified as fitting to the positive class by the different classifier. Additionally, a TPR of 0.98 indicates that 98% of the datasets are accurately divided into the positive class by the current classifier. The Python code for ROC-AUC (area under the curve) comparison of SVM & W-KNN classifier is displayed.

Table 10.3 offers valuable insights into the classifier's ability to differentiate between classes, aiding in the evaluation and optimization of the classification process. By analyzing TPR, FNR, and class-specific summaries, informed decisions can be made, allowing for the fine-tuning with multiple classifiers to accomplish best results across various defined classes.

10.6 Challenges of Utilizing ML with Python for PEs Applications

The use of ML in PEs has several advantages, including improved system monitoring, defect identification, and efficiency improvement. However, different obstacles occur as a result of the unique properties of PEs systems, as well as the constraints of employing Python for these tasks as given below:

- **Data Quality and Availability**: PE systems generate vast amounts of data, but this data is not always labeled or structured for ML applications. Additionally, noise, missing values, and anomalies are common in sensor data, which can interfere with model accuracy and reliability.
- **Real-Time Processing Constraints**: PEs applications, such as defect detection and system control, frequently demand real-time or near-real-time processing. However, Python's Global Interpreter Lock (GIL) and interpreted nature might impede real-time performance, particularly when working with high-frequency data from converters, inverters, or other power components.
- **High Computational Demands of ML Models**: Advanced ML models, often require substantial computational resources, particularly for training. PE engineers may struggle to optimize these models on standard processing units, especially when handling complex datasets or time-series data over long

periods. Using high-performance libraries (e.g. Tensor Flow or PyTorch) with GPU acceleration can mitigate this issue.

- **Limited Availability of Domain-Specific Libraries**: While Python's ML ecosystem is robust, there are limited libraries and tools specifically tailored for PE applications. Engineers must often modify or extend existing libraries to address domain-specific requirements, which can be time-consuming and may require extensive testing. Python's flexibility is to create custom libraries and functions to meet their unique needs, although this requires expertise in both PE and Python.
- **Data Privacy and Security**: PEs systems, particularly those in critical infrastructure like smart grids, often involve sensitive data that must be protected from unauthorized access and tampering. Using cloud-based ML solutions may introduce security and privacy risks, as data transferred between devices and servers is vulnerable to interception.

10.7 Conclusion and Future Scope

The suggested ML technique, characterized by Python's high processing accuracy, minimal communication needs, and suitability for islanding detection in microgrids, emerges as a resilient and effective option for enhancing the reliability of PEs systems. These findings underscore the significance of ML techniques for effective condition monitoring and fault detection in Pes-based grid system. This chapter effectively trained and evaluated a three-phase grid system utilizing SVM and W-KNN ML methods. The suggested system determined that the output from the Fine SVM is more precise than that of the W-KNN, and the ML algorithms can forecast fault-tolerant circumstances throughout diverse grid operations.

In recent times, ML has emerged as a potent solution for addressing various challenges in PEs fault analysis using Python Also, the aging or degradation impact on PEs devices can be caused by PEs faults as additional problem to take into account. In the same way, we can replicate particular parametric errors in a PE converter so that our machine-learning model can identify trends. This makes it easier to anticipate which parts of our circuit may fail, giving us the ability to perform preventative maintenance. Advanced techniques leveraging artificial intelligence (AI) and ML require extensive datasets for training, yet they possess the capability to derive valuable insights, resulting in exceptional outcomes, through the utilization of various algorithms such as SVM, ANN, fuzzy logic, KNN, and expert systems within AI frameworks. It is clear that fault classification/detection methods in PEs can be greatly improved by optimization library tool under real-time validation.

Exercises

10.1 How can PYTHON-based ML algorithms enhance the performance and efficiency of power electronic converters?

10.2 Explain the use of supervised and unsupervised learning in identifying anomalies within PE systems. Write the practical example, advantages, and limitations?

10.3 What are some common challenges in implementing ML algorithms in PYTHON for PE applications?

10.4 Discuss the key steps for data pre-processing in ML models.

10.5 How does feature selection impact the accuracy of ML models in PE applications?

10.6 Explain how PYTHON libraries like Scikit-Learn and TensorFlow can be used for ML tasks in PE.

10.7 What are the benefits of using a confusion matrix to evaluate classifier performance in PE applications? Provide an example.

10.8 Explain the concept of Low-Voltage Ride-Through (LVRT) and High-Voltage Ride-Through (HVRT) in power systems. How can ML models be used to simulate and improve these fault classification features?

10.9 How do evaluation metrics such as True-Positive Rate (TPR) and False-Negative Rate (FNR) provide insight into classifier performance?

10.10 Describe the concept of the Region of Convergence (ROC) curve and its significance in comparing model performance for PE.

10.11 Discuss the advantages of PYTHON-based visualization tools, such as Matplotlib and Seaborn, for representing PE data.

10.12 Write PYTHON code for predicting 24-hour energy consumption using ML models for a PV rooftop system in a home application.

References

1 J. Verma, L. Sandys, A. Matthews, and S. Goel, "Readiness of artificial intelligence technology for managing energy demands from renewable sources," *Eng. Appl. Artif. Intell.*, vol. 135, p. 108831, Sep. 2024. doi: 10.1016/j.engappai.2024.108831.

2 A. Haque, K. V. S. Bharath, M. Amir, and Z. Khan, "Role and applications of power electronics, renewable energy and IoT in smart cities," in *Smart Cities: Power Electronics, Renewable Energy, and Internet of Things*. Boca Raton: CRC Press, 2023, pp. 66–95. doi: 10.1201/9781032669809-4.

3 P. Venkata, V. Pandya, K. Vala, and A. V. Sant, "Support vector machine for fast fault detection and classification in modern power systems using quarter cycle data," *Energy Rep.*, vol. 8, pp. 92–98, Dec. 2022. doi: 10.1016/j.egyr.2022.10.279.

4 K. Chen, C. Huang, and J. He, "Fault detection, classification and location for transmission lines and distribution systems: A review on the methods," *High Volt.*, vol. 1, no. 1, pp. 25–33, Apr. 2016. doi: 10.1049/hve.2016.0005.

5 B. Ravikumar, D. Thukaram, and H. P. Khincha, "Application of support vector machines for fault diagnosis in power transmission system," *IET Gener. Transm. Distrib.*, vol. 2, pp. 119–130, 2008. doi: 10.1049/iet-gtd:20070071.

6 B. Taşcı, A. Omar, and S. Ayvaz, "Remaining useful lifetime prediction for predictive maintenance in manufacturing," *Comput. Ind. Eng.*, vol. 184, p. 109566, Oct. 2023. doi: 10.1016/j.cie.2023.109566.

7 V. Vita, G. Fotis, V. Chobanov, C. Pavlatos, and V. Mladenov, "Predictive maintenance for distribution system operators in increasing transformers' reliability," *Electronics*, vol. 12, no. 6, p. 1356, Mar. 2023. doi: 10.3390/electronics12061356.

8 A. Theissler, J. Pérez-Velázquez, M. Kettelgerdes, and G. Elger, "Predictive maintenance enabled by machine learning: Use cases and challenges in the automotive industry," *Reliab. Eng. Syst. Saf.*, vol. 215, p. 107864, Nov. 2021. doi: 10.1016/j.ress.2021.107864.

9 S. Ayoub, A. Haque, M. Amir, and V. S. B. Kurukuru, "Intelligent islanding classification with optimal k-nearest neighbors technique for single phase grid integrated PV system," in *2022 IEEE 3rd Global Conference for Advancement in Technology (GCAT)*, IEEE, Oct. 2022, pp. 1–6. doi: 10.1109/GCAT55367.2022.9972088.

10 Z. M. Çınar, A. Abdussalam Nuhu, Q. Zeeshan, O. Korhan, M. Asmael, and B. Safaei, "Machine learning in predictive maintenance towards sustainable smart manufacturing in industry 4.0," *Sustainability*, vol. 12, no. 19, p. 8211, Oct. 2020. doi: 10.3390/su12198211.

11 S. Stavrev and D. Ginchev, "Reinforcement learning techniques in optimizing energy systems," *Electronics*, vol. 13, no. 8, p. 1459, Apr. 2024. doi: 10.3390/electronics13081459.

12 N. Mazaheri, D. Santamargarita, E. Bueno, D. Pizarro, and S. Cobreces, "A deep reinforcement learning approach to DC-DC power electronic converter control with practical considerations," *Energies*, vol. 17, no. 14, p. 3578, Jul. 2024. doi: 10.3390/en17143578.

13 N. Shah, A. Haque, M. Amir, and A. Kumar, "Investigation of renewable energy integration challenges and condition monitoring using optimized tree in three phase grid system," in *2023 7th International Conference on Computing Methodologies and Communication (ICCMC)*, IEEE, Feb. 2023, pp. 1582–1588. doi: 10.1109/ICCMC56507.2023.10083636.

14 W. Choi and C. Ahn, "Deep Q-network-based controller for cabin cooling system of electric vehicles," *IEEE Access*, vol. 11, pp. 137274–137284, 2023. doi: 10.1109/ACCESS.2023.3339826.

15 M. Alkhodari and L. Fraiwan, "Convolutional and recurrent neural networks for the detection of valvular heart diseases in phonocardiogram recordings," *Comput. Methods Programs Biomed.*, vol. 200, p. 105940, Mar. 2021. doi: 10.1016/j.cmpb.2021.105940.

16 M. Abdelmaksoud, M. Torki, M. El-Habrouk, and M. Elgeneidy, "Convolutional-neural-network-based multi-signals fault diagnosis of induction motor using single and multi-channels datasets," *Alexandria Eng. J.*, vol. 73, pp. 231–248, Jul. 2023. doi: 10.1016/j.aej.2023.04.053.

17 L. Alzubaidi *et al.*, "Review of deep learning: Concepts, CNN architectures, challenges, applications, future directions," *J. Big Data*, vol. 8, no. 1, p. 53, Dec. 2021. doi: 10.1186/s40537-021-00444-8.

18 L. S. Maraaba, A. S. Milhem, I. A. Nemer, H. Al-Duwaish, and M. A. Abido, "Convolutional neural network based inter-turn fault diagnosis in LSPMSMs," *IEEE Access*, vol. 8, pp. 81960–81970, 2020. doi: 10.1109/ACCESS.2020.2991137.

19 D. Velasquez *et al.*, "A hybrid machine-learning ensemble for anomaly detection in real-time industry 4.0 systems," *IEEE Access*, vol. 10, pp. 72024–72036, 2022. doi: 10.1109/ACCESS.2022.3188102.

20 L. Rajaoarisoa, M. Kuk, S. Bobek, and M. Sayed-Mouchaweh, "Hybrid and co-learning approach for anomalies prediction and explanation of wind turbine systems," *Eng. Appl. Artif. Intell.*, vol. 133, p. 108046, Jul. 2024. doi: 10.1016/j.engappai.2024.108046.

21 S. Raschka, J. Patterson, and C. Nolet, "Machine learning in Python: Main developments and technology trends in data science, machine learning, and artificial intelligence," *Information*, vol. 11, no. 4, p. 193, Apr. 2020. doi: 10.3390/info11040193.

22 B. Gomatheeshwari and J. Selvakumar, "Appropriate allocation of workloads on performance asymmetric multicore architectures via deep learning algorithms," *Microprocess. Microsyst.*, vol. 73, p. 102996, 2020. doi: 10.1016/j.micpro.2020.102996.

23 S. K. H. Shah, A. Hellany, M. Nagrial, and J. Rizk, "Influence of environmental changes on power quality disturbances in hybrid renewable energy system," *Energy Rep.*, vol. 9, pp. 164–173, Oct. 2023. doi: 10.1016/j.egyr.2023.08.047.

24 I. A. Saifi, A. Haque, M. Amir, and V. S. Bharath Kurukuru, "Intelligent islanding classification with MLPNN for hybrid distributed energy generations in microgrid system," in *2023 International Conference on Intelligent and Innovative Technologies in Computing, Electrical and Electronics*, Jan. 2023, pp. 982–987. doi: 10.1109/IITCEE57236.2023.10091089.

25 H. Rahimighazvini, Z. Khashroum, M. Bahrami, and M. H. Masali, "Power electronics anomaly detection and diagnosis with machine learning and deep learning methods: A survey," *Int. J. Sci. Res. Arch.*, vol. 11, no. 2, pp. 730–739, Mar. 2024. doi: 10.30574/ijsra.2024.11.2.0428.

26 L. Lei, B. Wu, X. Fang, L. Chen, H. Wu, and W. Liu, "A dynamic anomaly detection method of building energy consumption based on data mining technology," *Energy*, vol. 263, p. 125575, Jan. 2023. doi: 10.1016/j.energy.2022.125575.

27 D. Kim, S. Hong, S. Cui, and I. Joe, "Deep reinforcement learning-based real-time joint optimal power split for battery–ultracapacitor–fuel cell hybrid electric vehicles," *Electronics*, vol. 11, no. 12, p. 1850, Jun. 2022. doi: 10.3390/electronics11121850.

28 M. J. Salehpour and M. J. Hossain, "Leveraging machine learning for efficient EV integration as mobile battery energy storage systems: Exploring strategic frameworks and incentives," *J. Energy Storage*, vol. 92, p. 112151, Jul. 2024. doi: 10.1016/j.est.2024.112151.

29 Z. Jia, J. Li, X.-P. Zhang, and R. Zhang, "Review on optimization of forecasting and coordination strategies for electric vehicle charging," *J. Mod. Power Syst. Clean Energy*, vol. 11, no. 2, pp. 389–400, 2023. doi: 10.35833/MPCE.2021.000777.

30 H. Wang, M. Liserre, and F. Blaabjerg, "Toward reliable power electronics: Challenges, design tools, and opportunities," *IEEE Ind. Electron. Mag.*, vol. 7, no. 2, pp. 17–26, Jun. 2013. doi: 10.1109/MIE.2013.2252958.

31 M. Massaoudi, H. Abu-Rub, S. S. Refaat, I. Chihi, and F. S. Oueslati, "Deep learning in smart grid technology: A review of recent advancements and future prospects," *IEEE Access*, vol. 9, pp. 54558–54578, 2021. doi: 10.1109/ACCESS.2021.3071269.

32 A. Baliyan, K. S. Kaswan, and J. S. Dhatterwal, "An empirical analysis of Python programming for advance computing," in *2022 2nd International*

Conference on Advance Computing and Innovative Technologies in Engineering (ICACITE), IEEE, Apr. 2022, pp. 1482–1486. doi: 10.1109/ICACITE53722.2022.9823643.

33 I. Stancin and A. Jovic, "An overview and comparison of free Python libraries for data mining and big data analysis," in *2019 42nd International Convention on Information and Communication Technology, Electronics and Microelectronics (MIPRO)*, IEEE, May 2019, pp. 977–982. doi: 10.23919/MIPRO.2019.8757088.

34 X. Cai, H. P. Langtangen, and H. Moe, "On the performance of the Python programming language for serial and parallel scientific computations," *Sci. Program.*, vol. 13, no. 1, pp. 31–56, Jan. 2005. doi: 10.1155/2005/619804.

35 J. Yu *et al.*, "CONGO^2: Scalable online anomaly detection and localization in power electronics networks," *IEEE Internet Things J.*, vol. 9, no. 15, pp. 13862–13875, Aug. 2022. doi: 10.1109/JIOT.2022.3143123.

36 W. Saad, E. Hegazy, and M. Shokair, "Maximum power point tracking based on modified firefly scheme for PV system," *SN Appl. Sci.*, vol. 4, no. 4, p. 94, Apr. 2022. doi: 10.1007/s42452-022-04976-3.

37 L.-F. Gaona-Cárdenas, N. Vázquez-Nava, O.-F. Ruíz-Martínez, A. Espinosa-Calderón, A.-I. Barranco-Gutiérrez, and M.-A. Rodríguez-Licea, "An overview on fault management for electric vehicle onboard chargers," *Electronics*, vol. 11, no. 7, p. 1107, Mar. 2022. doi: 10.3390/electronics11071107.

38 I. A. Saifi and P. Saravana Prakash, "Power quality improvement using hexagon connected autotransformer based 24-pulse AC-DC converter," *IEEE Trans. Transp. Electrif.*, vol. 11, pp. 1–1, 2024. doi: 10.1109/TTE.2024.3406965.

39 M. S. Hossain Lipu *et al.*, "Power electronics converter technology integrated energy storage management in electric vehicles: Emerging trends, analytical assessment and future research opportunities," *Electronics*, vol. 11, no. 4, p. 562, Feb. 2022. doi: 10.3390/electronics11040562.

40 I. Mizher Baht, F. F. Al Sarhan, P.-M. Nicolae, and N. Bahat, "Micro grid operation mode with different MPPT controller," *TEM J.*, vol. 10, no. 1, pp. 158–164, Feb. 2021. doi: 10.18421/TEM101-19.

41 N. Kongtrakul, W. Wangdee, and S. Chantaraskul, "Comprehensive review and a novel technique on voltage unbalance compensation," *IET Smart Grid*, vol. 6, no. 4, pp. 331–358, Aug. 2023. doi: 10.1049/stg2.12106.

42 R. Khan *et al.*, "Energy sustainability–survey on technology and control of microgrid, smart grid and virtual power plant," *IEEE Access*, vol. 9, pp. 104663–104694, 2021. doi: 10.1109/ACCESS.2021.3099941.

43 R. Jain, Y. N. Velaga, K. Prabakar, M. Baggu, and K. Schneider, "Modern trends in power system protection for distribution grid with high DER penetration," *e-Prime Adv. Electr. Eng. Electron. Energy*, vol. 2, p. 100080, 2022. doi: 10.1016/j.prime.2022.100080.

44 Y. Geng, K. Yang, Z. Lai, P. Zheng, H. Liu, and R. Deng, "A novel low voltage ride through control method for current source grid-connected photovoltaic inverters," *IEEE Access*, vol. 7, pp. 51735–51748, 2019. doi: 10.1109/ACCESS.2019.2911477.

45 T. S. Tran, D. T. Nguyen, and G. Fujita, "The analysis of technical trend in islanding operation, harmonic distortion, stabilizing frequency, and voltage of islanded entities," *Resources*, vol. 8, no. 1, p. 14, Jan. 2019. doi: 10.3390/resources8010014.

46 H. Haes Alhelou, M. Hamedani-Golshan, T. Njenda, and P. Siano, "A survey on power system blackout and cascading events: Research motivations and challenges," *Energies*, vol. 12, no. 4, p. 682, Feb. 2019. doi: 10.3390/en12040682.

47 I. A. Saifi, M. Amir, A. Haque, and A. Iqbal, "Investigation of condition monitoring system for grid connected photovoltaic (GCPV) system with power electronics converters using machine learning techniques," *e-Prime Adv. Electr. Eng. Electron. Energy*, vol. 9, p. 100722, Sep. 2024. doi: 10.1016/j.prime.2024.100722.

48 S. R. Mohamed, P. A. Jeyanthy, D. Devaraj, M. H. Shwehdi, and A. Aldalbahi, "DC-link voltage control of a grid-connected solar photovoltaic system for fault ride-through capability enhancement," *Appl. Sci.*, vol. 9, no. 5, p. 952, Mar. 2019. doi: 10.3390/app9050952.

11

Integration of AI in Power Electronics in Real-time

Kurukuru Varaha Satya Bharath[1] *and Ahteshamul Haque*[2]

[1] *Silicon Austria Labs, Villach, Austria*
[2] *Advance Power Electronics and Research Laboratory, Jamia Millia Islamia (A Central University), New Delhi, India*

11.1 Overview

The primary objective of this chapter is to develop an Advanced Active Thermal Control (ATC) strategy to enhance the performance and extend the operational reliability of Silicon Carbide (SiC) Metal–Oxide–Semiconductor Field-Effect Transistors (MOSFETs). SiC MOSFETs are favored in power electronics for their superior capabilities in managing large currents and high voltages with high efficiency, enabled by their wide bandgap semiconductor properties. These devices exhibit lower on-resistance, higher breakdown voltages, and faster switching speeds compared to traditional silicon-based MOSFETs, making them ideal for high-power applications.

However, the advantages of SiC MOSFETs come with the challenge of rapid heating due to their high-power density. The thermal conductivity of SiC is significantly better than that of silicon, allowing for more efficient heat dissipation, but the high-power densities involved can still lead to rapid temperature increases. This rapid heating can compromise device efficiency, induce thermal stress, accelerate material degradation, and eventually lead to premature failure [1, 2].

To address these thermal management challenges, the proposed ATC strategy involves several key components:

- **Real-time Temperature Monitoring:** Integration of precise temperature sensors, such as thermocouples or infrared sensors, at critical points on the MOSFET die and package to continuously monitor the device's temperature. This data is essential for real-time assessment of thermal conditions.

Artificial Intelligence for Power Electronics, First Edition.
Edited by Ahteshamul Haque, Saad Mekhilef, and Azra Malik.

- **Dynamic Thermal Management:** Implementation of dynamic control algorithms that adjust operating parameters such as switching frequency, duty cycle, and gate drive strength based on real-time temperature data. This adaptive approach helps prevent thermal runaway by modulating power dissipation in response to temperature fluctuations.
- **Thermal Modeling and Prediction:** Development of accurate thermal models that predict temperature rise based on electrical input parameters and operating conditions. These models use finite element analysis (FEA) or computational fluid dynamics (CFD) to simulate heat distribution and identify potential hotspots.
- **Active Cooling Techniques:** Incorporation of active cooling methods such as liquid cooling systems, thermoelectric coolers, or heat pipes. These systems actively remove excess heat from the MOSFET, maintaining optimal operating temperatures even under high power conditions.
- **Integrated Control Systems:** Creation of a control system that combines sensor feedback with predictive models to dynamically adjust thermal management strategies. This system utilizes digital controllers or field-programmable gate arrays (FPGAs) for high-speed processing and response.
- **Reliability Enhancements:** Focus on enhancing the long-term reliability of SiC MOSFETs by reducing thermal cycling stresses through controlled ramping of power levels and by ensuring even heat distribution across the device.

The ATC strategy aims to maintain the SiC MOSFETs within a safe temperature range under varying operational loads, thereby maximizing their performance and lifespan. This involves continuous monitoring and adjustments to prevent thermal runaway – a critical requirement for maintaining system integrity and reliability in demanding applications such as renewable energy converters, electric vehicles, and advanced power supply systems [3, 4].

The integration of these advanced thermal management techniques not only mitigates the risk of overheating but also promotes improved energy efficiency and extended device longevity. The effectiveness of this approach is supported by empirical data and simulations from previous research, which highlight the thermal challenges specific to SiC MOSFETs and demonstrate the innovative solutions developed to address them.

11.2 Control Development

11.2.1 Electro-Thermal Modeling

Deriving a power loss model for a SiC MOSFET power module involves considering various factors that contribute to power dissipation in the device. The power loss in a SiC MOSFET can be categorized into conduction losses, switching losses, and gate drive losses [5].

- **Conduction Losses:**
  Conduction losses occur during the on-state operation of the MOSFET when current flows through the device, and they are primarily influenced by the on-state resistance ($R_{ds(on)}$) and the root-mean-square current (I_{rms}) [5, 6]. The I_{rms} represents the effective or equivalent steady-state current flowing through the MOSFET during its on-state. This is crucial because MOSFETs operate in switching applications where the current waveform is typically non-constant. The RMS current is calculated using the following equation:

$$I_{rms} = \sqrt{\frac{1}{T}\int_0^T i^2(t)dt} \tag{11.1}$$

  where $i(t)$ is the instantaneous current and T is the period of one complete cycle. Further, the $R_{ds(on)}$ is a pivotal parameter in determining conduction losses. It signifies the resistance of the MOSFET when it is fully conducting. The conduction losses ($P_{conduction}$) can be expressed by the following equation:

$$P_{conduction} = I_{rms}^2 \cdot R_{ds(on)} \tag{11.2}$$

  This highlights the quadratic relationship between conduction losses and both RMS current and on-state resistance. Usually, the choice of MOSFET with lower on-state resistance is critical for minimizing conduction losses, enhancing the overall efficiency of the power module. But, in practical applications, the on-state resistance can vary with temperature and other operating conditions. Moreover, the temperature dependence of the on-state resistance can be modeled using temperature coefficients provided by the datasheet. To further investigate into the complexities of conduction losses, it's essential to consider the impact of variations in RMS current and on-state resistance on power dissipation. Increasing the RMS current or on-state resistance results in higher conduction losses, emphasizing the need for careful consideration of these parameters during the design phase. The total conduction losses contribute to the overall power dissipation in the SiC MOSFET power module. These losses form the foundation of the broader power loss model, which includes both conduction losses and switching losses.
- **Switching Losses:**
  Switching losses in a MOSFET power module arise from the dynamic switching transitions between the on-state (conduction) and off-state. Switching losses can be further divided into two main components: turn-on (T_{on}) losses and turn-off (T_{off}) losses. The turn-on losses are associated with the energy dissipated during the switching transition from the off-state to the on-state. The primary contributors are the charging of the input capacitance (C_{iss}) and the associated voltage

across the MOSFET. The C_{iss} represents the capacitance between the gate and source terminals of the MOSFET [7–9]. During turn-on, the gate voltage (V_{in}) rises, causing C_{iss} to charge. The energy (E_{on}) dissipated during this charging process is given by:

$$E_{on} = \frac{1}{2} \cdot C_{iss} \cdot V_{in}^2 \tag{11.3}$$

The corresponding T_{on} is the energy dissipation per switching cycle and is calculated using the switching frequency (f_{sw}):

$$T_{on} = E_{on} \cdot f_{sw} \tag{11.4}$$

Furthermore, the turn-off losses are associated with the energy dissipated during the transition from the on-state to the off-state. The primary contributors are the discharging of the output capacitance (C_{oss}) and the associated voltage across the MOSFET. The C_{oss} represents the capacitance between the drain and source terminals of the MOSFET. During turn-off, when the gate voltage is removed, C_{oss} discharges. The energy (E_{off}) dissipated during this discharging process is given by:

$$E_{off} = \frac{1}{2} \cdot C_{oss} \cdot V_{in}^2 \tag{11.5}$$

The corresponding turn-off loss (T_{off}) is the energy dissipation per switching cycle and is calculated using the switching frequency (f_{sw}):

$$T_{off} = E_{off} \cdot f_{sw} \tag{11.6}$$

From the above equations, it is seen that the total switching losses encompass both T_{on} and T_{off} losses. These losses arise during the dynamic transition of the MOSFET between its on-state and off-state. The total switching losses per switching cycle ($P_{switching}$) are given by:

$$P_{switching} = T_{on} + T_{off} \tag{11.7}$$

The total switching losses in SiC MOSFET power modules are directly proportional to the f_{sw}, where higher frequencies lead to more frequent transitions, resulting in increased losses. Besides, the shape and characteristics of voltage and current waveforms during switching transitions significantly impact total switching losses, with well-controlled transitions minimizing losses and phenomena like ringing and overshoot contributing to additional losses. Moreover, dynamic effects, including parasitic elements like stray inductance and capacitance, should be considered in total switching loss calculations, and advanced simulation tools are valuable for accurately capturing these effects. Additional factors, such as body diode conduction losses during turn-off and dynamic effects, temperature effects, influencing $R_{ds(on)}$ and capacitance characteristics, play crucial roles in understanding the switching losses in SiC MOSFET power modules.

- **Gate Drive Losses:**
  In addition to the conduction and switching losses, there are gate drive losses in a power module which are associated with the energy dissipated during the charging and discharging of the MOSFET gate capacitance [7, 10]. In the analysis of gate drive losses, the gate charge (Q_g), as specified in the MOSFET datasheet, embodies the total charge required to transition the gate from one voltage level to another. This encompasses the charge needed for both turning the MOSFET on $Q_{g(on)}$ and off $Q_{g(off)}$. The V_{in}, serving as the input to the gate driver circuit, significantly influences the energy necessary for gate capacitance charging and discharging. The f_{sw}, representing the rate at which the MOSFET transitions between states, determines the frequency of gate charging and discharging cycles. The gate drive power loss (P_{gate}) can be precisely calculated as:

  $$P_{gate} = Q_g \cdot V_{in} \cdot f_{sw} \tag{11.8}$$

  Additionally, the impact of a gate resistor (R_g) on gate drive losses is crucial; higher gate resistors may reduce losses but result in slower switching speeds. The power waveform during gate charging and discharging follows a triangular shape, with the area under the curve representing the energy dissipated as gate drive losses. Finally, it is essential to consider temperature dependence, as gate drive losses can be influenced by temperature variations, requiring accurate consideration of temperature-dependent MOSFET characteristics, such as gate charge, in power loss calculations.
  From the above descriptions, the power loss model is defined as the sum of conduction, switching, and gate drive losses.

11.2.2 Thermal Model

Once the power loss model is established, the next step involves determining a thermal model for the power MOSFET to characterize its transient behavior and estimate the junction temperature under various heat pulse conditions. This approach eliminates the necessity for costly and time-consuming experimental testing for each configuration. To emphasize a crucial aspect, this section delves into the creation of an analytical model for a temperature-dependent RC network applicable to a power module featuring SiC MOSFETs. This concept has been explored in several publications, with Gorecki et al. [11] and Szekely [12] presenting methodologies for determining the thermal RC values of a structure through experimental transient heating or cooling temperature curves. Typically, these curves demonstrate a step function response. Bagnoli et al. [13] introduced the TRAIT analysis, wherein the heating or cooling curve of a device is represented using an exponential term. Expanding on these methodologies, researchers in [14, 15] applied the Variable Angle Method (VAM) to construct a Cauer RC network for a physical structure, enabling the dynamic modeling of thermal behavior in semiconductor packages. Considering the insights gathered from the literature, an analytical approach has been formulated for the calibration of the transient thermal network within a power module. The effective heat propagation path serves as the basis for constructing an equivalent RC network that captures the diverse layers of materials within the package. In the Cauer network configuration, each lump corresponds to a layer, and its thermal resistance is established based on the volume of that layer. The determination of the heat source and its projection onto the boundary of the subsequent layer involves the utilization of spreading angles, namely α_1, α_2, β_1, and β_2. These angles are contingent on the thermal conductivity and specific heat capacity of each layer and undergo variations due to the asymmetric placement of the heat source in relation to the substrate. The thermal resistance within the defined volume is computed using the following equation:

$$R_{th} = \frac{1}{4kl_x} \frac{1}{\gamma_e \tan\alpha - \tan\beta} \ln \frac{l_x + w\tan\alpha}{l_x + w\tan\frac{\beta}{\gamma_e}} \tag{11.9}$$

where γ_e is the aspect ratio of the heating element, and tan α and tan β are the spreading angles. Further, the thermal capacitance is calculated as follows:

$$C_{th} = V \cdot \rho \cdot c_\rho \tag{11.10}$$

where V is the active volume, ρ is the medium density, and c_ρ is the heat capacity.

Due to the distributed nature of the power MOSFET, acquiring straightforward dynamic models connecting thermal mass to the heat flow path proves challenging. Nevertheless, formulating the problem with a specified heat propagation path allows for the computation of thermal resistance within a precisely defined volume. This volume, based on the thermal resistance approximation, encompasses the heat flow, assuming adiabatic limiting walls, despite the partial inaccuracy of this assumption. Nevertheless, this model remains valid since the outer volume's influence is minimal in the overall heat conduction process. Consequently, the same volume employed for calculating thermal resistance is also utilized for determining thermal capacitance, as outlined by the equation [14, 15]:

$$V = 4\left[l_x l_y \omega + (l_x \tan\beta + l_y \tan\alpha)\frac{w^2}{2} + \tan\beta\tan\alpha\frac{w^3}{3}\right] \tag{11.11}$$

The inclusion of spreading angles considers the dynamics of transient thermal modeling, offering a more precise depiction of semiconductor packages. A brief schematic of the developed approach is shown in Figure 11.1. Further, to aid the analytical approach in accurate estimation of the T_{vj}, Multiphysics simulations are conducted to understand the high-temperature behavior of the SiC MOSFET as

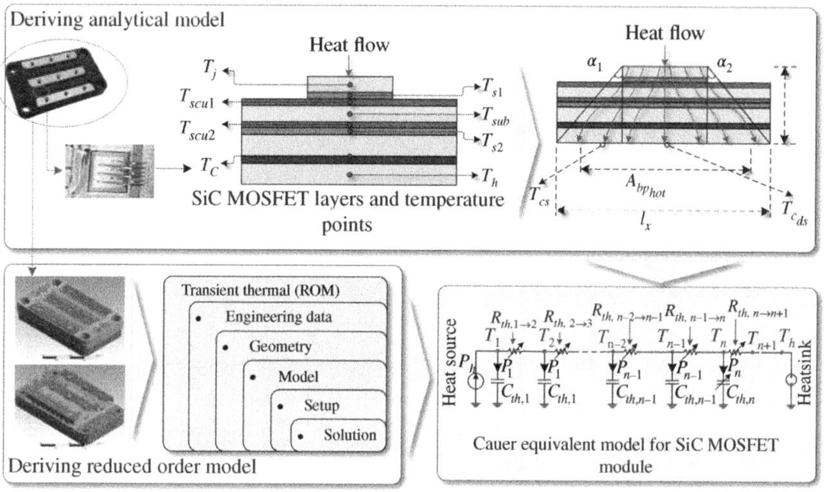

Figure 11.1 Thermal model derived from transient thermal characterization.

shown in Figure 11.1. This approach represents each layer of the packaged module as a thermal capacitance C_{th}, with thermal resistances R_{th} denoting the interconnections between the layers. To establish the temperature dependence at each layer, finite element method (FEM) simulations are employed on the modeled SiC MOSFET. These simulations consider the intricate interplay between electrical conduction, heat transfer, and material properties, providing a comprehensive analysis of the device's thermal behavior. The geometry and material properties of the SiC MOSFET have a significant influence on its thermal distribution. Therefore, defining temperature measurement points (TMPs) within individual layers becomes essential. TMPs serve as critical locations for monitoring the temperature response within the device. Analyzing the temperature response at each TMP obtained from the FEM simulations enables the estimation of R_{th} and C_{th}, as described in Eq. (11.9):

$$R_{th,n-1\rightarrow n+1} = \frac{\left(T^{ss}_{n-1} - T^{ss}_{n+1}\right)}{P_h} \tag{11.12}$$

where T^{ss}_{n+1} represents the steady-state temperatures at $n+1$ *TMPs* as identified in Figure 11.1, and P_h is the constant heating power.

$$C^{th}_n = \frac{P_n}{\frac{dT_n}{dt}} \tag{11.13}$$

where P_n is the transient power flowing through the capacitor at TMP_n, and T_n is the transient temperature response at TMP_n. P_n should be defined for each Cauer

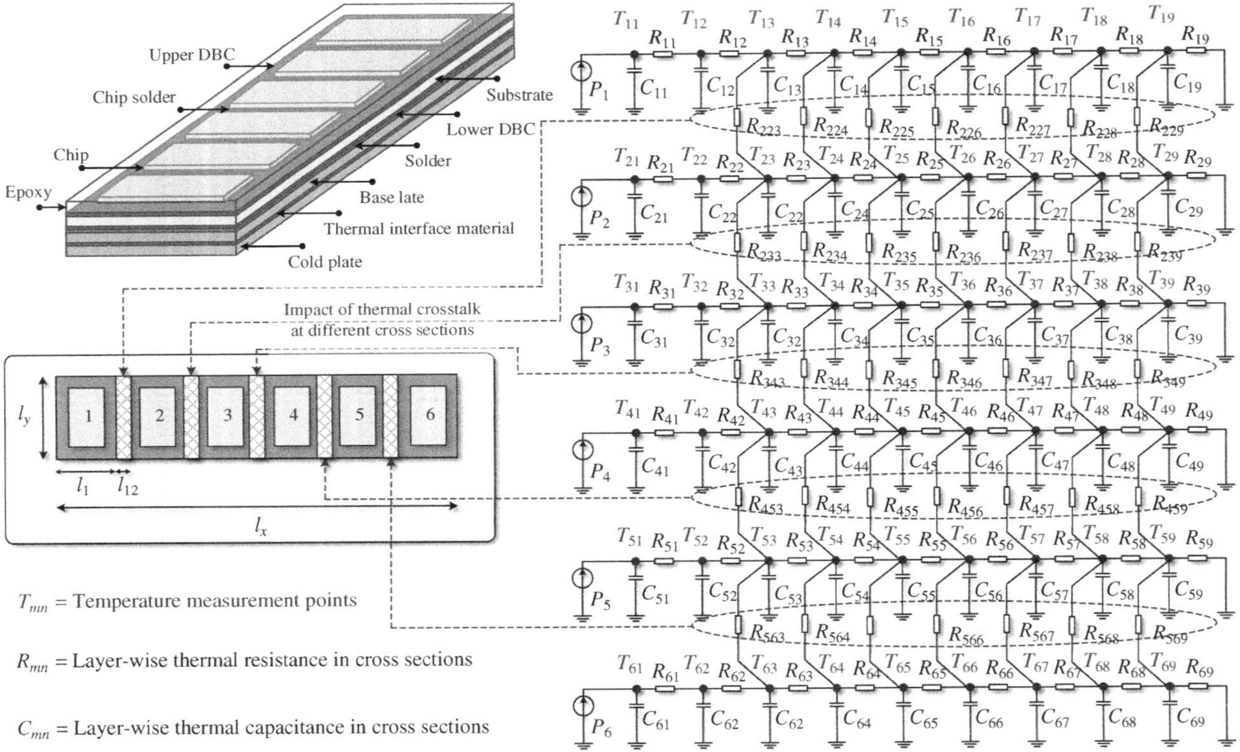

Figure 11.2 Impact of thermal cross talk in a power module identified through transient thermal characterization.

cell, as the power, which is injected in the capacitance node. Further, P_n is estimated as

$$P_n = \frac{T_{n-1} - T_n}{R_{th,n-1\to n}} - \frac{T_n - T_{n+1}}{R_{th,n\to n+1}} \tag{11.14}$$

This development for the transient thermal characterization of a power MOSFET involves a comprehensive process that combines simulations and modeling techniques. Beyond its suitability for single-chip setups, the above-discussed analytical approach can be expanded to analyze multi-chip configurations, facilitating the evaluation of thermal crosstalk within a semiconductor package. In instances where multiple chips are integrated into a single package, intricate thermal interactions arise among the chips and their surrounding materials. By integrating the notion of thermal crosstalk – referring to the impact of one chip's temperature on neighboring chips as shown in Figure 11.2, a more thorough comprehension of the thermal behavior of the entire package can be attained.

11.2.3 Thermal Observer

The data obtained from transient thermal characterization is utilized to estimate temperatures through the training of Support Vector data descriptor. The proposed method of data description involves the initialization of a hypersphere defined by a radius R and center a. The primary objective is to minimize the hypersphere's volume, achieved by minimizing R^2. An error function $F(R, a)$ is introduced, incorporating a penalty term using slack variables ξ_i, leading to a formulation that balances the hypersphere's size with the inclusion of data points. Constraints are imposed to ensure that all data points x_i lie within the defined hypersphere. The introduction of slack variables ξ_i allows for the penalization of distances greater than R^2, addressing the potential presence of outliers in the training set. Further, the Lagrange multiplier formulation is employed to incorporate these constraints into the objective function. This results in a dual formulation that maximizes a set of Lagrange multipliers (α_i), identifying support vectors crucial for the data description. The proposed method draws parallels with Support Vector Machines (SVM), particularly in the formulation of an optimization problem that seeks to minimize errors while considering a user-defined parameter controlling the fraction of data to be isolated. The approach is further extended through kernelization, enhancing the system's flexibility by replacing inner products with kernels. The optimization problem is transformed, resulting in an equivalent formulation to SVM, demonstrating the versatility and adaptability of the proposed method. More details of the approach are thoroughly discussed in [16]. Furthermore, labeling the characterization data with various temperature instances representing expected and exceeded

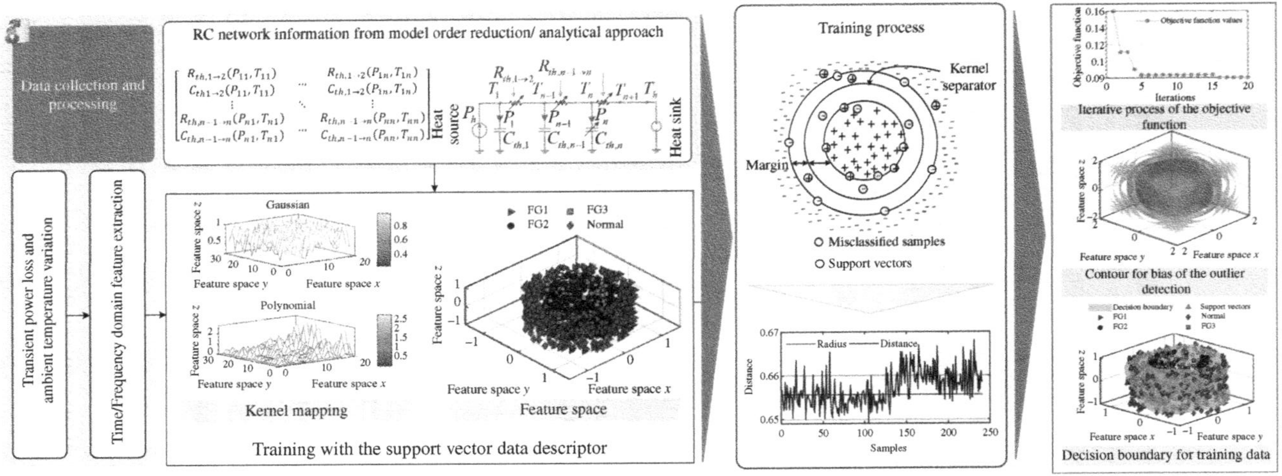

Figure 11.3 Thermal observed developed using support vector data descriptor.

temperature thresholds, an SVDD model with an appropriate kernel function encapsulates the "normal" data within the smallest hypersphere while maximizing the margin to detect anomalies. Once validated, this model estimates the junction temperature for given power loss data, and if the estimated temperature rise surpasses a predefined safety threshold, it serves as a warning for potential overheating, allowing timely preventive actions. The training process of the proposed approach is shown in Figure 11.3. Continuous model updating ensures adaptability to evolving operational conditions, making it a valuable tool for real-time thermal management and safety control in MOSFET applications. Subsequently, the trained model will be integrated into the thermal control system to provide estimated temperatures for each MOSFET in a closed-loop manner. The control module will then work to achieve minimally invasive thermal control based on these estimations.

Parallelly, to validate the estimated temperature values, the trained model along with the power loss model is subjected to sinusoidal excitation approach [17–19] based on the 3ω method [20, 21] which is widely established for measuring the thermal conductivity of dielectric solids. This approach determines the thermal conductivity by analyzing the radial flow of heat from the device under test (DUT). Further, by analytically solving the temperature oscillations in the line, an expression is established between the slope of the in-phase temperature oscillations versus frequency, and the thermal conductivity. Moreover, the temperature oscillations of the DUT are measured using the third harmonic component of the voltage across the line. A current at the angular frequency ω heats the sample as 2ω and produces a temperature oscillation at 2ω. The resistance of the DUT increases with increasing temperature and therefore exhibits a small AC component at 2 if a linear relationship between current and resistance is considered. The resistance oscillation (2ω) times the original driving current (ω) produces a small oscillation of the voltage across the line at 3ω. The advantage with this approach is that it derives a non-linear model of the DUT and then linearizes it around a certain operating condition to obtain a closed-form solution of the temperature estimation by considering the 3ω component. Generally, this estimation process considers several key parameters, including power dissipation, thermal resistance, and thermal capacitance.

Power dissipation is affected by various factors, including the on-state resistance $R_{ds(on)}$, applied input current, and energy losses during turn-on and turn-off processes. These elements directly contribute to the heat generated by the MOSFET during its operation, and higher power dissipation results in an elevated junction temperature. Additionally, the thermal resistance (R_{th}) plays a crucial role in dissipating heat from the MOSFET and is influenced by the device's structure and material properties, such as the number of layers, layer thickness, and thermal

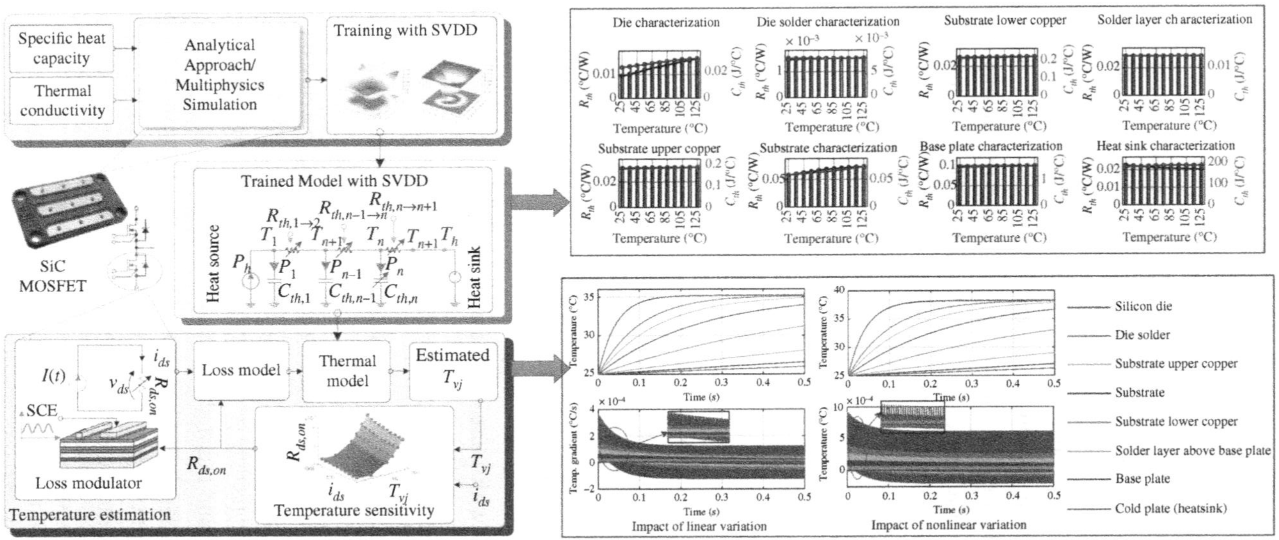

Figure 11.4 Transient thermal characterization using sinusoidal excitation for junction temperature estimation.

conductivity. A higher thermal resistance impedes efficient heat transfer, leading to higher junction temperatures for a given power dissipation. In contrast, thermal capacitance (C_{th}) is pivotal in the MOSFET's ability to store and dissipate heat efficiently, depending on the specific characteristics and composition of the device. A MOSFET with greater thermal capacitance excels at absorbing and releasing heat, contributing to the maintenance of lower junction temperatures. The schematic illustrating the transient thermal characterization using a sinusoidal current excitation approach for junction temperature estimation is depicted in Figure 11.4. The corresponding results in the figure show the variation of R_{th} and C_{th} at different layers for a given power loss and estimated temperature at each layer under the operation of transient thermal network.

11.2.4 Methods for Active Loss Manipulation

The method for active loss manipulation for achieving ATC utilizes the thermal observer discussed in Section 11.2.3. The control scheme is developed for the 3-phase dual inverter system, driving a Permanent Magnet Synchronous Motor (PMSM) through Field-Oriented Control (FOC). Typically, space-vector modulation is employed in the control scheme, and this is replaced with discontinuous pulse-width modulation (DPWM) for active loss manipulation [22]. DPWM is a modulation technique that, during specific intervals, confines the phase leg to either the positive or negative terminals of the DC bus by introducing a zero-sequence component u_z^* to the original references u_{jo}^*. The generation of the zero-sequence component is based on the information derived from the output voltage references. For a three-phase converter, the normalized output voltage references can be expressed as follows:

$$\begin{cases} u_{a0}^* = \dfrac{1}{2}m \cdot \sin(\omega_0 t) \\ u_{b0}^* = \dfrac{1}{2}m \cdot \sin(\omega_0 t - 120^\circ) \\ u_{c0}^* = \dfrac{1}{2}m \cdot \sin(\omega_0 t + 120^\circ) \end{cases} \tag{11.15}$$

Given that m represents the modulation index and ω_0 denotes the angular frequency of the AC voltage, u_{max} and u_{min} as the maximum and minimum values of the normalized three-phase output voltage references, respectively, using the following definitions:

$$\begin{cases} u_{max} = \max\left(u_{a0}^*, u_{b0}^*, u_{b0}^*\right) \\ u_{min} = \min\left(u_{a0}^*, u_{b0}^*, u_{b0}^*\right) \end{cases} \tag{11.16}$$

In discontinuous modulation, the maximum clamping interval within one fundamental period remains constant at 120° for one phase, irrespective of whether it is clamped to the positive or negative terminals of the DC bus. The DPWM introduces a new variable, α, to determine the positive interval θ_p, representing the clamping interval over one fundamental period when the phase leg is clamped to the positive terminal of the DC bus, and the negative interval θ_n, indicating the clamping interval during one fundamental period when the phase leg is clamped to the negative terminal. To minimize switching losses, the zero-sequence component is designed to achieve the maximum clamping interval $\left(\theta_p + \theta_n = \frac{2\pi}{3}\right)$. The expression for the zero-sequence component is given as follows:

$$u_{z1}^* = \begin{cases} \frac{1}{2} - u_{max}, u_{max} + u_{min} \geq \alpha \\ -\frac{1}{2} - u_{min}, u_{max} + u_{min} < \alpha \end{cases} \tag{11.17}$$

While α can take values larger than 0.25 m or smaller than −0.25 m, it will exert an identical impact on u_{z1}^* when α equals 0.25 m or −0.25 m. Therefore, α is restricted within the range [−0.25 m, 0.25 m]. In addition to these positive and negative intervals, the phase angle becomes another determinant for the zero-sequence component. To enable flexible generation of the zero-sequence component using the generalized DPWM, a phase angle control variable, β, is introduced. β is constrained to the interval $\left[0, \frac{2\pi}{3}\right]$ and is utilized to generate a trigger signal for the zero-sequence component. The modification of normalized three-phase output voltage references is necessary to obtain the trigger signal:

$$\begin{cases} u_{a0t}^* = \frac{1}{2} m \cdot \sin(\omega_0 t - \beta) \\ u_{b0t}^* = \frac{1}{2} m \cdot \sin(\omega_0 t - 120^\circ - \beta) \\ u_{c0}^* = \frac{1}{2} m \cdot \sin(\omega_0 t + 120^\circ - \beta) \end{cases} \tag{11.18}$$

u_{maxt} and u_{mint} values are determined by the modified normalized three-phase output voltage references, representing the upper and lower limits, respectively.

$$\begin{cases} u_{maxt} = \max\left(u_{a0t}^*, u_{b0t}^*, u_{b0t}^*\right) \\ u_{mint} = \min\left(u_{a0t}^*, u_{b0t}^*, u_{b0t}^*\right) \end{cases} \tag{11.19}$$

Given α, the trigger signal can be formulated as follows:

$$u_{tri} = \begin{cases} 1, u_{maxt} + u_{mint} \geq \alpha \\ 0, u_{maxt} + u_{mint} < \alpha \end{cases} \tag{11.20}$$

Using the trigger signal u_{tri}, the modification of the zero-sequence component in Eq. (11.17) is as follows:

$$u_z^* = \begin{cases} \frac{1}{2} - u_{max}, u_{tri} = 1 \\ -\frac{1}{2} - u_{min}, u_{tri} = 0 \end{cases} \tag{11.21}$$

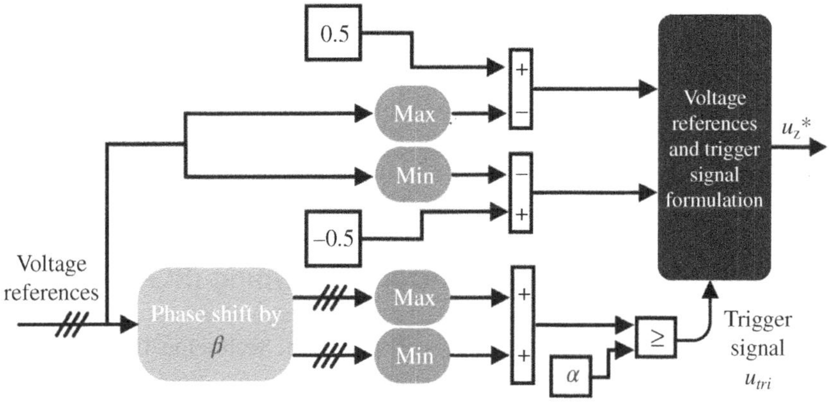

Figure 11.5 Zero-sequence component generation block diagram.

Utilizing variables α and β, the developed generalized DPWM can dynamically generate diverse zero-sequence components. The schematic in Figure 11.5 offers an overview of the zero-sequence component generator within the DPWM. Furthermore, the values of α and β are selected in such a way as to minimize switching loss when applying the generalized DPWM [23].

Hence, the selection of α and β is mapped to the temperature variation of each MOSFET in the converter. Mapping temperature variations to α and β involves establishing a dynamic relationship with the thermal conditions of the MOSFETs in the power module, aiming to optimize the modulation strategy based on changing temperatures and ensuring the MOSFETs operate within safe temperature limits. Temperature mapping begins by acknowledging that higher temperatures induce increased power losses due to elevated conduction and switching losses in the MOSFETs. The clamping interval control variable (α), governing positive and negative clamping intervals during one modulation period, can be dynamically adjusted in response to temperature changes. For higher temperatures, where MOSFET switching losses are prominent, increasing α reduces the switching frequency, thereby mitigating losses. Conversely, in lower temperatures, α is decreased to exploit improved switching characteristics. The phase angle control variable (β), associated with generating the zero-sequence component in DPWM, is also responsive to temperature. Higher temperatures suggest a need to decrease the zero-sequence component to minimize losses, prompting an adjustment of β to reduce its contribution. Conversely, lower temperatures may permit a higher zero-sequence component, encouraging an increase in β [22]. This mapping, complex and requiring detailed understanding, is crucial for balancing power losses and thermal conditions in the power module.

An adaptive control strategy enhances this process by dynamically adjusting α and β based on real-time feedback from temperature sensors or estimates obtained through the SVDD approach. Continuous monitoring of temperature enables the adaptive control algorithm to optimize α and β in real-time, ensuring an effective modulation strategy tailored to the prevailing thermal conditions.

11.3 Overview of Rapid Control Prototyping (RCP)

Rapid Control Prototyping (RCP) is a development technique that allows to quickly test and implement control strategies using real-time systems. This method is particularly important in scenarios where the control demands both precision and adaptability under variable operational conditions. RCP is crucial because it provides a dynamic testing environment where theoretical control strategies can be applied to physical hardware immediately after their conception [24]. This instant feedback loop is invaluable for testing the viability and effectiveness of control strategies under real-world conditions. It significantly reduces the time and resources spent on developing prototypes, allowing engineers to fine-tune controls and algorithms much faster than traditional development cycles would permit.

Figure 11.6 depicts the systematic workflow of developing and validating a control system for power electronics applications. The process begins with

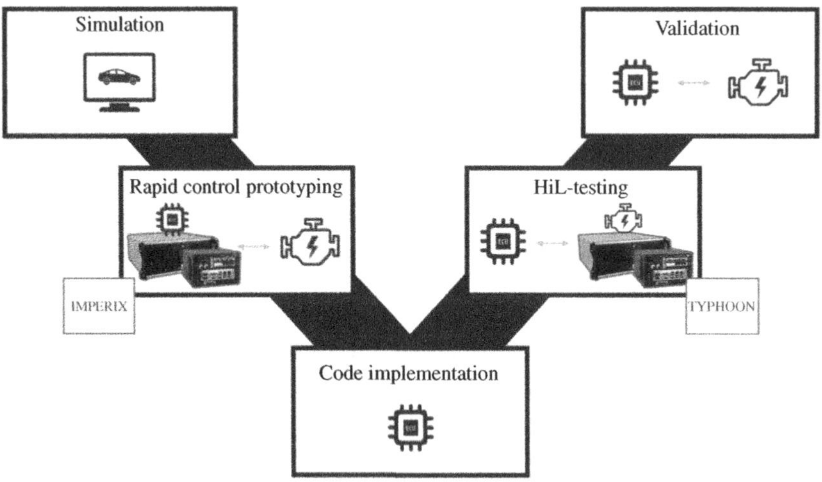

Figure 11.6 Workflow integration of simulation, rapid control prototyping, HIL testing, and code implementation.

the simulation phase, where models are created to represent the system's behavior. These models are then used within an RCP framework, facilitated by an Imperix B-Box, which allows for real-time testing and adjustment of control algorithms. Following RCP, the Hardware-in-the-Loop (HIL) testing phase uses a Typhoon HIL setup to simulate real-world physical processes and validate the control strategies under various scenarios, ensuring that the system responds correctly to simulated inputs. The final phase involves code implementation, where the validated control algorithms are coded into an electronic control unit (ECU) for deployment in the actual hardware system. This integrated approach ensures that the developed system is thoroughly tested and validated at each stage, resulting in a robust and reliable control strategy ready for real-world application [25].

In the context of a converter which involves managing the thermal performance of SiC MOSFETs, RCP enables the integration and testing of advanced thermal management techniques that are critical for preventing overheating and ensuring optimal performance. The ability to rapidly iterate and modify these controls in response to actual hardware performance data helps in accurately managing the rapid thermal dynamics associated with SiC MOSFETs. This is especially critical in high-power applications where the thermal stability of MOSFETs can directly impact the reliability and efficiency of the entire system. Furthermore, RCP allows for the integration of complex, multi-variable control algorithms that can react in real-time to changes in the system's environment or operational status. This adaptability is key in converters where the performance and safety of power electronic systems cannot be compromised. By using RCP, engineers can ensure that the developed thermal management strategies are not only theoretically sound but also practically feasible and robust against the unpredictable nature of real operational environments.

11.3.1 Imperix B-Box

The Imperix B-Box is an advanced controller specifically designed for the rapid prototyping of power electronics systems. Its core functionality revolves around its ability to execute complex control algorithms at high speeds, which is essential for managing the fast dynamics typically found in SiC MOSFET-based power systems. The B-Box is equipped with high-resolution analog and digital inputs and outputs, which facilitate precise measurement and control of electrical parameters. A key feature of the B-Box is its real-time execution environment, which allows it to respond to changes in the system within microseconds, crucial for applications that require immediate reaction to prevent system failures, such as in thermal management of power modules [26]. Its modular design supports a variety of plug-in cards, enabling customization to fit specific

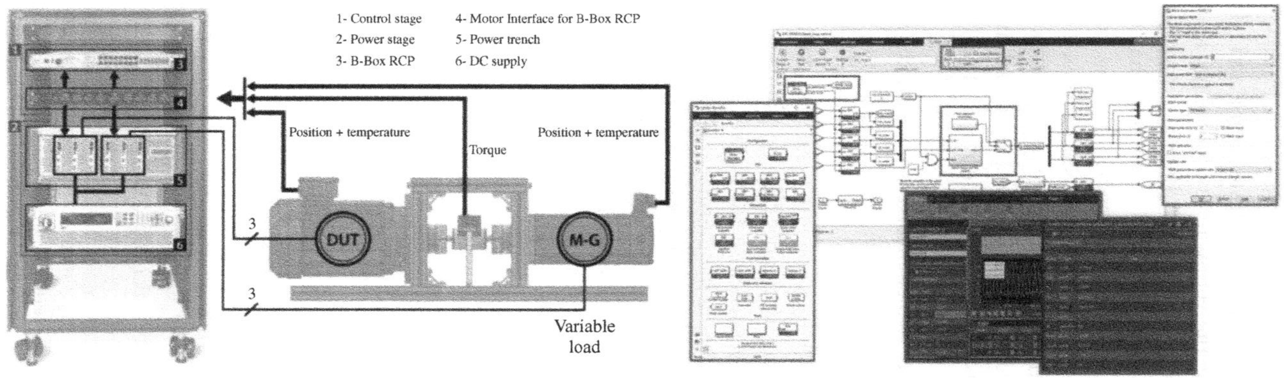

Figure 11.7 Integrated system setup for HIL-RCP validation and control strategy implementation.

project requirements, from basic experiments to complex industrial systems. The programming interface of the B-Box is designed to be user-friendly, supporting graphical programming via its own software or integration with other tools like MATLAB/Simulink. This ease of use accelerates the development process, allowing engineers to focus more on innovation and less on setup and troubleshooting.

Figure 11.7 illustrates an integrated setup for validating a control strategy through HIL and RCP. On the left side of the image is the physical setup featuring the Imperix B-Box for RCP, a power stage for executing power conversion tasks, and a DC power supply unit, all of which are interconnected. This setup is used to implement and test control strategies developed in the simulation phase. The central part of the image represents a schematic of a test bench with a DUT connected to a motor–generator (M–G) unit simulating a variable load, highlighting the points for position, temperature, and torque measurements. On the right, the Automated Code Generation Software Development Kit (ACG-SDK) Simulink interface shows the control logic that includes PWM generation and various control parameters, interfacing with the B-Box. This interface enables real-time analysis and tuning of the system, completing the loop from simulation to actual deployment.

11.3.2 ACG-SDK in MATLAB

The ACG-SDK (Automated Code Generation Software Development Kit) for MATLAB is a critical tool for engineers using MATLAB/Simulink for the development of control systems. This SDK facilitates the direct translation of Simulink models into executable code that can run on controllers like the Imperix B-Box. This seamless integration is invaluable for model-based design, where systems are first designed and tested virtually in MATLAB. Using the ACG-SDK, engineers can develop their control strategies in the familiar MATLAB/Simulink environment and then automatically generate C code that is optimized for real-time performance on the B-Box [27]. This capability not only speeds up the prototyping process but also enhances the reliability of the control systems by allowing for extensive pre-deployment testing in a simulated environment. Additionally, the ACG-SDK supports real-time parameter tuning and logging, features that are essential for iterative development processes. Engineers can adjust parameters on the fly and immediately see the effects of these changes, enabling fine-tuning of control strategies based on real-world data. This real-time interaction between software and hardware through the ACG-SDK minimizes development risks and helps ensure that the final product performs as expected under actual operating conditions.

11.4 System Configuration

11.4.1 Description of the 3-Phase Dual Converter System

The converter topology utilized in this research is a 3-phase dual inverter configuration, specifically engineered to maximize power output while optimizing the physical layout for efficiency and thermal management. This configuration shown in Figure 11.8 and Table 11.1 comprises two interconnected half-bridges per phase. Employing a dual half-bridge per phase effectively doubles the current carrying capacity relative to a single half-bridge setup. This enhancement is critical as it significantly reduces the voltage drops across the semiconductor switches and diminishes power dissipation in each switch. Lower power dissipation is pivotal in maintaining high efficiency and stability, as it directly impacts the thermal load managed by the cooling system and the overall reliability of the power converter.

In this specific configuration, the converter drives a PMSM using FOC and uses a 1200 V 480 A SiC MOSFETs. The SiC MOSFET allows for an enhanced design that supports high load currents without the adverse effects of increased thermal burden typically associated with high current densities.

11.4.2 Integration with FOC and DPWM

The FOC technique, also known as vector control, decomposes the motor's stator currents into two orthogonal components that can be controlled independently: one aligned with the rotor's magnetic field (d-axis component) and one perpendicular to it (q-axis component). These components represent the magnetic flux and torque production, respectively. The implementation of FOC involves measuring the rotor position and using this information to calculate the required current

Table 11.1 Technical Specifications for the 3-Phase Dual Inverter Configuration.

Parameter	Value
DC voltage range	40–900 V
DC link capacitor value	C = 370 μF per Stack
Switching frequency range	2–100 kHz
Transversal inductance	Approximately 300 μH
Longitudinal inductance	Less than 500 nH
Nominal current (RMS)	315 A/half-bridge
Nominal power ratings	440 kVA/Stack

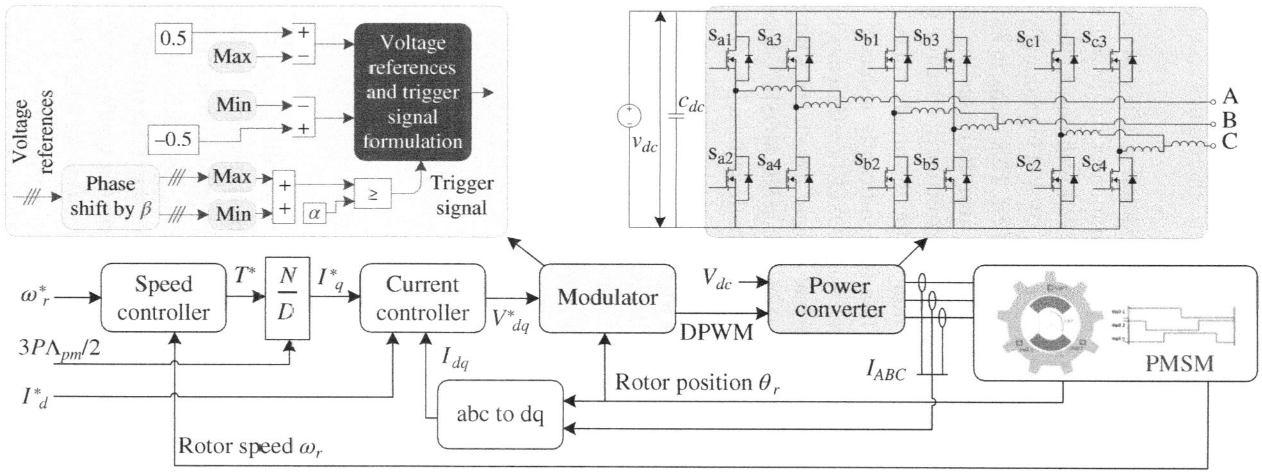

Figure 11.8 Schematic of a 3-phase dual inverter configuration.

components that will produce the desired torque with minimal losses. By continuously adjusting these currents in response to changes in motor load and speed, FOC can significantly improve the efficiency and dynamic response of a motor. This precise control is essential in applications where quick acceleration and high torque accuracy are necessary.

Further, Discontinuous Pulse Width Modulation (DPWM) is a modulation scheme used to control the timing and duration of the power switches in the inverter driving the motor. Unlike standard PWM techniques which keep switches active throughout the entire electrical cycle, DPWM intentionally creates periods where all switches in a phase leg are turned off. This break in switching reduces the overall switching frequency, which in turn decreases the switching losses in the power semiconductors. DPWM is particularly effective in reducing the thermal load on SiC MOSFETs. By minimizing the number of switch transitions, the heat generated by the MOSFETs is significantly reduced. This reduction in thermal stress not only enhances the MOSFETs' efficiency but also extends their lifespan by reducing the thermal fatigue associated with high-frequency switching. By integrating FOC and DPWM in a SiC MOSFET-based 3-phase dual converter system utilizes the strengths of both approaches to manage thermal dynamics effectively. FOC optimizes the current paths for energy efficiency, reducing unnecessary power dissipation which can cause heat build-up. Meanwhile, DPWM decreases the frequency of thermally stressful events, such as switch transitions that generate heat. Together, these strategies reduce the operational temperature of the SiC MOSFETs, maintaining them within safe operational limits even under high load conditions. This careful control of temperature is critical in preventing thermal runaway situations where excessive heat could lead to device failure.

11.5 Development Process

11.5.1 System Design in HIL

The development of the converter model was a critical step in accurately representing the real-world behavior of a 3-phase dual converter system equipped with SiC MOSFETs. This model was intricately designed within the Typhoon HIL 606 simulation environment as shown in Figure 11.9 to capture the complex dynamics of the system with high precision. One of the most critical aspects captured in the converter model was the detailed thermal response of SiC MOSFETs. The simulation incorporated the specific heat capacity (J/kg K) and thermal resistance (K/W) parameters intrinsic to Silicon Carbide, which dictate the thermal behavior of these devices under operational stress. In the process of modeling the thermal behavior of Silicon Carbide (SiC) MOSFETs, two parameters are pivotal: specific

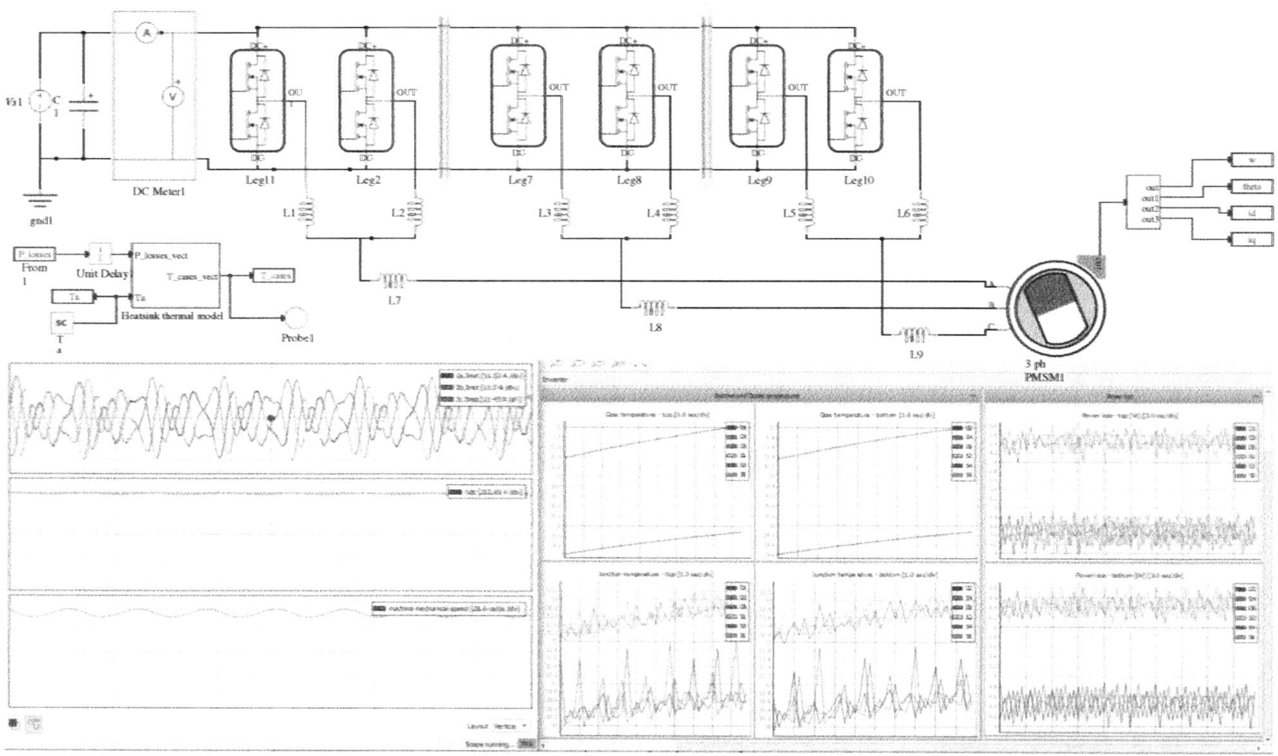

Figure 11.9 Implementation layout of a 3-phase dual converter driving a PMSM in Typhoon HIL environment.

heat capacity and thermal resistance. The specific heat capacity of a SiC MOSFET is a measure of the energy required to increase its temperature by 1 °C. Modeling this characteristic accurately is essential for understanding the thermal energy dynamics of the device, especially as it undergoes state changes during operational activities like switching events or when handling different loads. The thermal resistance, on the other hand, quantifies the MOSFET's ability to transfer heat away from its junction to the case and the surrounding environment. A lower value of thermal resistance means that the MOSFET is more effective at dissipating heat, which is a critical factor in preventing excessive temperature rise at the junction – known as thermal runaway – and ensuring the device operates within safe temperature limits.

Further, dynamic thermal modeling forms a core part of assessing and predicting the performance of SiC MOSFETs under various operational conditions. In this context, the simulation tracks the junction temperature in real-time, which is a crucial indicator of the MOSFET's thermal state. The model incorporates a heat transfer analysis that accounts for the heat generated by both switching and conduction losses, alongside the inherent thermal properties of the device. Using principles such as Fourier's law of heat conduction and Newton's law of cooling, the simulation provides insight into the effectiveness of heat dissipation from the junction to the heat sink and the ambient environment. The simulation also responds to varying load conditions by adjusting current and voltage levels to see how they influence the thermal behavior of the MOSFETs. This dynamic response helps identify the operational limits and how deviations from these limits, due to changes in load or external temperature, could potentially compromise the MOSFETs' performance. By modeling these factors, engineers can determine the critical thresholds that, if exceeded, could result in thermal overload, leading to device degradation or outright failure. Such predictive capabilities are indispensable for designing MOSFETs that are not only efficient but also reliably durable across their operational lifespan.

The thermal modeling for SiC MOSFETs explored various cooling techniques, such as passive air and active liquid cooling, to maintain operational temperatures and prevent thermal runaway. Simulations tested the efficacy of thermal interface materials and heat sink designs to enhance heat dissipation. Assessing the switching behavior, the simulation focused on gate charge, threshold voltage, and energy per switching event, which are critical for high-frequency efficiency and stability. It also examined dynamic switching and conduction losses, considering gate drive voltage, switching frequency, and di/dt effects on power loss. By integrating these factors, the simulation provided insights for creating a balanced system with optimal switching frequency and heat management. This detailed analysis guides the selection of MOSFETs, circuitry, and cooling strategies, ensuring efficient and thermally stable system operations.

11.5.1.1 Integration of Efficiency Models into System Simulation

The integration of efficiency models into system simulations has allowed for a comprehensive understanding of SiC MOSFET performance under various load conditions and the effect of both intrinsic and extrinsic parasitic elements on their efficiency [28]. This simulation was extensive, involving scenario analysis where different operating conditions were scrutinized. This process aimed to find the optimal conditions that ensure peak MOSFET performance while mitigating reactive and switching losses. To this end, the simulation helped in pinpointing suitable filter capacitors that minimize voltage overshoots, crafting layout strategies to cut down inductance, and selecting operating frequencies that strike a balance between efficiency and performance.

Further enhancing the system's capabilities, the simulation facilitated the development of adaptive control strategies. These strategies dynamically tweak operating parameters in real time, reacting to the current load conditions and system efficiency metrics. By adopting this flexible approach, the system consistently operates at high efficiency, regardless of the range of conditions it encounters, which is instrumental in prolonging the life and reliability of SiC MOSFET-based systems.

To support this sophisticated simulation, resource allocation was meticulously planned for the 3-phase dual converter system within the Typhoon HIL platform, as detailed in the following Table 11.2:

The use of multiple cores ensures the simulation can run complex and interactive models in real time, critical for analyzing the performance and efficiency of power electronic systems under varying operational conditions. Core couplings are crucial for maintaining high-fidelity synchronization between computational tasks that simulate interdependent physical phenomena, such as thermal effects coupled with electrical dynamics [29]. Adequate memory and enough I/O channels are essential to handle the extensive data and connectivity requirements of this simulation, facilitating a comprehensive evaluation of the converter's performance and its control strategies.

11.5.2 Implementation in Imperix B-Box

The Imperix B-Box setup marks a crucial stage in the deployment of real-time control strategies for power electronics, serving as an advanced controller to expedite prototyping and the evaluation of power conversion and motor control concepts. The B-Box's integration with the Typhoon HIL environment is strategic, allowing control algorithms to be pre-validated under simulated conditions, which markedly curtails the risks and duration of development. In the interfacing phase, the Imperix B-Box was physically linked to the Typhoon HIL through interfaces adept at managing both input and output signals, spanning analog and digital I/O

Table 11.2 Resource Allocation for SiC MOSFET-Based 3-Phase Dual Converter Simulation on Typhoon HIL System.

Resource Type	Utilization Detail	Description
Cores used	4 (of 8 available)	Four cores are dedicated to simulating the dynamic behavior of the SiC MOSFETs and handling the complex interactions in the power conversion process.
Core couplings	2 pairs coupled	Two pairs of cores are tightly coupled to manage real-time simulation of electromagnetic and thermal interactions between the phases and the control logic.
Memory allocation	2 GB (of 4 GB available)	Two gigabytes are allocated for real-time data processing and logging simulation results, ensuring smooth operation and data integrity.
Input/output channels	48 used (of 64 available)	48 I/O channels are utilized to handle signals from the power stages, sensors, and control inputs/outputs, reflecting the high connectivity demands of the system.
Simulation components	SiC MOSFET models, inductances, capacitors, control logic	Detailed component models include SiC MOSFETs, associated passive elements like inductors and capacitors, and the control algorithms necessary for FOC and DPWM techniques.

ports that required precise mapping to their counterparts on the Typhoon HIL. The robustness of communication between the B-Box and Typhoon HIL is critical, involving the establishment of Ethernet or USB connections to enable swift data exchange. Communication protocols such as CAN or Modbus TCP/IP are meticulously configured to ensure synchronized operations, facilitating a seamless flow of commands and data between the simulation and the real-time controller.

A significant hurdle during setup was synchronizing the real-time clock of the B-Box with the Typhoon HIL's simulated timeframe. Accurate synchronization is essential; discrepancies can result in phase shifts or signal delays, compromising control signal integrity and system response. To overcome this, a synchronization protocol using a shared time reference was implemented. Typically, a master clock signal from the Typhoon HIL dictates the pace, with the B-Box adjusting its internal clock accordingly, employing methods like the Network Time Protocol (NTP) or Precision Time Protocol (PTP) as needed for the desired precision. Signal levels and timings also underwent meticulous calibration. Aligning the signal levels between the Typhoon HIL outputs, and the B-Box inputs is pivotal to prevent unintended signal alteration due to mismatches. This adjustment is achieved through software settings or external conditioning circuits tailored to the B-Box's parameters. Timing calibration is just as crucial, especially for high-frequency operations, necessitating a thorough examination of signal propagation delays and processing intervals. The B-Box software includes calibration tools that help measure and fine-tune the timing of inputs and outputs to preclude any potential discrepancies [30].

The successful configuration of the Imperix B-Box with the Typhoon HIL thus encompasses not only the physical and logical setup but also an exact calibration process, ensuring that the system functions according to specifications in actual applications. This comprehensive approach allows engineers to simulate, experiment, and refine their power systems within a secure environment before proceeding to the final implementation, significantly bolstering the reliability and operational efficiency of the developed system. Table 11.3 outlines possible numerical values associated with different aspects mentioned in the setup and configuration of the Imperix B-Box and its integration with the Typhoon HIL environment. These values are examples and can vary based on the specific requirements of the system being developed.

11.5.2.1 Control Strategy Implementation

The implementation of the ATC strategy within the MATLAB/Simulink environment involved utilizing the ACG-SDK to create a seamless transition from high-level simulation models to executable code that can run on the Imperix B-Box. This process began with the design of a sophisticated control algorithm aimed at managing the thermal properties and operational efficiency of SiC MOSFETs. The control model was developed to continuously monitor and react to real-time temperature data gathered from sensors positioned at critical points within the MOSFET array. By integrating these temperature measurements, the system could dynamically adjust the operating parameters such as gate drive voltage, switching frequency, and duty cycle to optimize thermal management. This capability is essential in preventing the SiC MOSFETs from reaching temperatures that could lead to thermal runaway, thus safeguarding against efficiency losses and potential hardware failures. Using the ACG-SDK, this high-fidelity model was translated into code that the B-Box could execute [31]. The SDK facilitates this by generating C code that is compatible with the real-time operating system running on the B-Box, thereby ensuring that the simulation fidelity is maintained in the actual hardware implementation. This translation involves not only the primary control logic but also includes the integration of various fail-safe measures and diagnostic routines to enhance system reliability and safety.

The Simulink model shown in Figure 11.10 represents a FOC system for motor management. It processes inputs like bus voltage and machine currents through analog-to-digital conversion blocks, essential for accurate digital control. The

Table 11.3 Key Technical Parameters and Example Values for the Setup and Configuration of the Imperix B-Box and Its Integration with Typhoon HIL.

Parameter	Example Value	Description
Ethernet communication speed	100 Mbps	Speed at which data is transferred between the Imperix B-Box and Typhoon HIL via Ethernet connection.
USB communication speed	480 Mbps (USB 2.0)	Speed for USB connections, if used instead of Ethernet, for data transfer between devices.
Synchronization precision	<1 ms	The precision required for synchronizing the real-time clocks of the B-Box and Typhoon HIL.
Signal voltage levels	0–10 V	The range of signal voltage levels for analog inputs/outputs between the Typhoon HIL and the Imperix B-Box.
Signal adjustment tools	Software settings, External circuits	Tools used to match signal levels between Typhoon HIL outputs and B-Box inputs.
Network Time Protocol (NTP) precision	1–50 ms	Typical synchronization precision for NTP used in less precise applications.
Precision Time Protocol (PTP) precision	<1 ms	Typical synchronization precision for PTP, used for highly precise applications.
Modbus TCP/IP transfer rate	Variable, typically <100 ms	Latency and transfer rate for Modbus TCP/IP, depending on network conditions and system configuration.
Gate drive voltage	15–20 V	Typical gate drive voltages used to control the MOSFETs in the B-Box.
Switching frequency range	2–100 kHz	Frequency range over which the MOSFETs can switch in power electronic applications.
di/dt (Rate of current change)	Up to 1000 A/μs	Maximum rate of current change during switching events, affecting voltage spikes and EMI.
Maximum load current	Up to 100 A (per channel)	Typical maximum current per channel that the B-Box can handle during operations.
Thermal management	Active cooling (fans), passive (heat sinks)	Methods used to manage the thermal load of the B-Box and attached electronics.
Operational temperature range	−40 to 85 °C	Temperature range within which the B-Box can operate reliably.

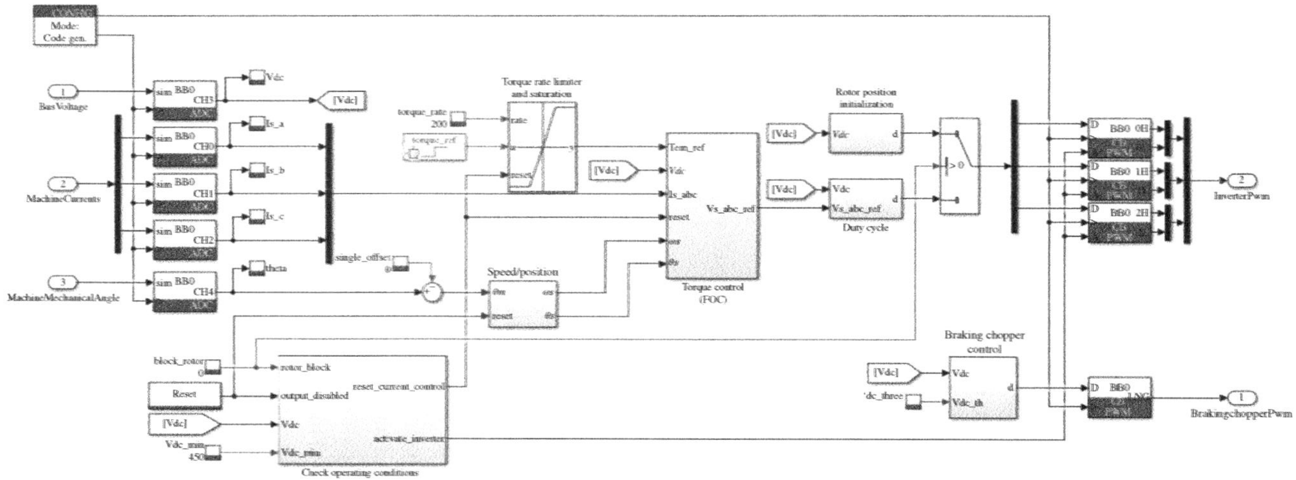

Figure 11.10 Simulink schematic of FOC-based motor control: integrating ADC inputs, torque management, and PWM inverter control with safety monitoring.

system also tracks the mechanical angle, vital for rotor positioning. Control strategy components within the model ensure motor command limits and calculate torque references based on motor dynamics. The FOC block is central to the model, manipulating current to control motor torque and power delivery. Inverter control then converts these calculations into PWM signals to drive the motor, while a braking chopper controller is in place for energy dissipation during deceleration. Real-time monitoring blocks oversee system safety, and rotor position initialization is crucial for the FOC's precision. Though not explicitly shown, ACG-SDK integration with the Imperix B-Box suggests a seamless transition from the model to actual hardware implementation. This setup allows for dynamic adjustment based on live motor feedback, aiming to maintain efficiency and prevent thermal overload, which is pivotal for SiC MOSFET longevity in the motor control application.

Once the initial code was deployed to the B-Box, the system underwent a series of iterative testing phases. These tests were designed to validate the effectiveness of the thermal management strategy under a variety of operating conditions, ranging from low-load scenarios to peak thermal stress situations. Adjustments were made to the control parameters based on the outcomes of these tests, which were crucial in refining the model's accuracy and the system's responsiveness.

11.6 Hardware-in-the-Loop (HIL) and RCP Interface

11.6.1 HIL–RCP Integration

The integration process of the HIL system with RCP is multi-layered technical task that ensures the simulated and physical components communicate and function together as a cohesive unit. This process begins with establishing a robust system connection. A communication link is set up between the HIL simulator and the RCP system – often facilitated by Ethernet, CAN, or USB interfaces – equipped with high-bandwidth capabilities to handle the swift exchange of data necessary for real-time simulation. This establishes the foundational infrastructure needed for seamless data transmission. In the data exchange phase, specific communication protocols such as TCP/IP, UDP, or proprietary protocols supported by the HIL hardware are configured to orchestrate the flow of information. These protocols are vital as they dictate how the RCP system transmits control commands to the HIL simulator and, just as importantly, how it interprets and reacts to feedback from the simulated environment. The RCP system uses this data exchange through a interface cockpit as shown in Figure 11.11 to send a PWM command to the HIL, which simulates the corresponding effects on a motor model.

The real-time simulation is a pivotal aspect of the integration where the HIL simulator is synchronized with the RCP to mirror the timing and behavior of the actual

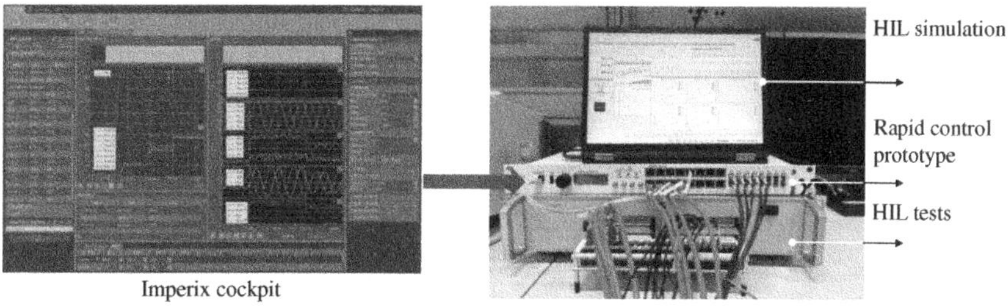

Figure 11.11 Control system development workflow: from Imperix cockpit visualization to HIL simulation, rapid control prototyping, and HIL testing setup.

physical system. This synchronization ensures that the model within the HIL operates at a timestep that matches real-world operations, often in the microsecond range, to accurately reflect the dynamic responses of the physical system to various stimuli. Achieving such fidelity requires the HIL simulator to have robust computational power and real-time operating systems that can keep pace with the demands of a high-speed RCP system. Feedback mechanisms are then intricately woven into the system architecture. These mechanisms allow the RCP system to perform on-the-fly adjustments based on the real-time data received from the HIL simulator. This creates a closed-loop system where the simulated inputs – such as temperature readings, current measurements, or voltage levels – are fed back into the RCP system, prompting it to modify its control outputs accordingly. The feedback loop is critical for the ATC strategy, enabling it to make informed decisions based on simulated thermal conditions and adjust parameters like fan speeds, heat sink operations, or switch timing to maintain optimal temperatures. The detailed technical orchestration of system connection, data exchange, real-time simulation, and feedback loops is fundamental to the validation of the ATC strategy. This complex interplay ensures that the simulated environment closely mirrors real-world conditions, allowing engineers to refine and validate the ATC strategy under controlled yet realistic scenarios, paving the way for deployment in actual operational contexts.

11.6.2 Results and Observations

The integration and validation of the HIL and RCP systems were undertaken to evaluate the implemented ATC strategy, focusing specifically on its responsiveness to control actions and its impact on power loss mitigation. In an HIL–RCP environment created, the dual inverter topology, key to driving a PMSM, was subjected to a variety of dynamic load conditions that imitated real-world torque demands on the motor. The overall workflow for the implemented process starting from the developments made is shown in Figure 11.12.

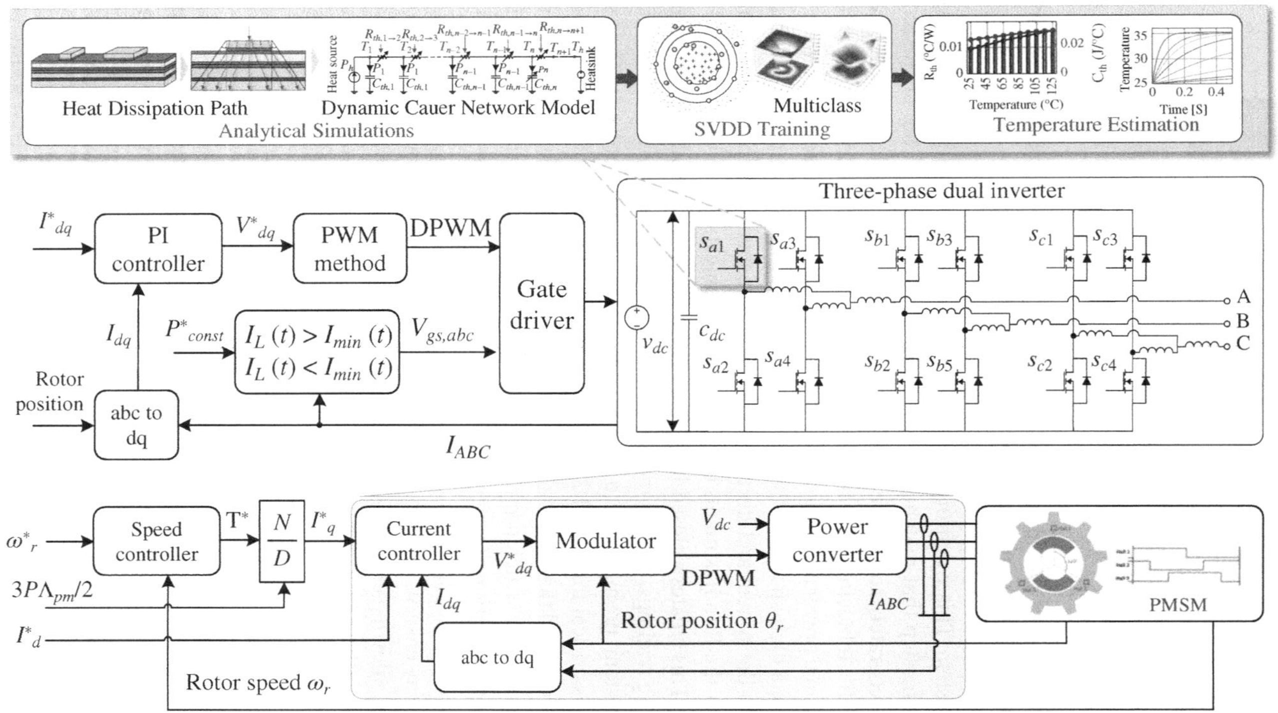

Figure 11.12 Integrated control and thermal management system for SiC MOSFETs in PMSM drive applications.

The simulation capitalized on the FOC technique for managing the PMSM's torque and speed, while DPWM was instrumental in optimizing the switching events to minimize thermal strain on the SiC MOSFETs. A notable point in the process was setting the switching frequency of the converter to 20 kHz, a parameter critical to balancing the trade-off between efficiency and performance. Thermal behavior modeling took a forefront in the analysis, employing a temperature-dependent Cauer network to simulate the heat dynamics within the SiC MOSFET module. This simulation produced a thermal map that identified potential overheating hotspots, which is essential for preemptive thermal management. The Support Vector Data Description (SVDD) model, trained on the data from these thermal simulations, proved critical in encapsulating normal operation patterns and detecting deviations, thereby ensuring robust anomaly detection. Real-time power loss measurements from each MOSFET were processed by the SVDD model within the Simulink framework, translated through a custom MATLAB Function Block. The temperature estimates generated by this block were crucial in adjusting the DPWM control signals, enabling on-the-fly thermal regulation. Simulation outcomes shown in Figures 11.13 and 11.14 demonstrated a marked improvement in the system's ability to manage thermal loads, maintain operational efficiency, and consequently prevent thermal overload – a testament to the ATC strategy's effectiveness.

From Figure 11.13, it can be observed that the phase currents remained balanced, torque response accurately tracked reference values, and rotor speed exhibited stability, even in the face of rapid load changes. The d-axis and q-axis current profiles adhered to their reference trajectories, highlighting the system's capability in maintaining vector control despite fluctuations. Moreover, a comparative analysis of inverter power loss in Figure 11.9 emphasized the superior performance of the ATC system, which substantially contained power loss as compared to standard control strategies.

The integration of HIL-RCP, therefore, not only validated the ATC strategy but also highlighted the enhanced performance capabilities and reliability of SiC MOSFET-based systems under variable load conditions. These advancements in thermal management, validated through rigorous simulation and testing, are poised to make significant contributions to the design and operation of advanced power electronic systems.

11.7 Conclusion

In conclusion, this result successfully demonstrated the advancement in the management of SiC MOSFET temperatures within a 3-phase dual inverter system by integrating transient thermal characterization with SVDD. The approach has led

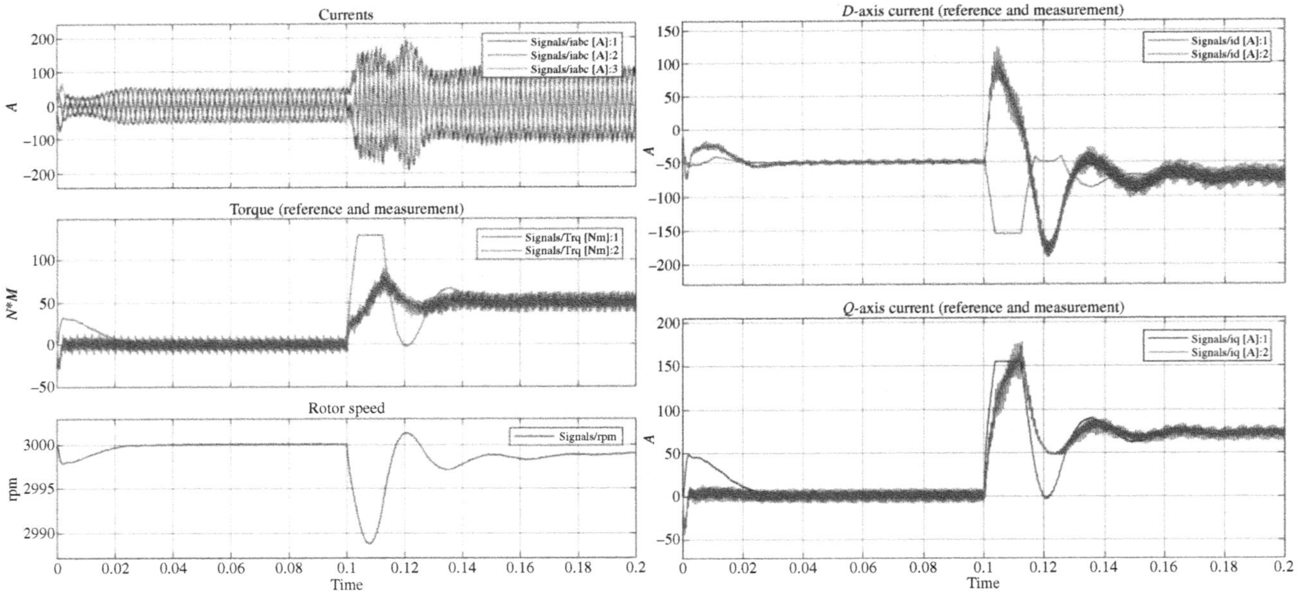

Figure 11.13 Comprehensive evaluation of motor drive performance with active thermal control and *D*-axis and *Q*-axis current profiles under field-oriented control.

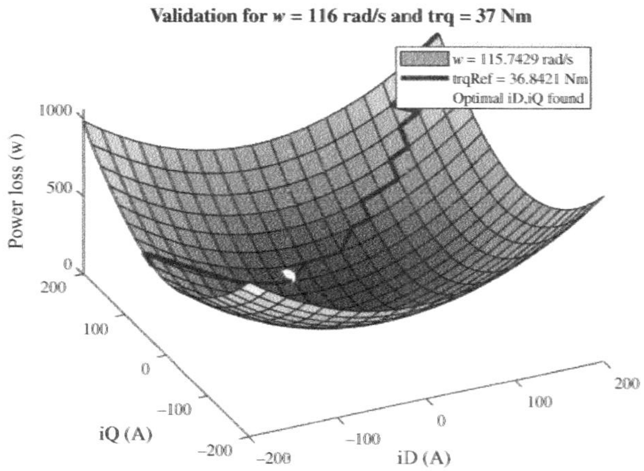

Optimization for Power Loss Minimization

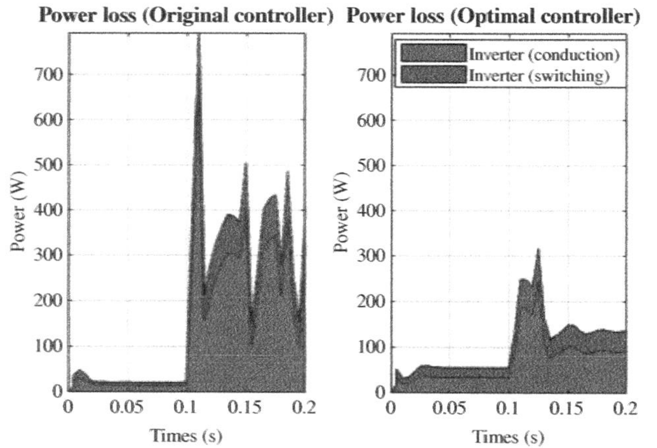

Power loss reduction achieved through optimized FOC in response to a step change in loa torqued

Figure 11.14 Inverter power loss analysis demonstrating the efficacy of developed active thermal control.

to a responsive ATC system that accurately estimates junction temperatures, a critical factor for reliable and efficient performance in power electronic applications. Simulations confirm that this method substantially lowers power losses and enhances system robustness against thermal fluctuations. This research provides a pathway for implementations of machine-learning-assisted thermal controls in high-demand industrial settings, with the potential to significantly improve the operational lifespan of power conversion systems.

Exercises

11.1 What advantages do SiC MOSFETs offer in high-power applications compared to traditional silicon-based MOSFETs?

11.2 Elaborate the associated thermal management challenges when using SiC MOSFETs.

11.3 How does the Adaptive Thermal Control (ATC) strategy, including real-time temperature monitoring and dynamic thermal management, help address the rapid heating issues in SiC MOSFETs?

11.4 What role do predictive thermal models and active cooling techniques play in maintaining the reliability and efficiency of SiC MOSFETs under high-power conditions?

11.5 What are the primary components of power loss in a SiC MOSFET power module, and how do conduction losses depend on the RMS current and on-state resistance?

11.6 What causes switching losses in SiC MOSFETs, and how do turn-on and turn-off losses contribute to the total switching losses?

11.7 How does the capacitance between gate and source, and drain and source, impact the energy dissipation during the switching process in a MOSFET?

11.8 What role does gate drive loss play in power dissipation in a SiC MOSFET?

11.9 What is the significance of thermal modeling in SiC MOSFETs, and how does an RC network approach help in estimating junction temperature?

11.10 How does the Variable Angle Method (VAM) help in constructing a Cauer RC network for modeling thermal behavior in semiconductor packages like SiC MOSFETs?

11.11 How does active loss manipulation through DPWM reduce switching losses in a three-phase inverter system?

11.12 What is Rapid Control Prototyping (RCP), and how does it enhance the testing and implementation of control strategies under variable operational conditions?

References

1 J. Wang and X. Jiang, "Review and analysis of SiC MOSFETs' ruggedness and reliability," *IET Power Electron.*, vol. 13, no. 3, pp. 445–455, Feb. 2020. 10.1049/iet-pel.2019.0587.

2 Y. Zheng, R. Potera, and T. Witt, "Characterization of early breakdown of SiC MOSFET gate oxide by voltage ramp tests," in *2021 IEEE International Reliability Physics Symposium (IRPS)*, Mar. 2021, pp. 1–5. 10.1109/IRPS46558.2021.9405196.

3 B. Yu and L. Wang, "Online accurate measurement of steady-thermal resistance of SiC MOSFETs for DC solid-state power controller," *IEEE Trans. Power Electron.*, vol. 36, no. 5, pp. 5006–5021, May 2021. 10.1109/TPEL.2020.3028844.

4 Z. Zhang, L. Liang, and H. Fei, "Investigation on safe-operating-area degradation and failure modes of SiC MOSFETs under repetitive short-circuit conditions," *Power Electron. Devices Components*, vol. 4, p. 100026, Mar. 2023. 10.1016/j.pedc.2022.100026.

5 M. H. Ahmed, M. Wang, M. A. S. Hassan, and I. Ullah, "Power loss model and efficiency analysis of three-phase inverter based on SiC MOSFETs for PV applications," *IEEE Access*, vol. 7, pp. 75768–75781, 2019. 10.1109/ACCESS.2019.2922741.

6 S. Hazra, S. Madhusoodhanan, G. K. Moghaddam, K. Hatua, and S. Bhattacharya, "Design considerations and performance evaluation of 1200-V 100-A SiC MOSFET-based two-level voltage source converter," *IEEE Trans. Ind. Appl.*, vol. 52, no. 5, pp. 4257–4268, Sep. 2016. 10.1109/TIA.2016.2587098.

7 U. Jadli, F. Mohd-Yasin, H. A. Moghadam, P. Pande, J. R. Nicholls, and S. Dimitrijev, "Measurement of power dissipation due to parasitic capacitances of power MOSFETs," *IEEE Access*, vol. 8, pp. 187043–187051, 2020. 10.1109/ACCESS.2020.3030269.

8 R. K. Williams, M. N. Darwish, R. A. Blanchard, R. Siemieniec, P. Rutter, and Y. Kawaguchi, "The trench power MOSFET—Part II: Application specific VDMOS, LDMOS, packaging, and reliability," *IEEE Trans. Electron Devices*, vol. 64, no. 3, pp. 692–712, Mar. 2017. 10.1109/TED.2017.2655149.

9 R. K. Williams, M. N. Darwish, R. A. Blanchard, R. Siemieniec, P. Rutter, and Y. Kawaguchi, "The trench power MOSFET: Part I—history, technology, and prospects," *IEEE Trans. Electron Devices*, vol. 64, no. 3, pp. 674–691, Mar. 2017. 10.1109/TED.2017.2653239.

10 G. Lakkas, "MOSFET power losses and how they affect power-supply efficiency," *Analog Appl.*, vol. 10, pp. 22–26, Apr. 2016.

11 K. Gorecki, P. Gorecki, and J. Zarebski, "Measurements of parameters of the thermal model of the IGBT module," *IEEE Trans. Instrum. Meas.*, vol. 68, no. 12, pp. 4864–4875, Dec. 2019. 10.1109/TIM.2019.2900144.

12 V. Székely, "A new evaluation method of thermal transient measurement results," *Microelectron. J.*, vol. 28, no. 3, pp. 277–292, Mar. 1997. 10.1016/S0026-2692(96)00031-6.

13 P. E. Bagnoli, C. Casarosa, M. Ciampi, and E. Dallago, "Thermal resistance analysis by induced transient (TRAIT) method for power electronic devices thermal characterization. I. Fundamentals and theory," *IEEE Trans. Power Electron.*, vol. 13, no. 6, pp. 1208–1219, Nov. 1998. 10.1109/63.728348.

14 F. N. Masana, "A new approach to the dynamic thermal modelling of semiconductor packages," *Microelectron. Reliab.*, vol. 41, no. 6, pp. 901–912, Jun. 2001. 10.1016/S0026-2714(01)00013-0.

15 F. N. Masana, "A closed form solution of junction to substrate thermal resistance in semiconductor chips," *IEEE Trans. Components, Packag. Manuf. Technol. Part A*, vol. 19, no. 4, pp. 539–545, 1996. 10.1109/95.554935.

16 V. S. B. Kurukuru, M. A. Khan, and R. Singh, "Health monitoring framework for electric vehicle drive train in digital twin," in *2023 25th European Conference on Power Electronics and Applications (EPE'23 ECCE Europe)*, Sep. 2023, pp. 1–10. 10.23919/EPE23ECCEEurope58414.2023.10264291.

17 Y. Zhu, K. Ma, and X. Cai, "Thermal characterization method of power semiconductors based on H-bridge testing circuit," *IEEE Trans. Power Electron.*, vol. 34, no. 9, pp. 8268–8273, Sep. 2019. 10.1109/TPEL.2019.2900253.

18 C. Grosse *et al.*, "Rapid thermal characterization of materials with ultra-high resolution of droplet size specimens using the three-omega method," in *2018 17th IEEE Intersociety Conference on Thermal and Thermomechanical Phenomena in Electronic Systems (ITherm)*, May 2018, pp. 562–566. 10.1109/ITHERM.2018.8419617.

19 T. Tong and A. Majumdar, "Reexamining the 3-omega technique for thin film thermal characterization," *Rev. Sci. Instrum.*, vol. 77, no. 10, p. 104902, Oct. 2006. 10.1063/1.2349601.

20 D. G. Cahill, "Thermal conductivity measurement from 30 to 750 K: The 3ω method," *Rev. Sci. Instrum.*, vol. 61, no. 2, pp. 802–808, 1990. 10.1063/1.1141498.

21 C. Dames, "Measuring the thermal conductivity of thin films: 3 omega and related electrothermal methods," *Annu. Rev. Heat Transf.*, vol. 16, pp. 7–49, 2012. 10.1615/AnnualRevHeatTransfer.2012004678.

22 F. Liu, K. Xin, and Y. Liu, "An adaptive Discontinuous Pulse Width Modulation (DPWM) method for three phase inverter," in *2017 IEEE Applied Power Electronics Conference and Exposition (APEC)*, Mar. 2017, pp. 1467–1472. 10.1109/APEC.2017.7930892.

23 D. Nayak, S. Member, Y. R. Kumar, and S. Member, "Discontinuous modulation based active thermal control for enhancing the KVA limit of three-phase inverters discontinuous modulation based active thermal control for enhancing the KVA limit of three-phase inverters," *TechRxiv*. pp. 0–9, 2023. 10.36227/techrxiv.24591477.v1.

24 B. Schmitz-Rode *et al.*, "A modular signal processing platform for grid and motor control, HIL and PHIL applications," in *2022 International Power Electronics Conference (IPEC-Himeji 2022- ECCE Asia)*, May 2022, pp. 1817–1824. 10.23919/IPEC-Himeji2022-ECCE53331.2022.9807061.

25 J. Poon, P. Haessig, J. G. Hwang, and I. Celanovic, "High-speed hardware-in-the loop platform for rapid prototyping of power electronics systems," in *2010 IEEE Conference on Innovative Technologies for an Efficient and Reliable Electricity Supply*, Sep. 2010, pp. 420–424. 10.1109/CITRES.2010.5619770.

26 M. S. Vekić, S. U. Grabić, D. P. Majstorović, I. L. Čelanović, N. L. Čelanović, and V. A. Katić, "Ultralow latency HIL platform for rapid development of complex power electronics systems," *IEEE Trans. Power Electron.*, vol. 27, no. 11, pp. 4436–4444, Nov. 2012. 10.1109/TPEL.2012.2190097.

27 F. Behi and A. Kornecki, "Software development with automatic code generation: Observations from novice developer viewpoint," *Proceedings of the International Conference on Software Engineering Research & Practice, Conference Program Language Compilation SERP 2006*, May 2006, pp. 289–295.

28 J. Martins *et al.*, "Test platform for rapid prototyping of digital control for power electronic converters," in *IECON 2019 - 45th Annual Conference of the IEEE Industrial Electronics Society*, Oct. 2019, pp. 2056–2061. 10.1109/IECON.2019.8927836.

29 H. Vardhan, B. Akin, and H. Jin, "A low-cost, high-fidelity processor-in-the loop platform: For rapid prototyping of power electronics circuits and motor drives," *IEEE Power Electron. Mag.*, vol. 3, no. 2, pp. 18–28, Jun. 2016. 10.1109/MPEL.2016.2550239.

30 B. Lu, X. Wu, H. Figueroa, and A. Monti, "A low-cost real-time hardware-in-the-loop testing approach of power electronics controls," *IEEE Trans. Ind. Electron.*, vol. 54, no. 2, pp. 919–931, Apr. 2007. 10.1109/TIE.2007.892253.

31 B. Chabibi, A. Douche, A. Anwar, and M. Nassar, "Integrating SysML with simulation environments (Simulink) by model transformation approach," in *2016 IEEE 25th International Conference on Enabling Technologies: Infrastructure for Collaborative Enterprises (WETICE)*, Jun. 2016, pp. 148–150. 10.1109/WETICE.2016.39.

Index

Artificial Intelligence for Power Electronics, First Edition.
Edited by Ahteshamul Haque, Saad Mekhilef, and Azra Malik.

Printed and bound by CPI Group (UK) Ltd, Croydon, CR0 4YY

17/09/2026

14964332-0005